U0737702

河南省"十四五"普通高等教育规划教材

先进制造技术

（英汉对照）

任小中　薛玉君　钟相强　张志文
田晓光　张东明　苏永生　编著

机械工业出版社

本书是按照我国高等教育要与国际接轨、培养国际性复合型人才的要求，结合作者近年来在"本科教学工程"建设方面的实践与成果编著的英汉双语教材。这是目前国内少见的以英汉对照形式编写的"先进制造技术"双语教材。

本书在综合国内外最新研究成果和相关参考文献的基础上，从科学思维、学科综合和技术集成的角度，系统介绍了各种先进制造技术的理念、基本内容、关键技术和最新成果，旨在使学生了解国内外先进制造前沿技术，拓宽知识面，掌握先进制造技术的理念和方法，启发和培养学生的科学思维，培养他们科学创新和工程实践的能力。全书共6章，内容包括先进制造技术概论、先进设计技术、先进制造工艺、制造自动化技术、制造企业的信息管理技术和先进制造模式。各章后均附有复习题与习题。

本书内容综合性强，涉及的知识面广，编写形式新颖，可作为高等院校机械工程、航空航天制造工程、管理科学与工程等各类与制造技术有关的学科及专业的本科生和研究生教材或参考书，特别适合作为同类课程的双语教材，也可供机械制造业工程技术人员参考。

图书在版编目（CIP）数据

先进制造技术：英汉对照/任小中等编著. —北京：机械工业出版社，2021.11（2025.4重印）

河南省"十四五"普通高等教育规划教材

ISBN 978-7-111-69623-0

Ⅰ.①先… Ⅱ.①任… Ⅲ.①机械制造工艺-高等学校-教材-汉、英 Ⅳ.①TH16

中国版本图书馆 CIP 数据核字（2021）第 238993 号

机械工业出版社（北京市百万庄大街22号 邮政编码100037）

策划编辑：王勇哲 责任编辑：王勇哲 安桂芳
责任校对：徐鲁融 封面设计：张 静
责任印制：单爱军

北京虎彩文化传播有限公司印刷

2025 年 4 月第 1 版第 3 次印刷

184mm×260mm · 21 印张 · 518 千字

标准书号：ISBN 978-7-111-69623-0

定价：65.00 元

电话服务 网络服务

客服电话：010-88361066 机 工 官 网：www.cmpbook.com

010-88379833 机 工 官 博：weibo.com/cmp1952

010-68326294 金 书 网：www.golden-book.com

封底无防伪标均为盗版 机工教育服务网：www.cmpedu.com

前　　言

随着经济全球化的不断深化和国际交流活动的日益频繁，社会对人才外语水平和国际竞争能力的要求越来越高。双语教学是我国高等教育与国际高等教育接轨的一项重要举措。双语教学可以在传授学科知识的同时，提高学生的外语水平，开拓学生的国际视野，是培养高素质国际性人才的有效途径。

双语教材是开展双语教学的基础，合适的双语教材是保证双语教学质量的关键。尽管原版教材有其优势，但与国内教学大纲和教学体系不相适应。为此，编者结合近年来在"本科教学工程"建设方面取得的成果及在先进制造工程领域的科研实践，编写了这本英汉双语教材。本书是以"英汉对照"形式编著的"先进制造技术"双语教材，旨在为双语教学提供优质的教学资源，提高双语教学质量。

当前，我国虽已成为世界上的制造大国，但还不是制造强国。我国制造业要想在激烈的国际市场竞争中求得生存和发展，就必须掌握和科学运用最先进的制造技术。这就要求培养一大批掌握先进制造技术、具有科学思维、创新意识及工程实践能力的高素质专业人才。为了帮助学生掌握先进制造技术的理念和内涵，了解先进制造技术的最新发展，促进先进制造技术在我国的研究和应用，本书从"大制造"的角度讲述了机械制造业的前沿技术及制造企业的信息管理技术，并介绍了一些先进制造模式及其在现代制造企业中的推广应用。

全书以先进制造工艺技术为核心，以先进工程设计技术、先进制造工艺和制造自动化技术为主干，以先进生产管理技术和先进制造模式为软环境，系统地构建了先进制造工程的知识体系。本书具有以下特色：

（1）知识体系完整　从"大制造"角度构建产品设计—制造工艺—制造自动化—企业信息管理—先进制造模式的层次架构，把新颖而实用的制造技术引入本书，注重学科的交叉融合。

（2）编写形式新颖，适用范围广泛　本书采用"英汉对照"形式编写，内容上保持一致，不仅适用于双语班的学生，也适用于非双语班的学生。

（3）内容全面，突出"先进"　不仅介绍先进制造工艺，而且介绍与制造密切相关的先进设计技术和自动化技术，还介绍了与之配套的先进管理技术和先进制造模式。中文内容聚集国内同类课程的优质资源，英文表述则吸取原版教材之语言精华。全书内容尽可能与国际先进水平接轨。

（4）英文内容的一致性、准确性和可读性强　这是编写双语教材对国内编者的最大挑战。本书并非一味地追求严格地按中文内容翻译，而是注重中、英文基本内容的一致性和英文词义的准确性，尽量采用国外同类教材中的专业词汇和表述方式，以及常用的语法和简单易懂的句子，从而保证内容易读、易懂。

（5）理论联系实际，注重综合素质培养　本书注重介绍先进制造的原理、方法和技术，强调实例教学，坚持理论联系实际，培养学生分析和解决实际问题的能力，为学生毕业后择业奠定宽泛的专业基础。

（6）配套齐全 本书分别配有中英文电子课件，以及章后的主要习题解答。凡是为学生订购了该教材的任课教师，均可向机械工业出版社免费索取。其中电子课件为开放式课件，任课教师可根据实际教学情况自行增删或改编，以满足个性化教学要求。

本书由黄河交通学院任小中、河南科技大学薛玉君、安徽工程大学钟相强、黄河交通学院田晓光、张东明、河南科技大学张志文、安徽工程大学苏永生共同编著。本书具体编写分工：任小中编写绪论，第1章，第2章的2.5~2.6节，第3章的3.4、3.6~3.7节；薛玉君编写第3章的3.3、3.5节，第6章的6.6~6.8节；钟相强编写第2章的2.1~2.4节；苏永生编写第3章的3.1~3.2节；张东明编写第4章；张志文编写第5章；田晓光编写第6章的6.1~6.5节。全书由任小中负责统稿。

江苏大学任乃飞教授担任本书主审，对教材进行了认真的审阅，提出了很多宝贵的建议和意见，编者在此表示由衷的感谢！

在本书编写过程中，我们参考了国内外出版的一些教材，谨此向有关作者表示诚挚的谢意，并向所有关心和帮助本书出版的人表示感谢！

本书的编写与出版得到了黄河交通学院"河南省智能制造技术与装备工程技术研究中心"建设经费的资助，在此表示衷心的感谢！

由于编者所及资料和水平有限，书中难免有错漏和不当之处，敬请广大读者批评指正。

编 者

Preface

With the deepening of economic globalization and the increase of international communication activities, requirements of society for the talents in foreign language proficiency and international competitive power are higher and higher. To carry out bilingual teaching is an important measure to make the domestic higher education connect international higher education. While the bilingual teaching is used to teach subject knowledge, it can improve the foreign language proficiency and open the international perspective of students. It is an effective way to cultivate international talents with high quality.

Bilingual textbook is the basis of bilingual teaching. Appropriate bilingual textbook is the key to ensure teaching quality. Original English textbooks have their own advantages, but they do not fit with domestic teaching program and curriculum system. Therefore, combining the achievements acquired in the construction of "Undergraduate Teaching Project" and scientific researches in the field of advanced manufacturing engineering, we have compiled this bilingual textbook written in both English and Chinese. It is a bilingual textbook on "Advanced Manufacturing Technology" written in both English and Chinese. The purpose of this book is to provide the excellent teaching resource for bilingual teaching and to improve bilingual teaching quality.

China has been a big manufacturing country but not a powerful one in the world. The manufacturing industry of China has to master and utilize scientifically the up-to-date advanced manufacturing technologies if it wants to survive and develop in the international market with intensive competition. It requires that China has to train a large number of high-quality professional talents armed with advanced manufacturing technology, scientific thinking and innovation consciousness, as well as engineering practice ability. In order to make the students master the idea and connotation of advanced manufacturing technology, learn its latest developments, and promote the research and application of advanced manufacturing technology in our country, from the view of "broad manufacturing", the editors introduce a lot of frontier technologies in manufacturing industry and information management technologies of manufacturing enterprises in this book. Some advanced manufacturing modes and their popularization and application in modern manufacturing enterprises are also introduced together.

Taking advanced manufacturing process as the core, advanced engineering design, advanced manufacturing process and manufacturing automation as the main part, advanced management technology and advanced manufacturing modes as the soft environment, the knowledge hierarchy of advanced manufacturing engineering is constructed in this book. This book has the features as follows:

(1) Integral knowledge hierarchy Hierarchical architecture composed of product design-manufacturing process-manufacturing automation-enterprise information management-advanced manufac-

turing mode is constructed from the view of broad manufacturing. The new and applicable manufacturing technologies are introduced into the book, paying attention to the overlapping and merging of different subjects.

(2) New compiling system and wide application　The book is written in both English and Chinese to keep the consistency of contents. It is suitable for the students both in bilingual class and in ordinary class.

(3) Comprehensive content, prominent "advanced"　The book introduces not only advanced manufacturing process, but also advanced design technology and automation technology having to do with advanced manufacturing. The advanced management technology and advanced manufacturing mode matching with advanced manufacturing are also introduced in this book. The content in Chinese collects the quality resources of domestic similar courses, the expression in English absorbs the language essence of the original textbook. Contents are kept in line with the international advanced level as far as possible.

(4) Consistency, accuracy and good readability　Writing bilingual textbook is the biggest challenge to domestic editors. Editors pay more attention to the consistency of English with Chinese in basic contents and the correctness of word meaning, instead of simply going in for strict translation. We try our best to use professional vocabularies and description manner from the same kind of foreign textbooks, widely used grammar and sentence patterns in the book so as to make the readers read and understand easily.

(5) Integrating theory with practice, stressing on the training of overall quality　Editors pay attention to the introduction of advanced manufacturing philosophy, methods and technologies, stress on the case teaching, insist on linking theory with practice so as to cultivate the ability to analyze and solve practical problems, and lay broad professional foundation for students to choose jobs after graduation.

(6) Complete supporting materials　The book is equipped with a set of courseware in English and a set of courseware in Chinese, and the solutions to primary exercises in every chapter. All teachers who have order this book for their students can get them for free. The electronic courseware is open type, so the teachers can add, delete, or recompose the contents in it, according to their own situation to satisfy various kinds of individualized teaching requirements.

The book is compiled jointly by Ren Xiaozhong (Huanghe Jiaotong University), Xue Yujun (Henan University of Science and Technology), Zhong Xiangqiang (Anhui Polytechnic University), Tian Xiaoguang and Zhang Dongming (Huanghe Jiaotong University), Zhang Zhiwen (Henan University of Science and Technology), and Su Yongsheng (Anhui Polytechnic University). The specific divisions of writing tasks are as follows: Ren Xiaozhong compiled Introduction, Chapter 1, Section 2.5 ~ 2.6 in Chapter 2, Section 3.4 and 3.6 ~ 3.7 in Chapter 3; Xue Yujun compiled Section 3.3 and 3.5 in Chapter 3, Section 6.6 ~ 6.8 in Chapter 6; Zhong Xiangqiang compiled Section 2.1 ~ 2.4 in Chapter 2; Su Yongsheng compiled Section 3.1 ~ 3.2 in Chapter 3; Zhang Dongming compiled Chapter 4; Zhang Zhiwen compiled Chapter 5; Tian Xiaoguang compiled Section 6.1 ~ 6.5 in Chapter 6. Ren Xiaozhong is in charge of overall compiling and editing the entire bilingual textbook.

The book is primarily reviewed by Prof. Ren Naifei (Jiangsu University) as reviser. He has provided thorough reviews and revisions to the book, and proposed many invaluable advises and suggestions. We extend our heartfelt thanks for his significant contributions.

The book has referred to some textbooks published at home and abroad as references. Hereon, we express a most cordially thank to the authors. At the same time, we also announce our sincere acknowledgement to all who have provided their help and kindness for the publication of the book.

The book has proudly acquired special financial support from Henan Engineering Technology Research Center on Intelligent Manufacturing Technology and Equipment of Huanghe Jiaotong University. We gratefully announce our sincere acknowledgement.

Due to various limitations, there may be some improper contents or even mistakes in the first edition. All authors of the book respectfully invite criticisms and corrections from all readers for further improvement of the book, so that the flaws and errors can be improved in future versions.

Editor

Contents
目　　录

Introduction

绪 论

0.1　Manufacturing Technology and Manufacturing System

0.1.1　Manufacturing and Manufacturing Technology

1. Meaning of Manufacturing

The word "manufacture" originated from two Latin roots "manu", meaning by hand, and "facere", meaning to make. This means that manufacturing was completed manually. With the development of society and the progress of manufacturing technology, the connotation of manufacturing shows a significant historical trend. Since the Industrial Revolution, machinery has played an increasingly important role. If you look up the dictionary, you may find that the definition of manufacturing is "making of articles by physical labor or machinery, especially on a large scale." With machine tools, humans can produce goods faster and better. Manufacturing means the whole procedure by which people, according to their purpose and applying their knowledge and skills, make original materials into valuable products, and put them into market by means of manual or available objective tools and facilities.

Meaning of manufacturing can be classified into narrow sense manufacturing and broad sense manufacturing. In narrow sense, manufacturing refers to the machining and assembly process of a product in workshop. In broad sense, manufacturing refers to a set of correlated operations and activities in the life cycle of product, which includes market analysis, product design, material selection, planning, quality control, production management, marketing of the products, after-sales service and even the disposal of waste. At present, broad manufacturing is widely accepted by more and more people.

The functions of manufacturing are realized through manufacturing process course, material flow course and information flow course. Manufacturing process course refers to the activities in which the shape, dimensions and performance of objects to be machined are directly changed. Material flow course refers to the activities, such as transporting, reserving and clamping of the objects to be machined, performed in manufacturing. Information flow course refers to the activities such as information acquisition, analysis processing and monitoring etc. in manufacturing.

2. Manufacturing Technology

Manufacturing technology is a generic term of all production technologies which are used in manufacturing industries to produce various necessities for national economic construction and people's life. It refers to a group of technologies which are used to change raw materials and other production elements economically and rationally into finished/semi-finished products which can be directly used with higher added value. These technologies include how to make use of certain knowledge and skills, how to control available materials and tools and how to adopt all kinds of effective strategies and methods.

0.1.2　Manufacturing System and Manufacturing Engineering

1. Manufacturing System

Manufacturing system refers to the input/output (I/O) system to convert manufacturing re-

0.1 制造技术与制造系统

0.1.1 制造与制造技术

1. 制造的含义

制造（manufacturing）一词来源于拉丁语词根 manu（手）和 facere（做），这说明制造是靠手工完成的。随着社会的发展和制造技术的进步，制造也在顺应历史潮流有着更深层次的内涵。自第一次工业革命以来，机器发挥着越来越重要的作用。字典中，制造的定义是"利用人力或机器大规模制作物品"。人类使用机床可以把商品做得既快又好。制造是指人们根据自己的意图，运用掌握的知识和技能，利用手工或一切可以利用的工具和设备把原材料制成有价值的产品并把这些产品投放市场的整个过程的总称。

制造的含义有广义和狭义之分。狭义制造是指生产车间的加工和装配过程。广义制造是指包括市场分析、产品设计、材料选择、工艺规划、质量控制、生产管理、营销、售后服务直至产品报废处理等在内的整个产品寿命周期的一系列相关操作和环节。目前，广义制造已为越来越多的人所接受。

制造的功能是通过制造工艺过程、物料流动过程和信息流动过程来实现的。制造工艺过程是指直接改变被制造对象的形状、尺寸、性能的行为活动。物料流动过程是指被制造对象在制造过程中运输、储存、装夹等活动。信息流动过程是指制造过程中的信息获取、分析处理、监控等活动。

2. 制造技术

制造技术是制造业为国民经济建设和人民生活生产各类必需物资所使用的一切生产技术的总称，是将原材料和其他生产要素经济合理地转化为可直接使用的具有较高附加值的成品、半成品和技术服务的技术群。这些技术包括运用一定的知识和技能，操纵可以利用的物质、工具，采取各种有效的策略、方法等。

0.1.2 制造系统与制造工程

1. 制造系统

制造系统是指由制造过程及其所涉及的硬件、软件和人员所组成的一个将制造资源转变为成品或半成品的输入/输出系统。其中，硬件包括厂房、生产设备、工具、刀具、计算机及网络等。软件包括制造理论、制造技术、管理方法、制造信息及其有关的软件系统等。制造资源主要是指物能资源，如原材料、毛坯、半成品、能源等。制造系统的功能结构如图0-1所示。

2. 制造工程

制造工程是一个以制造科学为基础、由制造模式和制造技术构成的、对制造资源和制造信息进行加工处理的有机整体。先进制造工程是传统制造工程与计算机技术、数控技术、信息技术、控制论及系统科学等学科相结合的产物。随着科学技术的进步，制造工程的概念有了新的含义，即除了设计和生产外，现代制造工程的功能还包括企业活动的其他方面，如产品的研究与开发、市场和销售服务等。制造工程随着国民经济的发展而在多学科的交叉渗透

source into finished or semi-finished products, which is composed of manufacturing process and hardware, software and personnel. Hardware includes factory building, production facility, tool, cutting tool, computer, network and so on. Software includes manufacturing theory, manufacturing technology, management method, manufacturing information and other related software system. Manufacturing resources refer primarily to material and energy resources, such as raw material, blank, semi-finished product, energy source and so on. The functional structure of manufacturing system is shown in Fig. 0-1.

2. Manufacturing Engineering

Based on manufacturing science, manufacturing engineering composed of manufacturing mode and manufacturing technology is an organized whole to process the manufacturing resource and manufacturing information. Advanced manufacturing engineering is the result formed by combining traditional manufacturing engineering with computer technology, NC technology, information technology, cybernetics and systematic science. With the progress of science and technology, a new meaning is brought to the manufacturing engineering. Besides design and production, the functions of modern manufacturing engineering include other activities, such as research and development of product, marketing and after-sale service and so on. With the development of national economy, manufacturing engineering is developing constantly in the interdisciplinary penetration, and it is used in the production practice mainly through manufacturing technology. Therefore, manufacturing engineering requires knowledge from other disciplines, such as electrical engineering, mechanical engineering, materials engineering, chemical engineering, and systems/information engineering.

0.2 Manufacturing Industry and Its Position in National Economy

0.2.1 Development of Manufacturing Industry

Manufacturing industry is the social production department whose task is to provide production materials for national economic departments and daily consumer goods for whole society. It involves various industries in national economy. There are more than 30 industries in manufacturing industries. As different manufacturing industries have different processing objects, there are big differences in manufacturing technology. This book involves mainly in the manufacturing issues in machine manufacturing fields.

The earliest human's manufacturing activities go back to the Stone Age. At that time, people made use of natural stones to make laboring tools which were used to hunt up natural resources for existence and survival. With the advent of the Bronze Age, and later the Iron Age, some primal manufacturing activities, such as spinning, smelting, forging, etc. , began to come forth in order to meet the needs of natural economy based on agriculture.

Industrial Revolution is the origin of modern civilization and the radical change of human production mode. Looking back to the history of human industry development, every innovation of science and technology is firstly embodied in the manufacturing industry and promotes enormously the

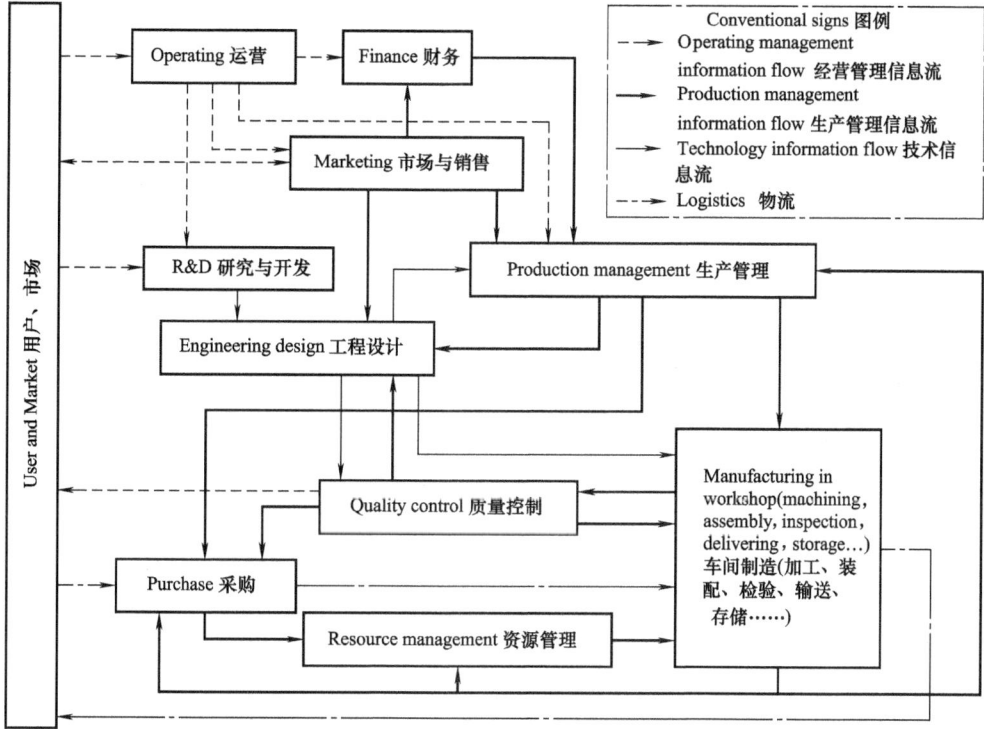

Fig. 0-1　Functional structure of manufacturing system 制造系统的功能结构

中不断地发展，且主要是通过制造技术应用于生产实践之中。因此，制造工程也需要其他学科的知识，如电气工程、机械工程、材料工程、化学工程及系统/信息工程等学科。

0.2　制造业及其在国民经济中的地位

0.2.1　制造业的发展历程

制造业是为国民经济各部门提供生产原料和为全社会提供日常消费品的社会生产部门。制造业涉及国民经济的各个行业，涵盖 30 多个行业。由于不同制造行业的加工对象不同，因此制造技术差异很大。本书主要涉及机械制造领域的制造问题。

最早的人类制造活动可追溯到石器时代。那时，人们为了生存，利用天然石料制作劳动工具以猎取自然资源。随着青铜时代和之后的铁器时代的到来，相继出现了纺织、冶炼、锻造等原始制造活动，以满足以农业为基础的自然经济发展的需要。

工业革命是现代文明的起点，是人类生产方式的根本性变革。回顾人类工业发展史，科学和技术的每一次革新，都首先体现在制造业上，都极大地促进了人类生产方式的改变和创新。自 18 世纪以来，制造业的发展经历了以下几个发展时期。

18 世纪末的第一次工业革命创造了机器工厂的"蒸汽时代"。蒸汽动力实现了制造活动的机械化，从此，人类进入工业 1.0 时代。

20 世纪初的第二次工业革命将人类带入大量生产的"电气时代"。电力的广泛运用促进

change and innovation of human production ways. Since 18th century, the development of manufacturing industry has gone through several development periods as follows.

The 1st Industrial Revolution at the end of 18th century created the "Steam Age" of machine factory. The steam power realizes the mechanization of manufacturing activities. Hence, human stepped into the time named as Industry 1.0.

The 2nd Industrial Revolution at the beginning of 20th century brought human into the "Electrical Age" with large volume production. The wide application of electrical power gave birth to production transfer line. Ford assembly line realized large volume production mode characterized by rigid automation technology. Hence, human stepped into the time named as Industry 2.0.

In the middle of 20th century, the invention of computer and the application of programmable logic controller (PLC) made the machine extend not only human physical strength, but also the human brain and opened a new era of digital control machine. The production mode with machine automation replaced gradually human work. This is the typical features of Industry 3.0 time at present.

In 21st century, Internet, new energy, new material and biotechnology are forming a huge industrial capacity and market at a great speed, and will upgrade the whole industrial production system to a new level and promote a new industry revolution. German Academy of Science and Technology (ACDTECH) and other organizations proposed a strategic plan called "4th generation industry—Industry 4.0". Its purpose is to ensure the competitive power of German manufacturing industry in the future and to lead the industry developing trend in the world. Essential difference between Industry 4.0 and former 3-time industry revolutions is that Industry 4.0 has a cyber physics system (CPS) with the ability of "man-machine" communication and "machine-machine" communication. It can be inferred that tomorrow's manufacturing industry will be a fenceless "intelligent plant".

0.2.2 Position of Manufacturing Industry in National Economy

Manufacturing industry is the main body of the national economy, the foundation of building country, the device of revitalizing state and the foundation of powerful nation. Looking at the countries of the world, all powerful countries in economy have their own developed manufacturing industry. The economy booms in many countries in which the manufacturing industry performs meritorious deeds never to be obliterated. The functions of manufacturing industry can be listed as follows:

(1) Manufacturing industry is the mainstay industry of national economy and the engine of economic growth In the developed countries, the manufacturing industry has created about 60% social wealth and 45% national economy income. In the United States, about 68% wealth comes from the manufacturing industry. In Japan, 49% GNP comes from the manufacturing industry. The production value of manufacturing industry in our country takes up about 45% of the total industrial output value.

(2) Manufacturing industry is the basic carrier to realize the industrialization of high technologies It is found by surveying industrialization history that numerous science and technology achievements are conceived in the development of manufacturing industry. At the same time, the manufacturing industry also provides scientific and technological means. A large number of high technologies

了生产流水线的出现。福特汽车装配生产线，实现了以刚性自动化技术为特征的大规模生产方式。由此人类进入了工业 2.0 时代。

20 世纪中期，计算机的发明、可编程控制器的应用使机器不仅延伸了人的体力，而且延伸了人的脑力，开创了数字控制机器的新时代。机械自动化生产方式逐步取代了人类作业，这正是目前工业 3.0 时代的典型特征。

进入 21 世纪，互联网、新能源、新材料和生物技术正在以极快的速度形成巨大的产业能力和市场，将使整个工业生产体系提升到一个新的水平，推动一场新的工业革命。德国技术科学院（ACDTECH）等机构联合提出"第四代工业——Industry 4.0"战略规划，旨在确保德国制造业的未来竞争力和引领世界工业发展潮流。工业 4.0 与前三次工业革命的本质区别在于其具有"人-机""机-机"相互通信能力的信息物理融合系统（Cyber Physics System，CPS）。由此推断，明天的制造业将是没有围墙的"智能工厂"。

0.2.2 制造业在国民经济中的地位

制造业是国民经济的主体，是立国之本、兴国之器、强国之基。纵观世界各国，任何一个经济强大的国家，无不具有发达的制造业。许多国家的经济腾飞，制造业功不可没。制造业的作用具体表现在以下几个方面：

（1）制造业是国民经济的支柱产业和经济增长的发动机　在发达国家中，制造业创造了约 60% 的社会财富、约 45% 的国民经济收入。其中美国 68% 的财富来源于制造业，日本 49% 的国民生产总值来源于制造业。我国制造业产值占工业总产值的比例为 45%。

（2）制造业是高技术产业化的基本载体　纵观工业化历史，众多的科技成果都孕育于制造业的发展之中。制造业也是科技手段的提供者，科学技术与制造业相伴成长。例如，20 世纪兴起的核技术、空间技术、信息技术、生物医学技术等高新技术无一不是通过制造业的发展而产生并转化为规模生产力的，其直接结果是诸如集成电路、计算机、移动通信设备、互联网、机器人、核电站、航天飞机等产品相继问世，并由此形成了制造业中的高新技术产业。

（3）制造业是吸纳劳动就业的重要行业　在工业国家中，约有 1/4 的人口从事各种形式的制造活动。在我国，制造业吸引了一半的城市就业人口，农村剩余劳动力转移也有近一半流入了制造业。

（4）制造业是国际贸易的主力军　近年来，国际贸易增长速度高于世界经济增长速度近两倍。由于初级产品的技术含量低，在国际市场的竞争力越来越弱，各国都千方百计扩大制成品的出口，以提高国际竞争力和附加价值。美、英、法、德、日等国家的制成品出口占全部出口比重的 90% 以上。20 世纪 90 年代以后，我国制造业的出口一直维持在 80% 以上，创造了接近 3/4 的外汇收入。

（5）制造业是国家安全的重要保障　现代战争已进入"高技术战争"的时代，武器装备的较量在很大意义上就是制造技术水平的较量。没有精良的装备，没有强大的装备制造业，一个国家不仅不会有军事和政治上的安全，而且经济和文化上的安全也会受到威胁。

0.3　先进制造技术的内容

先进制造技术是集机械、电子、信息、材料和管理等学科于一体的制造工程科学。本书

arisen in the 20th century, such as nuclear technology, space technology, information technology, biomedical technology etc. were all produced and converted into productive forces of scale. Its direct effectiveness was that many high-tech products, such as IC, computer, mobile communications equipment, Internet, robot, nuclear power station and space shuttle, etc. come out one after another, thereby, generating the high technology industries in manufacturing industry.

(3) Manufacturing industry is the key industry to recruit labor employment In industrialized countries, the people worked at manufacturing activities in various forms take up 1/4 of employers in the whole country. In China, one half of employed population of a city works at the manufacturing industry and about half of surplus labors in countryside transfers into manufacturing industry.

(4) Manufacturing industry is the main force in international trade In recent years, the growth rate of international trade is nearly two times more than that of the world economy. As the primary products have lower technology content, and its competitiveness in international market is getting weaker and weaker, countries of the world are enlarging the export of finished goods by all means to increase its competitiveness and added value in international market. The exports of finished goods in America, Britain, France, Germany, Japan, South Korea and Singapore have taken up above 90% of all exports. The exports in the manufacturing industry of China have kept over 80% and created about 3/4 foreign exchange earnings since 1990s.

(5) Manufacturing industry is an important assurance of national security Modern wars have come into the time of high-tech warfare. The competition in armaments is just the competition in manufacturing technology to a large extent. Without the excellent equipment and powerful equipment manufacturing industry, any country would have no safety not only in military and political affairs, but in economical and cultural activities.

0.3 Contents of Advanced Manufacturing Technology

Advanced manufacturing technology is a manufacturing science which integrates machine, electronics, information, material with management and other disciplines. From the view of "broad manufacturing", the frontier technologies in machine manufacturing industry and the information management technologies in modern manufacturing enterprises are expounded in the book. Main contents involve in advanced design technology, advanced manufacturing process, manufacturing automation and information management technologies in modern manufacturing enterprises. And several advanced manufacturing modes, such as computer integrated manufacturing, mass customization, concurrent engineering, lean production, agile manufacturing, network manufacturing, intelligent manufacturing and so on, and their applications in modern manufacturing enterprises are also introduced. It can be seen that advanced manufacturing technology is not a concrete technology, but a group of technologies integrated by multi-discipline high technologies. It should be emphasized that advanced manufacturing process is the core content of advanced manufacturing engineering. Without advanced manufacturing process, the CAX technologies, information technology and management technology etc. integrated with it will be like the water without source and like a tree without roots.

从"大制造"的角度，阐述了机械制造业的前沿技术以及现代制造企业的信息管理技术，主要内容涉及先进设计技术、先进制造工艺、制造自动化技术、现代制造企业的信息管理技术等方面，并介绍了计算机集成制造、大批量定制、并行工程、精益生产、敏捷制造、虚拟制造、网络制造、智能制造等先进制造模式在现代制造企业中的推广应用。由此可见，先进制造技术不是一项具体的技术，而是由多学科高新技术集成的技术群。应当强调的是，先进制造工艺是先进制造工程学的核心内容。离开了先进制造工艺，与之集成的计算机辅助技术、信息技术及管理技术等都将成为无源之水、无本之木。

0.4 本课程的特点和学习要求

0.4.1 本课程的特点

（1）关注度高 "先进制造技术"是高等院校机械工程、机械电子工程和工业工程等本科专业的专业选修（限选）课，也是很多院校硕士乃至博士研究生的学位课。

（2）内容丰富，综合性强 先进制造技术在传统制造工程的基础上融合了计算机技术、信息技术、数控技术及现代管理理念等，所涉及的内容非常广泛，学科跨度大，从产品的现代设计技术、超精加工技术、微纳加工技术、特种加工技术、快速原型制造技术等各种先进的主体技术，到绿色制造、柔性制造、计算机集成制造、大批量定制、精益生产、网络化制造和智能制造等先进制造理念，同时还涵盖了许多其他学科的知识，如管理科学、电子学、材料科学、信息科学等。

（3）理论基础性 先进制造技术涉及众多学科的理论知识，如超精密加工机理、高速加工理论，干切削理论，以及信息科学、材料科学、管理科学等，而机械类专业的本科生乃至研究生在学习本课程之前大都没有学过这些内容，或了解甚少。由于这门课程的教学学时有限，难以在规定时间内让学生掌握各种新技术的内涵与应用。

（4）技术先进性 先进制造技术的先进性主要体现在学科的集成性、工艺的动态性、设计制造的数字化，以及可持续发展的特性上。先进制造系统正朝着集成化、柔性化、数字化、网络化和智能化的方向发展。

（5）动态性强 由于各种先进制造技术不断吸收各种最新技术成果，课程内容还在不断地更新和充实。由于制造工艺的复杂性，课程在理论上和体系上正在不断完善和提高。

0.4.2 本课程的学习要求与学习方法

"先进制造技术"课程对当前高等工科院校的课程教学内容、人才培养体系以及学科间的交叉渗透提出了新的挑战。"先进制造技术"涉及众多先进的技术，这些技术的基本原理各不相同，许多技术原理深奥，理解和掌握都比较困难，往往需要单独开设一门课程才能讲清楚。这就需要构建先进制造工程的课程教学体系。

鉴于本课程的内容多，学时少，要求学生要学好铸造、锻压、焊接、热处理、机械加工等基础工艺，因为这些基础工艺经过优化而形成的优质、高效、低耗、清洁基础制造技术是先进制造技术的核心部分。要掌握数控技术及装备知识，同时要学习"CAD/CAM""现代设计方法""特种加工工艺""企业经营管理"等相关课程，从而培养学生广泛的收集与处

0. 4　Features of This Course and Requirements for Learners

0. 4. 1　Features of This Course

（1）High degree of concern　Advanced Manufacturing Technology is the professional elective（distributional electives）course for mechanical engineering, mechanical and electronic engineering, industrial engineering and other undergraduate specialties. It is also the postgraduate degree course or doctoral degree course in many universities.

（2）Rich content with high comprehensiveness　Advanced manufacturing technology integrated traditional manufacturing engineering with computer technology, information technology, NC technology and modern management ideas. It has a very wide content and large discipline span. It includes not only various advanced key technologies, such as modern design technology, ultra-precision machining technology, micro/nano fabrication technology, unconventional machining technology, rapid prototyping manufacturing（RPM）technology and so on, but also advanced manufacturing ideas, such as green manufacturing, flexible manufacturing, computer integrated manufacturing, mass customization, lean production, network manufacturing and intelligent manufacturing etc. It also contains many other subjects, such as management science, electronics, material science, information science etc.

（3）Strong theoretical basis　Advanced manufacturing technology involves theoretical knowledge of many subjects, such as ultra-precision machining mechanism, high-speed machining theory, dry cutting theory, as well as information science, material science, management science, etc. But the undergraduates and even graduates majored in mechanical engineering didn't learn these contents, or knew little before they learn this course. Because the course has a limit teaching hours, it is difficult for the students to master the connotations and applications of various new technologies within the allotted time.

（4）Progressiveness of technology　The progressiveness of advanced manufacturing technology embodied mainly in the integration of disciplines, dynamic process, digital design and manufacture, as well as sustainable development. Advanced manufacturing system is developing toward integration, flexibility, digitization, networking and intelligentializing.

（5）Dynamic characteristic　As various advanced technologies absorb constantly various up-to-date technological achievements, the contents of this course are being renewed and enriched constantly. As the complexity of manufacturing process, the contents of this course are being perfected and improved in theory and in system.

0. 4. 2　Requirements and Methods for Learning the Course

Advanced Manufacturing Technology puts forward new challenges to current teaching contents, cultivation system of talents and the cross penetration between subjects. Advanced Manufacturing Technology involves many advanced technologies which have different basic principles. As many

理信息的能力、获取新知识的能力、分析与解决问题的能力、组织管理能力、综合协同能力、表达沟通能力和社会活动能力等，尤其是不断增强的创新能力和工程实践能力。

要达到课程的培养目标，课堂教学固然是一个重要的学习途径，但要在有限的学时内掌握众多先进制造技术的基本原理、关键技术等内容是不现实的。自主学习是与传统的听课学习相对应的一种现代学习方式。自主学习的途径很多，如浏览课程网站或链接国内外有关先进制造技术的主要网站，参加有关先进制造的会议，聆听先进制造方面专家的讲座等。通过自主学习，学生可以按照自己的实际需要选择学习内容，突破时间和空间的限制，培养自身获取制造技术最新发展动态的能力。

technologies have a profound philosophy, they are difficult to understand and master. Often it is required to open a course separately in order to speak clearly. This asks for constructing the curriculum teaching system based on advanced manufacturing engineering.

Seeing that the course has many contents and fewer lesson hours, students are required to learn such basic processes as casting, forging, welding, heat treatment and machining well, because these basic manufacturing technologies with high quality, high efficiency, low consume and cleanness formed by optimizing the preceding basic processes are the core of advanced manufacturing technology. Students should master the knowledge of NC technology and equipment, at the same time, learn CAD/CAM, Modern Design Method, Unconventional Machining Technology, Enterprise Management and other related courses so that they have the abilities to collect and process information, obtain new knowledge, analyze and solve problems, and the abilities in organization and management, comprehensive coordination, expression and communication and social activities, especially have their innovation ability and engineering practice ability enhanced constantly.

In order to achieve the training objectives of the course, of course, classroom teaching is one of the important approaches to learning. But it is impractical to master the principles of many advanced manufacturing technologies and key technologies in limited lesson hours. The autonomous learning is a modern learning mode contrast to the traditional attending lectures. There are many approaches in autonomous learning, such as browsing the course website, or linking the primary websites related to advanced manufacturing technology at home and abroad, attending the conferences related to advanced manufacturing technology, listening the lectures made by the experts in advanced manufacturing. Through autonomous learning, students can choose learning contents in light of their practical need, thus breaking through the temporal and spatial constraints and training their ability to gain the latest development of manufacturing technology.

Chapter 1　Introduction to Advanced Manufacturing Technology

第1章　先进制造技术概论

1.1　Advent of Advanced Manufacturing Technology

1.1.1　Background of Advanced Manufacturing Technology

Advanced manufacturing is relative to the traditional manufacturing. The concept "advanced manufacturing technology (AMT)" is put forward explicitly for the first time by America at the end of 1980s. It is well known that manufacturing industry in the United States has got unprecedented development in the World War Ⅱ and a later period and become the overlord of the manufacturing industry in the world at that time. Later, the international environment was changed, and the United States began to emphasize basic research, health and national defense construction and ignored the development of manufacturing industry. In 1970s, a group of American scholars preached constantly that the U. S. had stepped into "post industrialized society", and thought that the manufacturing industry is "sunset industry", advocated that the economic center should turn from manufacturing industry to high-tech industry and the third industry. Many scholars paid attention only to the theoretical results and didn't pay attention to the practical application, thus causing the situation called "America made invention, Japan made a fortune". Besides, the U. S. government had not supported industrial technology for a long time, thus leading to the decline of manufacturing industry and competitiveness of product in market. Many products of America, such as cars, domestic appliances, machine tools, semiconductors which had the advantage originally, were defeated in the market competition. Under the pincer attack from products of Japan with high quality and the cheap goods made in other Asian and Latin American countries, the survival space of American goods in the world market was constantly shrinked.

The circumstances mentioned above attracted wide attention from academia, business circles and politicians. They asked for the government one after another to organize, and coordinate and support the development of industrial technology so as to revive the American economy. Meanwhile, Toshiba event which created a big sensation happened. The U. S. government began to recognize the seriousness of the problem. A White House report said, the economic recession in the United States has threatened national security. " And then, The U. S. government spent millions of dollars to organize a large number of experts and scholars to carry out investigation and study. Academia, the business community and the government formed a consensus, that is, "after all, economic competition is the competition of manufacturing technology and manufacturing capability". In 1988, the U. S. government started investing in a large scale to study "manufacturing enterprise strategy in 21th century", formulated and implemented the advanced technology plan (ATP) and the manufacturing technology center (MTC) and achieved significant effects. It is obvious that the purpose to implement advanced manufacturing technology in the United States is to enhance the competitiveness of American manufacturing industry, retake the advantage of the American manufacturing industry and promote the economic development of the United States.

As soon as advanced manufacturing technology was put forward, European countries, Japan and other Asian newly industrialized countries started one after another to carry out the theoretical research and application of advanced manufacturing technologies, such as "intelligent manufacturing system (IMS)" of Japan, EREKA plan of European Community, "advanced technology country program" of South Korea and so on. Thus, the research and development of advanced manufacturing

1.1　先进制造技术的产生

1.1.1　先进制造技术的产生背景

先进制造是相对于传统制造而言的，而"先进制造技术"（Advanced Manufacturing Technology，AMT）这一概念则是由美国于20世纪80年代末首次明确提出的。众所周知，美国制造业在第二次世界大战及稍后一段时期得到空前发展，美国成为当时世界制造业的霸主。后来，国际环境发生了变化，美国开始强调基础研究、卫生健康和国防建设，而忽视了制造业的发展。20世纪70年代，一批美国学者不断鼓吹美国已进入"后工业化社会"，认为制造业为"夕阳工业"，主张经济中心由制造业转向高科技产业和第三产业。许多学者只重视理论成果，不重视实际应用，造成所谓"美国发明，日本发财"的局面。再加上美国政府长期以来对产业技术不予支持，使美国制造业产生衰退，产品的市场竞争力下降。许多原来美国占优势的汽车、家用电器、机床、半导体等产品在市场竞争中也纷纷败北。美国商品在来自日本的高质量产品，以及其他亚洲和拉丁美洲国家廉价商品的夹击下，市场生存空间不断萎缩。

上述情况引起美国学术界、企业界和政界人士的广泛关注，纷纷要求政府出面组织、协调和支持产业技术的发展，重振美国经济。期间，又发生了轰动一时的"东芝事件"。美国政府开始认识到问题的严重性。白宫一份报告称"美国经济衰退已威胁到国家安全"。于是，美国花费数百万美元，组织大量专家、学者进行调查研究，学术界、企业界和政府之间达成了共识，即"经济的竞争归根到底是制造技术和制造能力的竞争"。1988年，美国政府开始投资进行大规模"21世纪制造企业战略"研究，制订并实施了先进技术计划（ATP）和制造技术中心（MTC）计划，取得了显著的效果。可见，美国实施先进制造技术的目的就是增强美国制造业的竞争力，夺回美国制造工业的优势，促进其国家的经济发展。

先进制造技术一经提出，欧洲各国、日本以及亚洲新兴工业化国家也相继展开各自国家先进制造技术的理论和应用研究，如日本的智能制造系统（IMS）、欧共体的尤里卡（ERE-KA）计划、韩国的"高级先进技术国家计划"等，由此把先进制造技术的研究和发展推向高潮。

1.1.2　先进制造技术的内涵和特征

1. 先进制造技术的内涵

先进制造技术是那些不断吸取信息技术和现代管理技术的成就，并将其综合应用于包括产品设计、规划、生产、检验、管理、销售和产品回收利用等整个制造过程的所有制造技术的总称。其目的是实现高质、高效、低耗、清洁、柔性生产，同时提高生产的适应性和市场竞争能力。其要点在于：①目标是提高制造企业对市场的适应能力和竞争力；②强调信息技术、现代管理技术与制造技术的有机结合；③信息技术、现代管理技术在整个制造过程中的综合应用。

2. 先进制造技术的特征

与传统制造技术相比，先进制造技术的特点如下：

technology are pushed to the climax.

1. 1. 2　Connotation and Characteristics of Advanced Manufacturing Technology

1. Connotation of Advanced Manufacturing Technology

Advanced manufacturing technology is a general name of manufacturing technologies, which keep on adopting the achievements in information technology and modern management technology, and utilize them synthetically to the whole manufacturing processes including product design, planning, production, inspection, management, marketing and recycling. Its purpose is to realize the production with high quality, high productivity, low cost, cleanness, and flexibility, and to enhance the adaptability and competitively.

2. Characteristics of Advanced Manufacturing Technology

Compared with the traditional manufacturing technology, advanced manufacturing technology has the characteristics as follows:

(1) Practicality　Advanced manufacturing technology does not pursue the advanced and new technology, but pay attention to the best practical effect. It centers on raising benefit and aims at increasing the competitivity of enterprise.

(2) Extensively　Advanced manufacturing technology is not limited in the machining technology, but covers the whole process from product design, planning to production, marketing and even recycling.

(3) Dynamic characteristic　Advanced manufacturing technology has not a fixed mode, but is developing dynamically. It has different characteristics, emphases, targets and contents in different time and different country or area.

(4) Integration　Advanced manufacturing technology stresses on inter-permeating and inter-merging among various disciplines or majors. The border line is generally dimmed and disappeared. Technologies involved tend to be systematic and integrated.

(5) Systematicness　Advanced manufacturing technology has become a systematic engineering which can control material-flow, energy-flow, and information-flow in manufacturing processes.

(6) Emphasis on the production with high quality, high efficiency, low cost, cleanness and flexibility.

1. 2　Architecture and Classification of Advanced Manufacturing Technology

1. 2. 1　Architecture of Advanced Manufacturing Technology

Because the development of manufacturing technology in different countries is unbalance, advanced manufacturing technology has different contents and hierarchies in different countries and different development stages. A kind of architecture of advanced manufacturing technology is shown in Fig. 1-1. Thereinto, main technology group is the core of advanced manufacturing technology; supporting technology group is the basic technology used for supporting design and manufacturing process to make progress; the infrastructure of manufacturing technology is a series of measures which make advanced manufacturing technology adapt specific enterprise application environment, give full play to its function and obtain the best benefit, and it is also the mechanism and the soil for

（1）实用性　先进制造技术不是一味追求先进或新颖，而是注重技术的应用效果，注重提高效益和企业的市场竞争力。

（2）广泛性　先进制造技术并不仅仅局限于加工技术，而是包括从产品设计、规划，到生产、销售甚至循环再利用的整个过程。

（3）动态性　先进制造技术没有固定模式，而是动态发展的技术。在不同时期、不同国家或地区，先进制造技术的特点、重心、目标和内容有所不同。

（4）集成性　先进制造技术强调各学科、各专业的相互渗透和融合。学科间的界限通常很模糊，技术趋于系统化和集成化。

（5）系统性　先进制造技术是一项系统工程，可控制制造过程中的物料、能量及信息流。

（6）其他　强调生产的优质、高效、低耗、清洁和灵活性等。

1.2　先进制造技术的体系结构和分类

1.2.1　先进制造技术的体系结构

由于世界各国制造技术发展的不均衡，在不同的国家、不同的发展阶段，先进制造技术具有不同的内容和体系。图1-1所示为一种先进制造技术的体系结构。其中，主体技术群是先进制造技术的核心；支撑技术群是指支持设计和制造工艺两方面取得进步的基础性技术；

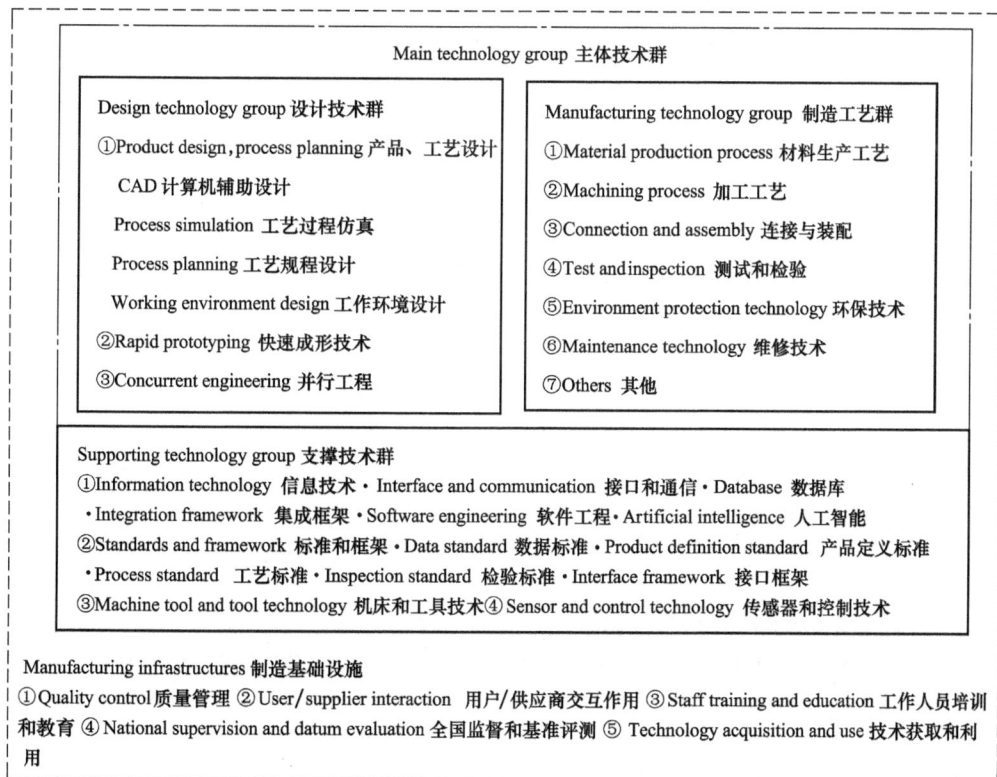

Main technology group 主体技术群

Design technology group 设计技术群
①Product design, process planning 产品、工艺设计
　CAD 计算机辅助设计
　Process simulation 工艺过程仿真
　Process planning 工艺规程设计
　Working environment design 工作环境设计
②Rapid prototyping 快速成形技术
③Concurrent engineering 并行工程

Manufacturing technology group 制造工艺群
①Material production process 材料生产工艺
②Machining process 加工工艺
③Connection and assembly 连接与装配
④Test and inspection 测试和检验
⑤Environment protection technology 环保技术
⑥Maintenance technology 维修技术
⑦Others 其他

Supporting technology group 支撑技术群
①Information technology 信息技术 · Interface and communication 接口和通信 · Database 数据库
· Integration framework 集成框架 · Software engineering 软件工程 · Artificial intelligence 人工智能
②Standards and framework 标准和框架 · Data standard 数据标准 · Product definition standard 产品定义标准
· Process standard 工艺标准 · Inspection standard 检验标准 · Interface framework 接口框架
③Machine tool and tool technology 机床和工具技术 ④Sensor and control technology 传感器和控制技术

Manufacturing infrastructures 制造基础设施
①Quality control 质量管理 ②User/supplier interaction 用户/供应商交互作用 ③Staff training and education 工作人员培训和教育 ④National supervision and datum evaluation 全国监督和基准评测 ⑤Technology acquisition and use 技术获取和利用

Fig. 1-1　Architecture of advanced manufacturing technology 先进制造技术的体系结构

the growth of advanced manufacturing technology. This hierarchy describes the components of advanced manufacturing technology and the role of each component in the manufacturing process mainly from the macro point of view rather than connotation of technical subjects.

(1) Main technology group It consists of product design technology and manufacturing technology.

1) Design technology group consists mainly of the designs of product, process and factory (such as CAD, CAE, design for machining and assembly, modular design, process simulation, CAPP, working environment design, design for environment protection etc.); rapid prototyping & manufacturing; concurrent engineering and other technologies.

2) Manufacturing technology group means the processes and equipment used for producing material products. Advanced manufacturing technology group consists mainly of the material production process, machining processes, connection and assembly, test and inspection; energy saving and clean production technology; maintenance technology and other technologies. Thereinto, the material production process includes smelting, rolling, die casting, sintering etc. Machining processes include cutting and grinding, unconventional machining, casting, forging, pressing, molding, heat treatment, surface coating and modification, precision and ultra-precision machining, lithography / deposition, composite processing etc. Connection and assembly includes welding, riveting, splicing, assembling, electronic packaging etc.

(2) Supporting technology group Supporting technology means the basis technologies used for supporting the design and manufacturing procedure to gain practical effects. Supporting technology group consists of information technology, standards and frameworks, machine tool and tool technology, sensor and control technology etc. Thereinto, information technology includes interface and communication, Internet and database, integration framework, software engineering, artificial intelligence, expert system, neural net, decision support system, multimedia technology, virtual reality technology etc. Standards and frameworks include data standard, product definition standard, process standards, inspection standard, interface framework etc.

(3) Manufacturing infrastructures These are the application environment, measures and mechanism of manufacturing in the enterprise. It involves chiefly in organization form and scientific management of new enterprises, just-in-time information system, marketing and user / supplier interaction, recruiting, employment, training and education of staff, total quality control, national supervision and datum evaluation, technology acquisition and use etc.

1. 2. 2　Classification of Advanced Manufacturing Technologies

It can be seen from Fig. 1-1 that advanced manufacturing technologies involve in a wide range of technical fields, which is made up of multi-discipline high technologies. In order to facilitate the reader to learn and master the basic system and primary coverage of advanced manufacturing technology, according to its research objects, advanced manufacturing technologies are summarized in the following categories:

1. Advanced Design Technology

Advanced design technology is the one which applies advanced technology and scientific knowledge, makes the design plan and puts the plan into practice according to the functional requirements of the product. Its importance is to design product on the basis of science, to promote the product from low-level to senior, to perfect the product function and to improve the product quality steadily. Advanced design technologies include the following contents:

制造基础设施则是使先进制造技术适用于具体企业应用环境，充分发挥其功能，取得最佳效益的一系列措施，是先进制造技术生长的机制和土壤。这种体系主要不是从技术学科内涵的层面来描述先进制造技术，而是着重从宏观的角度描述先进制造技术的组成以及各组成部分在制造过程中的作用。

（1）主体技术群　主体技术群包括产品的设计技术群和制造工艺群。

1）设计技术群主要包括产品、工艺过程和工厂设计（如计算机辅助设计、计算机辅助工程分析、面向加工和装配的设计、模块化设计、工艺过程仿真、计算机辅助工艺过程设计、工作环境设计、面向环保的设计等）；快速原型制造；并行工程和其他技术。

2）制造工艺群是指用于物质产品生产的过程和设备。先进制造工艺群的主要内容包括材料生产工艺、加工工艺、连接与装配、测试与检验、节能与清洁化生产技术、维修技术和其他技术等。其中，材料生产工艺包括冶炼、轧制、压铸、烧结等。加工工艺包括切削与磨削加工、特种加工、铸造、锻造、压力加工、模塑成形、热处理、表面涂层与改性、精密与超精密加工、光刻/沉积、复合材料工艺等。连接与装配包括焊接、铆接、粘接、装配、电子封装等。

（2）支撑技术群　支撑技术是指支持设计和制造过程取得实效的基础技术。支撑技术群包括信息技术、标准和框架、机床和工具技术、传感器和控制技术等。其中，信息技术包括接口和通信、网络和数据库、集成框架、软件工程、人工智能、专家系统、神经网络、决策支持系统、多媒体技术、虚拟现实技术等。标准和框架包括数据标准、产品定义标准、工艺标准、检验标准、接口框架等。

（3）制造基础设施　这是制造技术在企业的应用环境、措施和机制，主要涉及新型企业组织形式与科学管理，准时信息系统，市场营销与用户、供应商交换作用，工作人员的招聘、使用、培训和教育，全面质量管理，全局监督与基准评测和技术获取和利用等。

1.2.2　先进制造技术的分类

从图 1-1 可以看出，先进制造技术涉及的技术领域广泛，它是由多学科高新技术集成的制造技术。为便于读者学习与掌握先进制造技术的基本体系和主要内容，根据先进制造技术的研究对象，本书将先进制造技术归纳为以下几个大类：

1. 先进设计技术

先进设计技术是根据产品功能要求，应用先进技术和科学知识，制定设计方案并使方案付诸实施的技术。其重要性在于使产品设计建立在科学的基础上，促使产品由低级向高级转化，促进产品功能不断完善，产品质量不断提高。先进设计技术包含如下内容：

（1）先进设计方法　先进设计方法包括模块化设计、系统化设计、价值工程、模糊设计、面向对象的设计、反求工程、并行设计、绿色设计、工业设计等。

（2）产品可信性设计　产品的可信性是产品的可用性、可靠性和维修保障性的综合。可信性设计包括可靠性设计、安全性设计、动态分析与设计、防断裂设计、防疲劳设计、耐环境设计、健壮设计、维修设计等。

（3）设计自动化技术　设计自动化技术是指用计算机辅助完成设计任务和过程的技术。它包括产品的造型设计、工艺设计、工程图生成、有限元分析、优化设计、虚拟设计、工程数据库等内容。

（1）Advanced design methods　Advanced design methods include modular design, systematic design, value engineering, fuzzy design, object oriented design, reverse engineering, concurrent design, green design, industrial design and so on.

（2）Product creditability design　Product creditability design synthesizes the usability, reliability and maintenance supportability. It includes reliability design, security design, dynamic analysis and design, anti-fracture design, anti-fatigue design, environment resistant design, robust design, design for maintenance and so on.

（3）Design automation technology　Design automation technology means using computer to assist in completing design tasks and procedure. It includes the product modeling design, process planning, engineering drawing generation, finite element analysis, optimal design, virtual design, engineering data base etc.

2. Advanced Manufacturing Process

Advanced manufacturing process is the core and foundation of advanced manufacturing technologies, and the method and process to change different raw materials and semi-finished products into qualified products. Advanced manufacturing process includes primarily the precision forming technology with high efficiency, cutting process with high precision, unconventional machining technology, surface modification technology and so on.

（1）High-efficient precision forming technology　It is a general designation of all producing semi-finished products without machining allowance on all or local surfaces. It includes precision clean casting process, high-efficient precision plastic forming process, high quality and high efficiency welding and cutting technology, clean heat treatment with high quality and high efficiency, rapid prototyping & manufacturing technology and so on.

（2）High efficiency and high quality machining technology　It includes the precision and ultra-precision machining, high-speed cutting and grinding, NC machining of complex surface, efficient machining with loose abrasive and so on.

（3）Modern unconventional machining technology　Unconventional machining refers to the processes in which nontraditional energy transfer mechanism and/or nontraditional media for energy transfer are involved, such as high-energy beam machining (electron beam machining, ion beam machining, laser beam machining), electrical machining (electrochemical machining and electro-discharge machining), ultrasonic machining, high-pressure water jet machining, multi-energy compound machining, micro / nano fabrication and so on.

3. Manufacturing Automation Technology

Manufacturing automation means that the mechanical and electrical equipment is used to replace or magnify the manpower, even replace and extend a part of human intelligence and to complete automatically various production activities including storage, conveyance, machining, assembly and inspection etc. Manufacturing automation technology involves in NC technology, industrial robot technology, flexible manufacturing technology, sensor technology, automatic measurement technology, signal processing and recognition technology etc. Its purpose is to lighten the labor intensity of the operator, improve production efficiency, reduce the number of in-process products, save energy consumption and reduce production cost.

4. Information Management Technology in Modern Manufacturing Enterprise

Enterprise informatization means that applying advanced management ideas and methods, the production, operation, design, manufacturing and management existed in enterprise are integrated

2. 先进制造工艺

先进制造工艺是先进制造技术的核心和基础，是使各种原材料、半成品成为产品的方法和过程。先进制造工艺包括高效精密成形技术、高精度切削加工工艺、特种加工以及表面改性技术等内容。

（1）高效精密成形技术　它是生产局部或全部无余量半成品工艺的统称，包括精密洁净铸造工艺、精确高效塑性成形工艺、优质高效焊接及切割技术、优质低耗洁净热处理技术、快速原型制造技术等。

（2）高效高精度切削加工工艺　这种工艺包括精密和超精密加工、高速切削和磨削、复杂型面的数控加工、游离磨粒的高效加工等。

（3）现代特种加工工艺　特种加工是指那些采用非常规的能量转换机理或非常规介质进行能量转换的加工工艺，如高能束加工（电子束、离子束、激光束）、电加工（电解和电火花）、超声波加工、高压水射流加工、多种能源的复合加工、微纳加工等。

3. 制造自动化技术

制造自动化是指用机电装备取代或放大人的体力，甚至取代和延伸人的部分智力，自动完成特定作业，包括物料的存储、运输、加工、装配和检验等各个生产环节。制造自动化技术涉及数控技术、工业机器人技术、柔性制造技术、传感技术、自动检测技术、信号处理和识别技术等内容。其目的在于减轻操作者的劳动强度，提高生产效率，减少在制成品数量，节省能源消耗及降低生产成本。

4. 制造企业的信息管理技术

企业信息化就是运用先进的管理思想和方法，以计算机和网络技术为手段，整合企业现有的生产、经营、设计、制造和管理，为企业的决策提供及时、准确和有效的数据信息，以便对顾客要求做出快速反应。企业信息化的本质是加强企业的核心竞争力。企业信息化系统可分为企业资源计划系统、供应链管理系统、客户关系管理系统和产品生命周期管理系统。这4种信息系统的有机结合构成了企业信息化体系。

5. 先进制造模式

先进制造模式是将先进的信息技术与生产技术相结合的一种新思想和新哲理。其功能覆盖企业的生产预测、产品设计开发、加工装配、信息与资源管理直至产品营销和售后服务的各项生产活动，是制造业综合自动化的新模式。它包括计算机集成制造（CIM）、大批量定制（MC）、精益生产（LP）、并行工程（CE）、敏捷制造（AM）、智能制造（IM）等先进的生产组织管理模式和控制方法。

1.3　先进制造技术的发展趋势

当前，制造业正面临着新的挑战和机遇，制造技术正处于不断变化和完善之中。随着以信息技术为代表的高新技术的不断发展，为适应市场需求的多变性与多样化，先进制造技术正朝着数字化、集成化、精密化、极端化、柔性化、网络化、全球化、虚拟化、智能化、绿色化及管理创新的方向发展。

1. 数字化制造

数字化制造是先进制造技术发展的核心。它包含了以设计为中心的数字化制造、以控制

by means of computer and network technique to provide timely, accurate and effective data information so as to make a rapid response to the customer. The essence of enterprise informatization is to enhance the core competence of enterprise. Enterprise informatization systems can be classified into enterprise resource planning system, supply chain management system, customer relationship management system and product life cycle management system. The organic combination of these four kinds of information systems constitutes the enterprise informatization hierarchy.

5. Advanced Manufacturing Mode

Advanced manufacturing mode is a kind of new idea and new philosophy combined advanced information technology with production technology. Its functions include various production activities from the production forecast of enterprises, product design and development, machining and assembling, information and resource management up to product marketing and after-sale service. Advanced manufacturing mode is the new mode to realize integrated automation of manufacturing industry. Advanced manufacturing modes include many advanced production organization and management modes and control methods, such as computer integrated manufacturing (CIM), mass customization (MC), lean production (LP), concurrent engineering (CE), agile manufacturing (AM), intelligent manufacturing (IM) and so on.

1.3 Developing Tendency of Advanced Manufacturing Technology

Currently, manufacturing industry is encountering some new challenges and opportunities; manufacturing technology is constantly changing and improving. With the constant development of high technology represented by the information technology, advanced manufacturing engineering is developing in the direction of digitization, integration, precision, extramalization, flexible, networking, globalization, virtualization, intelligent, green and management innovation in order to meet the needs of the market variability and diversification.

1. Digitization Manufacturing

Digitization manufacturing is the core to develop advanced manufacturing technology. It comprises the digitization manufacturing centered on design, the digitization manufacturing centered on control, the digitization manufacturing centered on management. For manufacturing equipment, the control parameter is digital signal. For manufacturing enterprises, all kinds of information are transmitted in digital form through network in the enterprise so as to gather information rapidly according to the market conditions. In addition, product information, process information and resource information are analyzed, planed and recombined so as to complete the simulations of product design, machining process and production organization process, or the prototype manufacturing, and to realize the rapid reconstruction of production process and rapid response to the market. For the global manufacturing industry, the user releases information through network, different enterprises constitute dynamic alliance by network according to the needs to realize complementary advantages, resource sharing, rapidly collaborative design and to produce the corresponding products. In this way, under the digital manufacturing environment, a digital network is formed in a wide range of fields and even trans-regional or cross-border fields. In the process of research, design, manufacturing, sales and service, different enterprises interact with each other around the digital information given by the product, which becomes the most active factor in manufacturing activity.

2. Integration Manufacturing

Integration is reflected in three aspects: integration of technology, integration of management,

为中心的数字化制造和以管理为中心的数字化制造。对制造设备而言，其控制参数为数字信号。对制造企业而言，各种信息均以数字形式通过网络在企业内传递，以便根据市场情况迅速收集信息，并对产品信息、工艺信息与资源信息进行分析、规划与重组，实现对产品设计、加工过程与生产组织过程的仿真，或完成原型制造，从而实现生产过程的快速重组与对市场的快速响应。对全球制造业而言，用户借助网络发布信息，各类企业根据需求，通过网络形成动态联盟，实现优势互补，资源共享，迅速协同设计并制造出相应的产品。这样，在数字化制造环境下，在广泛领域乃至跨地区、跨国界形成一个数字化网；在研究、设计、制造、销售和服务的过程中，围绕产品所赋予的数字信息，成为驱动制造业活动最活跃的因素。

2. 集成化制造

集成化体现在 3 个方面：技术的集成、管理的集成、技术与管理的集成。先进制造技术就是制造技术、信息技术、管理科学与有关科学技术的集成。集成化的发展将使制造企业各部门之间以及制造活动各阶段之间的界限逐渐淡化，并最终向一体化的目标迈进。CAD/CAPP/CAM 系统的出现，使设计、制造不再是截然分开的两个阶段；FMC、FMS 的发展，使加工过程、检测过程、控制过程、物流过程融为一体；而计算机集成制造（CIM）的核心更是信息集成使一个个自动化孤岛有机地联系在一起，以发挥更大的效益；并行工程则强调产品及其相关过程设计的集成，这实际上是在一个更深层次上的集成。企业间的动态集成通过敏捷制造模式建立动态联盟，从而迅速开发出新产品，达到提升市场竞争力的目的。

加工技术的集成内容更多，如高能束加工、增材制造、生物加工制造等。

3. 精密化制造

精密化是指对产品、零件的加工精度要求越来越高，加工的极限精度正向纳米级、亚纳米级精度发展。精密加工与超精密加工技术是一个国家制造业水平的重要标志。它不仅能为其他高新技术产业提供精密装备，同时它本身也是高新技术的一个重要生长点。因而各工业发达国家均投入巨额资金发展该项技术。目前，超精密加工的尺寸误差已达到 $0.025\ \mu m$，表面粗糙度值达到 $Ra0.005\ \mu m$，所用机床定位精度达到 $0.01\ \mu m$，纳米级加工技术已接近实现。进一步的发展趋势是向更高精度、更高效率方向发展，向大型化、微型化方向发展，向加工检测一体化方向发展。超精密加工机理与应用的研究会更加广泛、更加深入。

4. 极端化制造

"极端"是指极端条件，或说要求很苛刻。如要求某种装置能在高温、高压、高湿、强磁场、强腐蚀等条件下正常工作，或要求某种材料或构件具有极高硬度、极高弹性等，或要求某种装置或零件在几何尺寸上极大、极小，甚至奇形怪状等。例如，原子存储器、芯片加工设备、微型飞机、微型卫星、微型机器人等都是"极小"的代表，而大飞机、航空母舰等属于"极大"产品。显然，这些产品都是科技前沿的产品。其中不得不提及"微机电系统（MEMS）"，它可以完成特种动作与实现特种功能，甚至可以沟通微观世界与宏观世界。

极端制造是指在极端条件下，制造极端尺度或极高功能的器件和系统。极端制造的内涵就是制造上的极端化、精细化。从表面上看，极端制造产品是机床尺度的变化，实质上则集中了众多的高新技术。极端制造是机床先进制造工程的发展趋势，它综合体现了机床设计与制造技术的创新能力。极端制造技术涉及现代设计、智能控制、超精密加工等多项高科技。因此，它需要发挥多学科优势进行联合攻关。

integration of technology and management. Advanced manufacturing technology is the integration of manufacturing technology, information technology, management science and other related scientific technologies. The development of integration will gradually fade the boundaries between departments in manufacturing enterprise and between stages in manufacturing activity and finally go towards the goal of integration. The advent of CAD/CAPP/CAM system make it practical that design and manufacturing is no longer the two completely separate phases. The development of FMC and FMS makes the machining, inspection, control and logistics and other processes as a whole. And the heart of computer integrated manufacturing (CIM) links individual islands of automation together through information integration to achieve greater benefits. Concurrent engineering stresses on the integration product design with other related process design, and this is an actual integration at the deeper level. The dynamic integration between enterprises forms dynamic alliance by means of agile manufacturing mode so as to develop new product rapidly and realize the goal to enhance the competitiveness of the market.

There are more contents in the integration of machining technology, such as high-energy beam machining, additive manufacturing, bio-manufacturing and so on.

3. Precision Manufacturing

Precision means that the requirement for the machining accuracy of product or workpiece is getting higher and higher, and the ultimate machining accuracy is developing towards nanometer grade and subnanometer grade. Precision machining and ultra-precision machining technologies are the important symbol of a national manufacturing industry level. These technologies cannot only provide precise equipment for other high-tech industries, they are also the important growing places of high technology. Therefore, every industrialized country invests huge amounts of money to develop the technology. At present, the dimensional error of ultra-precision machining has reached $0.025\mu m$, surface roughness Ra has reached $0.005\mu m$, positioning accuracy of machine tool used is $0.01\mu m$, nano scale machining technology has been closely achieved. Further development trend is in the direction of higher precision and higher efficiency, large size and miniaturization, integration of machining and inspection. The research on the mechanism and application of ultra-precision machining will be more extensive and more in depth.

4. Extramalization Manufacturing

"Extreme" means extreme condition, or very stringent requirement. For instance, a device is asked to work normally under the conditions of high temperature, high pressure, high humidity, strong magnetic field, strong corrosion and so on; or a material or component is asked to have an extremely high hardness, extremely high elasticity and so on; or a device is asked to be maximum, minimum in the geometric dimension, or even of odd shape. Atomic memory, chip processing equipment, micro aircraft, micro satellite, micro robot and so on all are the representative of the "minuteness". And large aircraft, aircraft carrier etc. belong to the "extreme large" product. Obviously, these products are in the frontiers of science and technology. That has to be mentioned is the "micro electro-mechanical systems (MEMS)". It can complete special actions and realize special functions. And even it can link up the micro world and the macro world.

Extreme manufacturing means to manufacture the component and system with extreme dimension or extremely high function under extreme conditions. The connotation of extreme manufacturing is the extramalization and the fine refinement. Superficially, extreme manufacturing product is the variation of machine tool dimension, but in fact, it gathers a large number of high and new technolo-

5. 柔性化制造

制造柔性化是制造企业对市场需求多样化的快速响应能力，即制造系统能够根据顾客的需求快速生产多样化新产品的能力。制造自动化系统从刚性自动化发展到可编程自动化，再发展到综合自动化，系统的柔性越来越大。模块化技术是提高制造自动化系统柔性的重要策略和方法。硬件和软件的模块化设计，不仅可以有效地降低生产成本，而且可以大大提高自动化系统的柔性。模块化产品设计可以有效改善设计工作的柔性，从而显著缩短新产品研制与开发周期；模块化制造系统可以极大提高制造系统的柔性，并可根据需要迅速实现制造系统的重组。并行工程（Concurrent Engineering，CE）和大规模定制生产（Mass Customization，MC）的出现，为制造系统柔性化提供了新的发展空间。

6. 网络化和全球化制造

制造网络化包括以下几个方面：①制造环境内部的网络化，实现制造过程的集成；②整个制造企业的网络化，实现企业中工程设计、制造过程、经营管理的网络化及其之间的集成；③企业与企业间的网络化，实现企业间的资源共享、组合与优化利用；④通过网络，实现异地制造；⑤网络化市场系统包括网络广告、网络销售、网络服务等。互联网（Internet）和内联网（Intranet）的出现，使企业之间的信息传输与信息集成以及异地制造成为可能。

计算机网络的问世和发展为制造全球化奠定了基础。随着经济全球化的出现，全球化制造的研究和应用迅速发展。制造全球化除了产品的跨国生产外，还包括产品设计与开发的国际化、制造产品与市场的分布与协调、市场营销的国际化、制造企业在全球范围的重组与整合、制造技术/信息和知识的全球共享、制造资源的跨国采购与利用等。制造全球化有利于生产要素在全球范围内的快速流动，最大规模地合理配置资源，追求最佳经济效益。

7. 虚拟化制造

虚拟制造（Virtual Manufacturing，VM）以系统建模技术和计算机仿真技术为基础，集现代制造工艺、计算机图形学、信息技术、并行工程、人工智能、多媒体技术等高新技术为一体，是一项由多学科知识形成的综合系统技术。虚拟制造将现实制造环境及制造过程，通过建立系统模型映射到计算机及相关技术所支持的虚拟环境中，在虚拟环境中模拟现实制造环境和产品制造全过程，从而对产品设计、制造过程及制造系统进行预测和评价。虚拟制造技术可以缩短产品设计与制造周期，提高产品设计成功率，降低产品开发成本，提高系统快速响应市场变化的能力。虚拟制造技术在制造自动化中将获得越来越多的应用。

8. 智能化制造

智能化是制造系统在柔性化和集成化基础上进一步的发展与延伸。智能制造技术是指在制造系统和制造过程的各个环节，通过计算机来实现人类专家的制造智能活动（分析、判断、推理、构思、决策等）的各种制造技术的总称。智能制造系统要求在整个制造过程中贯彻智力活动，使系统以柔性的方式集成起来，以在多品种、中小批量生产条件下实现"完善生产"。智能制造系统的特点是具有极强的适应性和友好性。具体表现如下：对于制造过程，要求实现柔性化和模块化；对于人，强调安全性和友好性；对于环境，要做到无污染、节能、资源回收和再利用；对于社会，则提倡合理的协作与竞争。当前的研究和应用进展集中在基于神经网络的智能检测、故障诊断、设计和优化，基于遗产算法的优化设计，基于框架的专家系统，基于 Agent 技术的智能制造系统等方面。由于在知识的表达与获取、人类学习与进化、自组织与创新机制等方面的研究还有待于深入，目前智能制造的应用水平距

gies. Extreme manufacturing is the developing trend of advanced manufacturing engineering; it embodies the innovation capability of machine tool design and manufacturing technology. Extreme manufacturing technology involves in many high technologies, such as modern design, intelligent control, ultra-precision machining and so on. Therefore, it requires to tackle key problems jointly by exerting multi-discipline advantages.

5. Flexible Manufacturing

Manufacturing flexibility is the rapid response ability of manufacturing enterprise to the market demand diversification, i. e. the ability of manufacturing system to manufacture rapidly new products with diversification according to the customer's demands. Manufacturing automation system has developed from rigid automation to programmable automation, and further to integrated automation, its flexibility is getting better and better. Modular technology is an important strategy and method to improve the flexibility of manufacturing automation system. The modular design of both hardware and software can not only reduce production cost effectively, but improve the flexibility of automation system significantly. Modular product design can improve the flexibility of design work effectively, and shorten the development cycle of new products significantly. The modular manufacturing system can greatly improve the flexibility of manufacturing system, and rapidly complete the reconstruction of manufacturing system according to requirement. The advent of concurrent engineering and mass customization provides a new development space for the flexibility of manufacturing system.

6. Networking and Globalization Manufacturing

Networking manufacturing includes the following aspects: ①the networking in the manufacturing environment can realize the integration of manufacturing process; ②the networking of entire manufacturing enterprise can realize the networking of engineering design, manufacturing process and operating management and the integration between them; ③the networking between enterprises can realize the resource sharing, combination and optical utilization; ④manufacturing in different places can be realized through networking; ⑤networked market system includes network advertisement, network sale, network service etc. The advent of Internet and Intranet makes the information transmission and information integration and the manufacturing in different places possible.

The advent and development of computer network lay the foundation for the manufacturing globalization. With the advent of economic globalization, the research and application of global manufacture develops rapidly. In addition to the transnational production of the products, manufacturing globalization includes internationalization of product design and development, distribution and coordination of manufacturing products and markets, internationalization of marketing, reorganization and integration of manufacturing enterprises in the world, global sharing of manufacturing technology/information and knowledge, transnational purchase and utilization of manufacturing resources etc. Manufacturing globalization is conducive to the rapid flow of production elements in the world and the rational allocation of resources in the largest scale so as to seek the best economic benefits.

7. Virtualization Manufacturing

Based on the system modeling technology and computer simulation technology, virtual manufacturing is an integrated system technology formed by multi-subject knowledge, which integrates modern manufacturing technology with computer graphics, information technology, concurrent engineering, artificial intelligence, multimedia technology and other high technologies. By the system modeling, virtual manufacturing puts the real manufacturing environment and manufacturing process into the virtual environment supported by the computer and related technologies, and simulates the real

人们的期望还相差甚远。

9. 绿色化制造

绿色制造（Green Manufacturing，GM），又称环境意识设计与制造（Environmentally Conscious Manufacturing，ECM），是指在保证产品的功能、质量、成本的前提下，综合考虑环境影响和资源效率的一种现代制造模式。其目标是使产品从设计、制造、包装、运输、使用到报废的整个寿命周期内对环境的负面影响最小，资源综合利用率最高。对于制造过程而言，绿色制造要求"绿色意识"渗透到从原材料投入到成品的全过程，包括节约原材料和能源，替代有毒原材料等，将一切排放物的数量与毒性削减在离开生产过程之前。对于产品而言，绿色制造覆盖构成产品整个生命周期的各个阶段，即从原材料提取到产品的最终处置，包括产品的设计、生产、包装、运输、消费及报废等，减少对人类和环境的不利影响。

当前，环境问题已成为世界各国关注的热点。各国政府部门和工业公司正在制定更加严格的法律法规以促进环保产品和工艺。不少国家的政府已经确立了正式的生态标志计划，旨在将环保产品告知客户。所有这些法律法规旨在使产品对环境的影响最小化。绿色制造的实施将带来 21 世纪制造技术的一系列重要变革。

10. 企业管理技术现代化

全世界都在经历着一个从前福特主义向后福特主义转化的过程，即生产由大规模批量生产转向大规模定制；由大企业垂直型的管理组织形式转向在生产过程中通过网络与其他企业相互协调的水平型组织形式；由死板封闭的刚性生产转向寻求其他企业创新合作的柔性生产；由寡头垄断型的市场结构向竞争合作型的市场结构转变。

以前的那种"金字塔式"的管理结构层级多，决策慢。处于这种管理结构中的员工没有横向合作意识，缺少交流和分享。而在这个互联网时代，企业的组织架构应该是扁平精简，决策能够迅速实行，员工之间有更多交流与创新合作。通过互联网，工业企业生产分工更加专业和明确，协同制造成为重要的生产组织方式，只有运营总部而没有生产车间的网络企业或虚拟企业开始出现。网络众包平台改变了企业的发包模式，发包和承包企业呈现网络虚拟化。电子商务的发展把企业营销渠道搬到了网上，拓宽了产品销售渠道，拓展了销售市场，降低了营销成本。供应链集成创新应用，使每个企业都演化成信息物理系统的一个端点。不同企业的原材料供应、机器运行、产品生产都由网络化系统统一调度和分派，产业链上下游协作日益网络化、实时化。

复习题与习题

1-1 简述先进制造技术提出的背景。

1-2 何谓先进制造技术？它有哪些特征？

1-3 先进制造技术的主体技术群包括哪些内容？

1-4 先进制造技术的支撑技术群包括哪些内容？

1-5 进制造技术的制造基础设施主要涉及哪些内容？

1-6 计技术群主要包括哪些内容？

1-7 先进制造工艺技术群主要包括哪些内容？

1-8 从广义制造的角度简述先进制造技术的组成。

1-9 简述先进制造技术的发展趋势。

manufacturing environment and the whole product manufacturing procedure in the virtual environment, thus forecasting and evaluating the product design, manufacturing process and manufacturing system. Virtual manufacturing technology is conducive to shortening product design and manufacturing cycle, improving the success rate of product design, reducing the product development cost and enhancing the ability of the system to respond quickly to market changes. Virtual manufacturing technology will get more and more applications in manufacturing automation.

8. Intelligentization Manufacturing

Intelligentization is the further development and extension of manufacturing system on the basis of flexibility and integration. Intelligent manufacturing technology is the generic name of various manufacturing techniques which are used through computer to realize the intelligent activities (analysis, judgment, reasoning, conception, decision making, etc.) of human experts in each link of manufacturing system and manufacturing process. Intelligent manufacturing systems require to implement the intellectual activity throughout the manufacturing process and to integrate the system in a flexible manner so as to realize "perfect production" under the condition of medium and small batch production conditions with multi-varieties. The characteristic of intelligent manufacturing system is that it has extremely strong adaptability and friendliness. Specific performances are as follows: for manufacturing process, it asks for flexibility and modularity; for human, it stresses on safety and friendliness; for environment, it asks to do without pollution, to save energy, to recover and reuse the resource; for society, it promotes reasonable cooperation and competition. Current research and application progress concentrates upon the intelligent detection, fault diagnosis, design and optimization based on neural network; the optimal design based on inheritance algorithm; the expert system based on framework; the intelligent manufacturing system based on agent technology and other aspects. Because these researches on the expression and acquisition of knowledge, the human learning and evolution and the self organization and innovation mechanism etc. have to be done further, the application level of intelligent manufacturing is far from people's expectations at present.

9. Green Manufacturing

Green manufacturing, also named as environmentally conscious manufacturing (ECM), is a kind of modern manufacturing mode considering comprehensively the environmental impact and resource efficiency under the premise of ensuring the product function, product quality and production cost. Its goal is to make the products have the least adverse influence on the environment, and to make the resource have the largest utilization in the life cycle from design, manufacturing, packaging, transportation, use to waste. For manufacturing process, green manufacturing requires the "green conscious" to penetrate into the whole process from raw materials to finished products, including saving raw materials and energy and replacing toxic raw materials, and reduces the number and toxicity of all emissions before leaving the production process. For the product, green manufacturing covers all stages of a product's life cycle from the extract of raw materials to the final disposal of the product, including product design, production, packaging, transportation, consumption and waste etc. so as to reduce the adverse effects on humans and the environment.

Currently, environmental issues have become the focus of attention payed by all countries in the world. Both the governments and industrial companies are making more strict regulations to promote environmental-friendly products and technology. Many governments have set up official eco-labeling schemes, intended to inform customers of environmental-friendly products. All of these regulations intend to minimize the environmental impact of products. The implementation of green manufacturing

1-10 企业信息化指的是什么？企业信息化系统由哪几部分组成？

1-11 数字化制造的内涵是什么？

1-12 集成化体现在哪几个方面？

1-13 何谓极端化制造？试举例说明。

1-14 制造网络化包括哪几个方面？

1-15 何谓虚拟制造？为什么要采用虚拟制造？

1-16 何谓智能制造技术？为什么要大力推行智能制造？

will bring a series of important changes in manufacturing technology in the 21th century.

10. Modernization of Enterprise Management Technology

All of the world is going through a process of transformation from the former Ford doctrine to the post Ford doctrine. That is, the production is transforming from mass production to mass customization; the vertical type of management organization in large enterprises is being changed into the horizontal type of organization coordinated with other enterprises through network in the production process; the stiff and closed inflexible production is being transformed into the flexible production seeking innovation and cooperation with other enterprises; the oligarch-monopoly type of market structure is being transformed into the competitive and cooperative market structure.

Previous "pyramid" management structure had many levels and slow decision. Employees in this management structure did not have a sense of horizontal cooperation and were lack of communication and sharing. In the Internet age, the organizational structure of the enterprise should be flat and simplified, decision can be implemented quickly, and there are more communication and innovation cooperation among employees. Through the Internet, the labor division of industrial production has become more professional and clear, collaborative manufacturing has become an important way of production organization, the internet enterprises or virtual enterprises which only have operational headquarters without manufacturing workshops begin to appear. Crowd/sourcing network platform has changed the mode of enterprises. Contract-issuing enterprise and contracting enterprise present in the network virtualization. The development of electronic commerce moves the enterprise marketing channel onto network, thus, broadening the product sales channels, expanding the sales market and reducing the marketing cost. The integration innovation and application of supply chain make each enterprise evolve into an endpoint of cyber-physical systems (CPS). The raw material supply, machine run and product production in different enterprises are all dispatched and allocated uniformly by network system. The networked and real-time cooperation between upstream and downstream of industry chain will be increasingly enhanced.

Review Questions and Problems

1-1　Briefly introduce the background of advanced manufacturing technology.

1-2　What is advanced manufacturing technology? What characteristics does it have?

1-3　What does the main technology group of advanced manufacturing technology include?

1-4　What does the support technology group of advanced manufacturing technology include?

1-5　What does the manufacturing infrastructure of advanced manufacturing technology mainly involve in?

1-6　What are the main contents of the design technology group?

1-7　What are the main contents of the advanced manufacturing process group?

1-8　State briefly the composition of advanced manufacturing technology from the view of generalized manufacturing.

1-9　State briefly the development trend of advanced manufacturing technology.

1-10　What does enterprise informatization mean? What is the enterprise information system composed of?

1-11　What is the connotation of digital manufacturing?

1-12　Which aspects does the integration reflect?

1-13　What is extreme manufacturing? Try to give several illustrations.

1-14　Which aspects does the manufacturing networking include?

1-15　What is virtual manufacturing? Why is the virtual manufacturing used?

1-16　What is intelligent manufacturing technology? Why should we vigorously carry out intelligent manufacturing?

Chapter 2   Advanced Design Technology

第2章　先进设计技术

2.1 Introduction to Advanced Design Technology

2.1.1 Connotation of Advanced Design Technology

Aiming at the demands of society and restricted by some design principles, product design is the process to create products by means of the design methods and tools. Product design is the key activity in the life cycle of a product and determines the congenital quality of a product. 75% of product quality accidents is caused by design error. The prevention in design is the most important and the most effective one. Therefore, the level of product design is directly related to the future and destiny of the enterprise.

Design technology refers to a variety of ways and means to solve specific design problems in the design process. Since the middle of twentieth century, with the development of science and technology and the advent of a variety of new materials, new processes and new technologies, the functions and structures of the products become more and more complex, and the market competition is increasingly intensive. Therefore, the traditional ways and means to develop products can hardly meet the needs of market and product design. With the development of computer science and application technology, a series of advanced design techniques have emerged in the field of engineering design.

Advanced design technology is the base of advanced manufacturing technology. It is a general term for the technology group that is used in the process of researching, improving, and creating product activities, taking meeting product quality, performance, time, comprehensive benefits of cost / price as the purpose, taking computer-aided design technology as the main body, taking a variety of scientific methods and techniques as a means. Moreover, its connotation is human, machine, environment compatible design concept, taking the market as driving force, taking knowledge acquisition as the center and taking product life cycle as the object.

2.1.2 Architecture of Advanced Design Technology

Advanced design technology has a lot of branch subjects. The basic system architecture and relationship with related subjects are as shown in Fig. 2-1.

1. Fundamental Technology

Fundamental technology refers to traditional design theory and method, including kinematics, statics, dynamics, material mechanics, thermodynamics, electromagnetics, and engineering mathematics. These fundamental technologies which are the source of modern design technology provide a solid theoretical basis for modern design technology.

2. Main Technology

Main technology refers to computer aided technology, such as computer aided x (x refers to product design, process design, NC programming, tooling design, etc.), optimization design, finite element analysis, simulation, virtual design, engineering database, etc. These technologies are becoming the backbone of advanced design technology groups due to their unique ability to deal with

2.1　先进设计技术概述

2.1.1　先进设计技术的内涵

产品设计是以社会需求为目标，在一定设计原则的约束下，利用设计方法和手段创造出产品结构的过程。产品设计是产品全寿命周期中的关键环节，它决定了产品的"先天质量"。产品的质量事故中有 75% 是设计失误造成的，设计中的预防是最重要、最有效的预防。因此，产品设计的水平直接关系着企业的前途和命运。

设计技术是指在设计过程中解决具体设计问题的各种方法和手段。自 20 世纪中期以来，随着科学技术的发展和各种新材料、新工艺、新技术的出现，产品的功能与结构日趋复杂，市场竞争日益激烈，传统的产品开发方法和手段已难以满足市场需求和产品设计的要求，计算机科学及应用技术的发展，促使工程设计领域涌现出了一系列先进的设计技术。

先进设计技术是先进制造技术的基础。它是以满足产品的质量、性能、时间、成本/价格综合效益最优为目的，以计算机辅助设计技术为主体，以多种科学方法及技术为手段，研究、改进、创造产品活动过程所用到的技术群体的总称。其内涵就是以市场为驱动，以知识获取为中心，以产品全寿命周期为对象，人、机、环境相容的设计理念。

2.1.2　先进设计技术的体系结构

先进设计技术的分支学科很多，其基本体系结构及其与相关学科的关系如图 2-1 所示。

1. 基础技术

基础技术是指传统的设计理论与方法，包括运动学、静力学、动力学、材料力学、热力学、电磁学、工程数学等。这些基础技术为现代设计技术提供了坚实的理论基础，是现代设计技术发展的源泉。

2. 主体技术

主体技术是指计算机辅助技术，如计算机辅助 X（X 是指产品设计、工艺设计、数控编程、工装设计等）、优化设计、有限元分析、模拟仿真、虚拟设计、工程数据库等，因其对数值计算和对信息与知识的独特处理能力，这些技术正成为先进设计技术群体的主干。

3. 支撑技术

支撑技术为设计信息的处理、加工、推理与验证提供多种理论、方法和手段的支撑，主要包括：①现代设计理论与方法，如模块化设计、价值工程、逆向工程、绿色设计、面向对象设计、工业设计、动态设计、疲劳设计、摩擦学设计、人机工程设计、可靠性设计等；②设计试验技术，如产品性能试验、可靠性试验、环保性能试验、数字仿真试验和虚拟试验等。

4. 应用技术

应用技术是针对实用目的解决各类具体产品设计问题的技术，如机床、汽车、工程机械、精密机械等设计的知识和技术。

2.1.3　先进设计技术的特征

先进设计技术具有以下特征：

numerical computation, information and knowledge.

3. Supporting Technology

Supporting technology provides a variety of theories, methods and means for design information processing, machining, reasoning and verification. It mainly includes the following technologies: ① Modern design theory and method, such as modular design, value engineering, reverse engineering, green design, object-oriented design, industrial design, dynamic design, fatigue design, tribological design, man-machine engineering design, and reliability design; ② Design and test technique, such as product performance test, reliability test, environmental performance test, digital simulation test and virtual test.

4. Application Technology

Application technology is the technology used for solving the specific product design issues for practical purposes, such as the knowledge and technology in the design of machine tool, automobile, engineering machinery, precision machinery and so on.

2.1.3 Characteristics of Advanced Design Technology

Advanced design technology has the following characteristics:

1) Advanced design technology is the inheritance, extension and expansion of traditional design theories and methods.

2) Advanced design technology is the intersection and synthesis of a variety of design techniques, theories and methods.

3) Advanced design technology realizes the computerization of design methods and the accuracy of design results.

4) Advanced manufacturing technology realizes the parallelization and intelligence of design process.

5) Advanced manufacturing technology realizes the feasibility design for the product lifecycle.

6) Advanced design technology is a comprehensive application of a variety of design experimental techniques.

2.2 Computer-Aided Design

2.2.1 Introduction

Computer-aided design (CAD) is used to assist or promote product design from conceptualization to documentation with the help of computer and graphics software. CAD technology can be used in all phases of the product design process. For example, computer-aided drafting is applied for the automatic drawing or the establishment of product documentation, while computer-aided design is used for increasing the productivity of product designers.

From the design process, computer-aided design process and traditional design methods and ideas are similar, but in terms of design cycle and quality, using CAD technology is more conven-

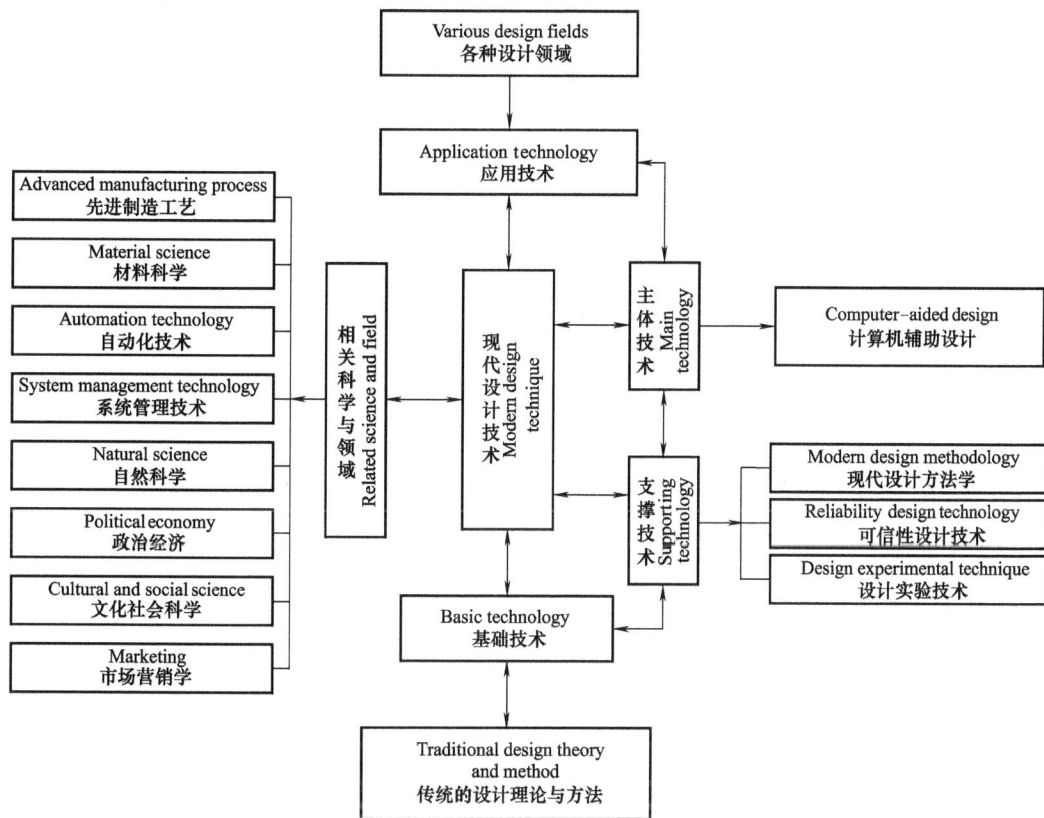

Fig. 2-1 Architecture of advanced design technology and its relationship with related subjects
先进设计技术的体系结构及其与相关学科的关系

1）先进设计技术是对传统设计理论与方法的继承、延伸与扩展。

2）先进设计技术是多种设计技术、理论与方法的交叉与综合。

3）先进设计技术实现了设计手段的计算机化与设计结果的精确化。

4）先进制造技术实现了设计过程的并行化、智能化。

5）先进制造技术实现了面向产品寿命周期全过程的可行性设计。

6）先进设计技术是对多种设计实验技术的综合应用。

2.2 计算机辅助设计

2.2.1 概述

CAD 是借助计算机和图形软件来帮助或促进包括从概念化到建立文档的产品设计。CAD 技术可被用于产品设计的各个阶段。例如，计算机辅助绘图用于自动绘图或产品文档建立过程，而计算机辅助设计用于提高产品设计者的生产力。

从设计过程看，计算机辅助设计过程与传统的设计方法和思路是相仿的，而在设计周期和质量方面，采用 CAD 技术要远比传统设计技术方便、灵活、有效得多。图 2-2 所示为计

ient, flexible, and effective than traditional design. Fig. 2-2 shows the process of computer-aided design. It can be seen that all phases of design are supported by powerful computer-aided tools, designer's job is simply to make creative thinking, conceptual design, and complete most work in the design by the computer.

2. 2. 2 Digital Product Model

The product refers to produce out the goods by use of modern processing methods and technology, for meeting certain functional requirements, which forms a body during different types of three-dimensional geometry bodies (aggregation).

The model is an expression of the object studied. The product model is an abstract simulation expressed for some expression forms of a body. The expression forms of these bodies include the design, manufacturing, management of products with shape information and non-shape information.

The product model under the computer environment is a computer numerical model which contains the information, such as design, manufacturing and management of products, also known as the digital model. The digital product model is divided into two-dimensional model and three-dimensional model.

The two-dimensional model mainly refers to the two-dimensional engineering drawing, as shown in Fig. 2-3. The advantages of the two-dimensional model are simple and practical. Two-dimensional model continues the traditional product expression form. The impact on the current production management system is the smallest. However, each view of the two-dimensional model is two-dimensional, people should understand the design shape of the product through the space imagination, so the two-dimensional model is not intuitive.

The three-dimensional model can be divided into wireframe model, surface model, solid model, feature model and parametric feature model, as shown in Fig. 2-4.

1. Wireframe Model

The wireframe model is a computer model that represents and builds a product with a finite set of vertices and edges that make up the product's three-dimensional entity.

The advantages of using the wireframe model to express the three-dimensional model are as follows:

1) It is easy to generate multi-view drawing, perspective and axonometric drawing.

2) When building the model, the operation is simple, the memory space is small, the operation response is fast.

3) The modeling of the three-dimensional model is consistent with people's design habits. It is intuitive and easy to learn.

The drawbacks of the wireframe model are as follows:

1) Because all crest lines are displayed, people can only understand the real shape of the object through the interpretation of the human brain. When the shape is complex, the crest line is too much, it will cause fuzzy understanding.

2) The contour of the surface body, such as the surface contour of the car and the aircraft,

算机辅助设计过程。可以看出，在设计的各个阶段都有功能强大的计算机辅助工具的支持，设计人员的工作仅仅是进行创造性思维，构思设计方案，并指挥计算机去完成设计中的大部分事务性工作。

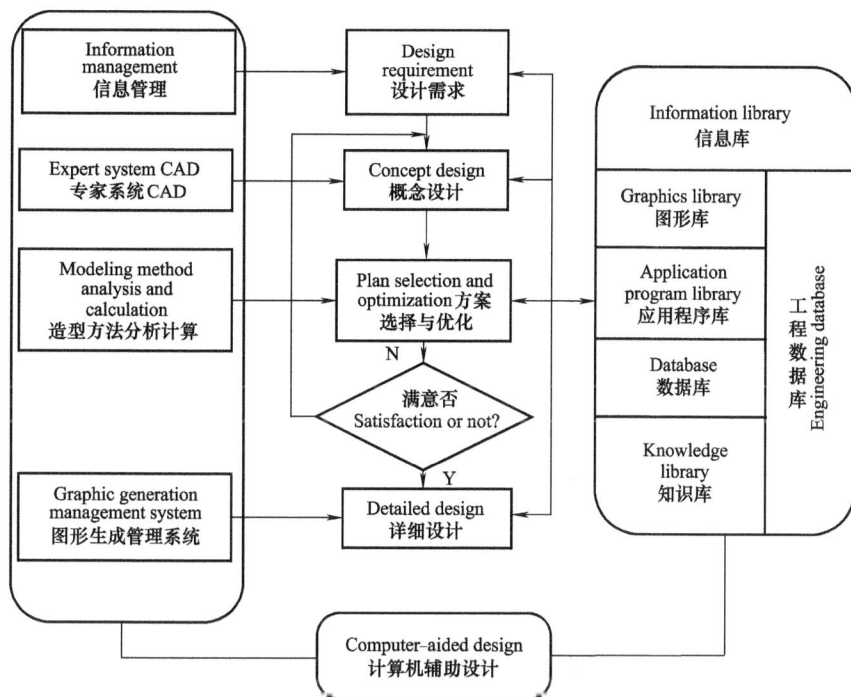

Fig. 2-2　Computer-aided design process 计算机辅助设计过程

2.2.2　数字化产品模型

产品是指用现代的加工方法和技术批量生产出来的物品，是满足一定的功能要求，由不同类型的三维几何形状构成的形体（集合体）。

模型是对所研究对象的一种表述。产品模型是对形体某种表现形式表示出来的抽象模拟。这些形体的表现形式包含了产品的设计、制造、管理等，既有形状信息，又有非形状的信息。

计算机环境下的产品模型是蕴含着产品的设计、制造、管理等信息的计算机数字模型，又称数字化模型。数字化产品模型分为二维模型和三维模型。

二维模型主要是指产品的二维工程图，如图 2-3 所示。二维模型的优点是简单、实用。二维模型延续了传统的产品表达形式，对当前的生产管理体制产生的冲击最小。但由于二维模型中的每一个视图都是二维的，人们要通过空间想象才能理解产品的设计形状，所以二维模型不直观。

三维模型可以分为线框模型、表面模型、实体模型、特征模型和参数化特征模型，如图 2-4 所示。

1. 线框模型

线框模型是指用构成产品三维实体的顶点、棱边的有限集合来表示和建立产品的计算机模型。

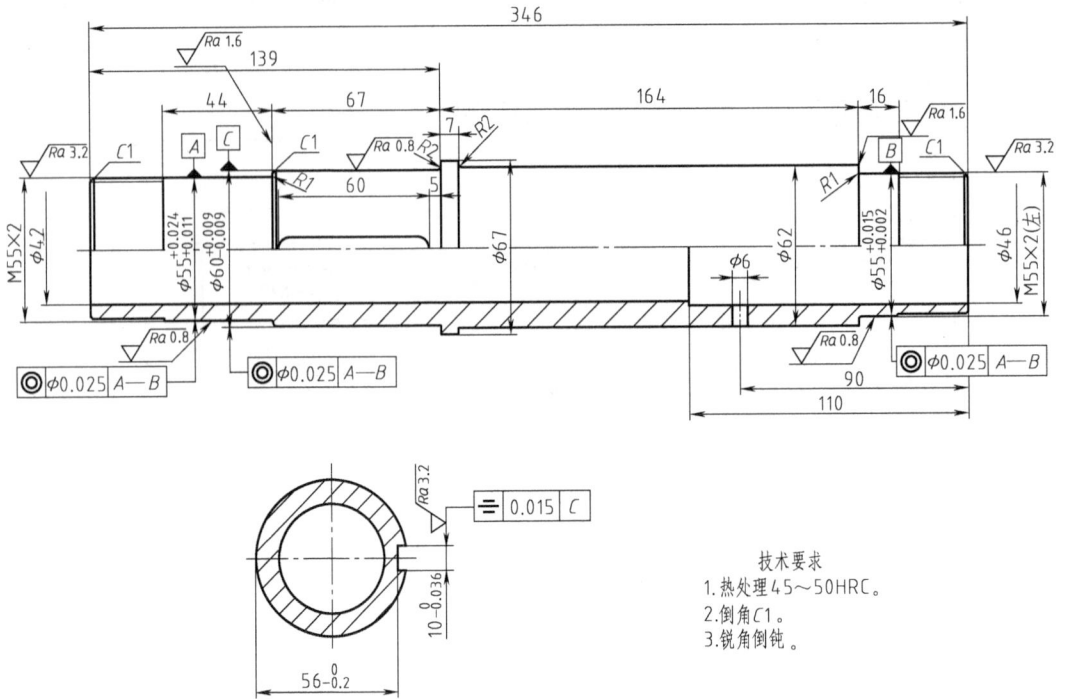

Fig. 2-3   2-D drawing of a shaft 轴的二维简图

can't be correctly expressed.

3) Be unable to achieve sectioning, to do the intersection calculation, to test the physical collision and interference, to generate CNC machining tool path, and to divide the finite element meshing automatically.

2. Surface Model

A surface model is a computer model that represents and builds a product with a finite set of vertices, edges, and surfaces which make up the 3D solid product.

The advantage of the surface model is that it can realize the blanking, coloring, surface area calculation, intersection of two surfaces, NC tool path generation, finite element meshing division and so on. In addition, the surface model is often used to construct complex surface objects, such as the mold, automobile and aircraft.

The disadvantage of the surface model is that it can only represent the surface of the object and its boundaries, is not the solid model. Therefore, the surface model can't be slit, can't calculate physical properties, can't check the collision and interference between objects.

3. Solid Model

A solid model is a computer model that uses the computer to store the basic solid primitives (such as cube, cylinder, sphere, etc.), and creates a complex shape through Boolean set operations.

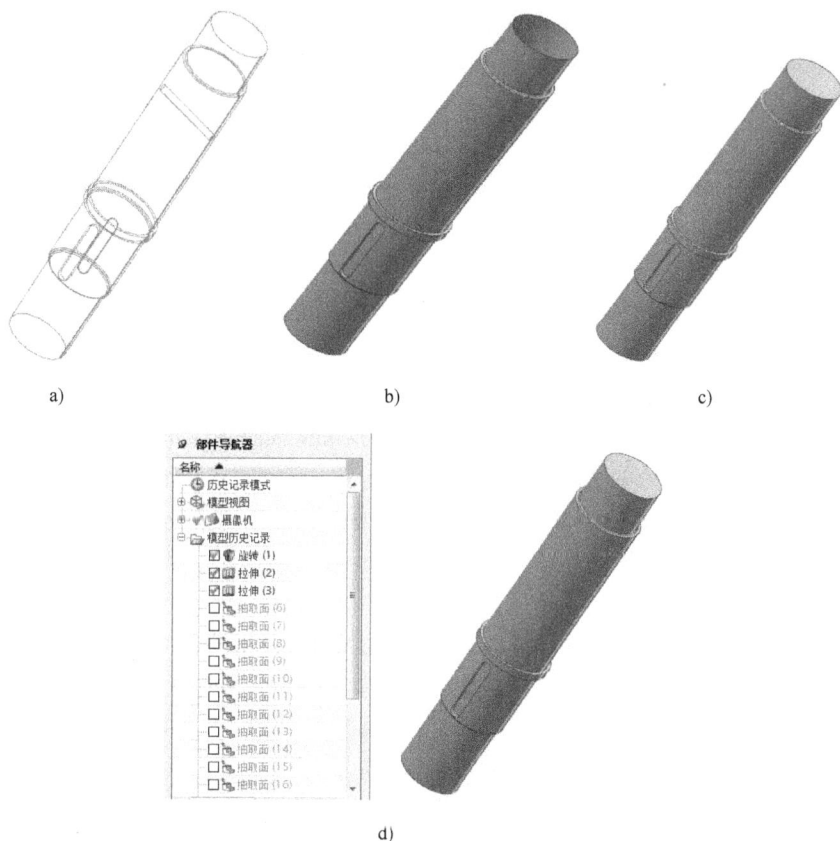

Fig. 2-4　Three-dimensional model　三维模型

a）Wireframe model 线框模型　b）Surface model 表面模型　c）Solid model 实体模型　d）Feature model 特征模型

应用线框模型表达三维模型的优点有：

1）方便生成多视图的工程图、透视图及轴测图。

2）构造模型时操作简便，内存空间小，操作响应快。

3）三维模型的建模符合人们的设计习惯，直观、简单易学。

线框模型的缺点有：

1）因为所有棱线全部都显示出来，物体的真实形状需通过人脑的解释才能理解，当形状复杂时，棱线过多，会引起模糊理解。

2）不能正确表达曲面形体的轮廓线，如汽车和飞机的表面轮廓线。

3）不能实现剖切，不能进行求交计算，无法检验实体的碰撞、干涉，无法生成数控加工刀具轨迹，不能自动划分有限元网格等。

2. 表面模型

表面模型是用构成产品三维实体的顶点、棱边、表面的有限集合来表示和建立产品的计算机模型。

表面模型的优点是能实现消隐、着色、表面积计算、两个曲面求交、数控刀具轨迹生成、有限元网格划分等功能。此外，表面模型常用于构造复杂的曲面物体，如模具、汽车、飞机等的表面。

The advantages of the solid model are as follows: blanking, slitting, finite element meshing division, NC tool path generation can be successfully achieved, and it has excellent visibility of the object because of coloring, lighting and texture processing technology. In addition, the solid model is also widely used outside CAD fields, such as computer art, advertisement, animation and so on. The solid model is the integration of product design and manufacturing automation and integration.

4. Feature Model

A feature model refers to the geometric model of the functional elements with engineering semantics obtained from highly generalized and abstracted engineering objects.

The feature model is used to establish the computer model of the product during defining and describing the part model by the feature and sets. The feature model is the product geometry model based on the solid model, which uses the feature with a certain design and processing function as the basic unit of the modeling. The feature model is the development direction of 3D model modeling technology.

5. Parametric Feature Model

A parametric feature model is a geometrical model with parametric features by using a dimension-driven or variable-driven design method to form the geometry model of the parametric feature, which applies the idea of parametric modeling on the feature model. Parametric feature model is also a development direction of 3D model modeling technology.

2. 2. 3　Examples of Digital Design

Digital design is the application of computer technology in product development, digital design requires specific and human-computer interactive computer environment for appropriate product design, namely three-dimensional modeling system. Currently, mainstream CAD system contains CATIA (Dassault Systems company in France), UG NX (Siemens company), Creo (PTC company in the United States), Inventor (Autodesk company in the United States), etc.

Basic steps for building part model in UG NX 9. 0 CAD system are as follows:

1) To build a new file and save it in the specified path.

2) To enter sketch design environment and draw sketch during selecting the appropriate reference plane.

3) To finish sketch drawing and generate a three-dimensional model with the appropriate tools.

4) To save the file.

Based on three-dimension design system, the part form can be understood as being made of a series of features with different semantics and relationships, or can be created one or more new features based on a basic character. The feature can be a positive space feature (a real part, such as boss, etc), it may be a negative space feature (subtracting or retracting part, such as a hole, slot, etc), or some basic features form an assembly body during union, intersection, difference set operations with several times. Time sequences of all features for forming a model are expressed by feature trees.

表面模型的缺点是只能表示物体的表面及其边界，还不是实体模型。因此，表面模型不能实行剖切，不能计算物性，不能检查物体间的碰撞和干涉。

3. 实体模型

实体模型是利用计算机内存储基本体素（如长方体、圆柱体、球体等），通过布尔集合运算生成复杂形体而建立产品的计算机模型。

实体模型的优点是：消隐、剖切、有限元网格划分、NC 刀具轨迹生成等都能顺利实现，而且由于着色、光照及纹理处理等技术的运用使物体有出色的可视性。此外，实体模型在 CAD 领域外也有广泛应用，如计算机艺术、广告、动画等。实体模型是产品设计与制造自动化及集成化的集成化。

4. 特征模型

特征模型是指从工程对象中高度概况和抽象后得到的具有工程语义的功能要素的几何模型。

特 2 征模型是通过特征及其集合来定义、描述零件模型从而建立产品的计算机模型。特征模型是以实体模型为基础的，用具有一定设计和加工功能的特征作为造型的基本单元建立起来的产品几何模型。特征模型是三维模型建模技术的发展方向。

5. 参数化特征模型

参数化特征模型是将参数化造型的思想应用于特征模型中，用尺寸驱动或变量驱动的设计方法对形体的特征进行参数化造型而形成的具有参数化特征的几何模型。参数化特征模型也是三维模型建模技术的发展方向。

2.2.3　数字化设计举例

数字化设计是计算机技术在产品开发中的应用，数字化设计要求具有适合产品设计工作的特定的人机交互式的计算机应用环境，即产品三维造型系统。目前，主流的产品三维造型系统主要有 CATIA（法国 Dassault Systems 公司）、UG NX（Siemens 公司）、Creo（美国 PTC公司）、Inventor（美国 Autodesk 公司）等。

在 UG NX 9.0 三维造型系统中建立零件模型的基本操作步骤如下：

1）建立新文件并且保存到指定的路径中。

2）选择适当的参考面进入草图设计环境，进行草图绘制。

3）完成草图绘制，用相应的工具生成三维模型。

4）保存文件。

在基于三维设计的体系中，零件的形状可理解为由具有不同工程语义并建立了关系的一系列特征构成，或是在一个基础特征上创建一个或多个新的特征。这些特征可以是正空间特征（指真实存在的部分，如凸台等），也可以是负空间特征（指减去或缩进去的部分，如孔、槽等），或是几个基础特征经过若干次并、交、差等集合运算形成的组合体。构成一个模型所有特征的时间顺序由特征树表达。

2.3　计算机辅助工程

2.3.1　CAE 概述

CAE 是指利用计算机技术对工程设计进行分析和评价，计算出产品的工作参数、性能

2. 3 Computer-Aided Engineering

2. 3. 1 Introduction to Computer-Aided Engineering (CAE)

CAE is the analysis and evaluation of the engineering design using computer-based techniques to calculate product operational, functional, and manufacturing parameters which are too complex for classical methods.

Study of the form, fit, and function characteristics of products is covered by the terms operational and functional in the definition, while an examination of the match between the design requirements and the production capability is included in the phrase "manufacturing parameters". The expression "too complex for classical methods" indicates that the process performed by CAE software could be performed manually, but the quantitative difficulty, number of the computations, or time required for analysis would be prohibitive.

CAE fits into the design process at the synthesis, analysis, and evaluation levels and is also consistent with the concurrent engineering (CE) principles described earlier. At the synthesis level, the primary CAE activity is focused on manufacturability using design for manufacturing and assembly principles. The output from the CAE operation at the analysis and evaluation levels is used by the CE team to determine the quality of the product design. Based on this CAE data, the product design is cycled through the top four steps in the design process until an optimum solution is generated.

Computer-aided engineering provides productivity tools to aid the production engineering area as well. Software to support group technology (GT), computer-aided process planning (CAPP), and computer-aided manufacturing (CAM), are grouped under the broad heading of CAE. As a result of these applications and the use of CAE in the design process. The level and variety of CAE software used, however, depends on the amount of concept and repetitive design practiced in the company and the type of manufacturing systems present. In the sections that follow, the type of CAE software commonly used through the design and in the production engineering area will be described.

2. 3. 2 Main Contents of CAE

CAE refers to the application of the computer in the modern production field, especially in the manufacturing industry; it mainly includes computer-aided design, computer-aided manufacturing and computer-integrated manufacturing system and so on.

1. Computer-Aided Design

By use of CAD in today's industrial manufacturing field, designers can draw various types of engineering drawings with the help of the computer and see a dynamic 3D stereogram on the display; during directly modifying the design drawing, the quality and efficiency of the drawing are greatly improved. In addition, designers can also use the methods of engineering analysis and simulation test, carry out computer logic simulation instead of the product test model (prototype) so that it can reduce product trial costs, shorten the product design cycle.

参数和制造参数，这些复杂参数若使用传统的人工方法将难以计算。

上述定义中的工作和性能两项参数覆盖了产品的形式、适应性及功能特征，而产品的设计指标与加工生产能力之间的匹配则通过制造参数来检验。使用人工方法将难以完成是指由CAE完成的处理过程虽然也可由人工方法完成，但因计算量太大而非常困难，或需要的时间太长而不能接受。

CAE在设计过程中适合应用于综合、分析和评价阶段，它也符合并行工程的原理。在综合阶段，CAE的主要工作是使用面向制造和装配的设计方法原理处理生产能力问题。在分析和评价阶段，CAE的工作结果被CE人员用来评定产品设计的质量。以CAE数据为基础，产品设计循环通过设计过程最顶层的四步，直至产生最优的设计方案。

CAE也是一种辅助生产工程领域的生产力工具。支持成组技术、计算机辅助工艺过程规划及计算机辅助制造的软件也囊括在广义的CAE概念之下。因为在整个设计过程中都要应用CAE和这些软件，但是，在哪个阶段使用及使用何种CAE软件则取决于企业实际的概念设计和重复设计的水平，还取决于所采用的制造系统类型。

2.3.2　CAE的主要内容

计算机辅助工程是指计算机在现代生产领域，特别是生产制造业中的应用，主要包括计算机辅助设计、计算机辅助制造和计算机集成制造系统等内容。

1. 计算机辅助设计

计算机辅助设计（CAD）是指在如今的工业制造领域，设计人员可以在计算机的帮助下绘制各种类型的工程图纸，并在显示器上看到动态的三维立体图后，直接修改设计图稿，极大地提高了绘图的质量和效率。此外，设计人员还可以通过工程分析和模拟测试等方法，利用计算机进行逻辑模拟，从而代替产品的测试模型（样机），降低产品试制成本，缩短产品设计周期。

目前，CAD技术已经广泛应用于机械、电子、航空、船舶、汽车、纺织、服装、化工以及建筑等行业，成为现代计算机应用中最为活跃的技术领域。

2. 计算机辅助制造

计算机辅助制造（CAM）是一种利用计算机控制设备完成产品制造的技术。例如，20世纪50年代出现的数控机床便是在CAM技术的指导下，将专用计算机和机床相结合后的产物。

借助CAM技术，在生产零件时只需使用编程语言对工件的形状和设备的运行进行描述后，便可以通过计算机生成包含加工参数（如走刀速度和切削深度）的数控加工程序，并以此来代替人工控制机床的操作。这样不仅提高了产品的质量和效率，还降低了生产难度，在批量小、品种多、零件形状复杂的飞机、轮船等制造业中备受欢迎。

3. 计算机集成制造系统

计算机集成制造系统（CIMS）是集设计、制造、管理三大功能于一体的现代化工厂生产系统，具有生产效率高、生产周期短等特点，是20世纪制造工业的主要生产模式。在现代化的企业管理中，CIMS的目标是将企业内部所有环节和各个层次的人员全都用计算机网络连接起来，形成一个能够协调统一和高速运行的制造系统。

2.3.3　CAE的关键技术——有限元技术

CAE的关键技术包含计算机图形技术、三维造型技术、数据交换技术、工程数据管理

At present, CAD technology has been widely used in machinery, electronics, aviation, ship, automotive, textile, clothing, chemical, building and other industries. It has become the most active technology field of computer application.

2. Computer-Aided Manufacturing

Computer-aided manufacturing (CAM) is a technology that uses computer control equipment to complete product manufacturing. For example, computer numerical control (CNC) machine tool under the guidance of CAM technology in the 1950s combines a dedicated computer with the machine tool.

With CAM technology, people only need to describe the shape of the workpiece and the operation of the device by use of the programming language. So people can generate a numerical control machining program that contains machining parameters (such as the speed and depth of cutter) by use of the computer and it can control machine operation with replacing the manual. This not only improves product quality and efficiency, but also reduces the difficulty of production. It is popular in manufacturing industries such as the aircraft manufacturing industry and shipbuilding industry with small batch, variety, complex parts.

3. Computer Integrated Manufacturing system

Computer integrated manufacturing system (CIMS) is a modern factory production system with three functions for design, manufacturing and management, which has the characteristics of high production efficiency and short production cycle and is the main production mode of the 20th century manufacturing industry. In modern enterprise management, the goal of CIMS is to connect all parts of the enterprise and all levels of personnel with the computer network to form a harmony and unification and high-speed operation manufacturing system.

2.3.3　Key Technologies of CAE—Finite Element Technology

Key technologies of CAE include computer graphics technology, 3D modeling technology, data exchange technology, engineering data management technology, simulation technology, optimization design and finite element technology, etc. The following content focuses on the finite element technique.

1. Basic Principles of Finite Element Analysis

Finite element method (FEM) is a discrete numerical analysis method based on variational principle. It is widely used because it is suitable for solving mechanical problems with any complicated structural shapes, boundary conditions and uneven material properties. So it almost can be applied to all the continuous medium and field for solving mechanical and mathematical physics equation, such as elastic mechanics, fatigue and fracture analysis, dynamic response analysis, fluid mechanics, heat transfer and electromagnetic fields and other issues.

The basic idea to solve the mechanical problem by finite element method is to discretize a continuous solution domain, that is, it is split into a finite number of elements that are connected to each other by nodes (discrete points), and a continuous elastomer is considered as a finite element combination. According to certain precision requirements, with a limited number of parameters to describe the mechanical properties of the element body, the mechanical properties of the continuum are the sum of the mechanical properties of all the elements. Based on this principle and the balance

技术、仿真技术、优化设计和有限元技术等。下面重点介绍其中的有限元技术。

1. 有限元分析基本原理

有限元方法（FEM）是一种根据变分原理进行求解的离散化数值分析方法。其由于适合求解任意复杂的结构形状和边界条件以及材料特性不均匀等力学问题而获得广泛应用，几乎可以应用于所有求解连续介质和场的力学及数学物理方程问题。例如，用于弹性力学、疲劳与断裂分析、动力响应分析、流体力学、传热学和电磁场等问题的求解。

有限元方法求解力学问题的基本思想是：将一个连续的求解域离散化，即分割成彼此用节点（离散点）相互联系的有限个单元。一个连续弹性体被看作是有限个单元体的组合。根据一定精度要求，用有限个参数来描述单元体的力学特性，而整个连续体的力学特性就是构成它的全部单元体的力学特性的总和。基于这一原理及各种物理量的平衡关系，建立起弹性体的刚度方程（即一个线性代数方程组）。求解该刚度方程，即可求出欲求的参量。

2. 有限元分析过程

有限元基本求解过程如图 2-5 所示。

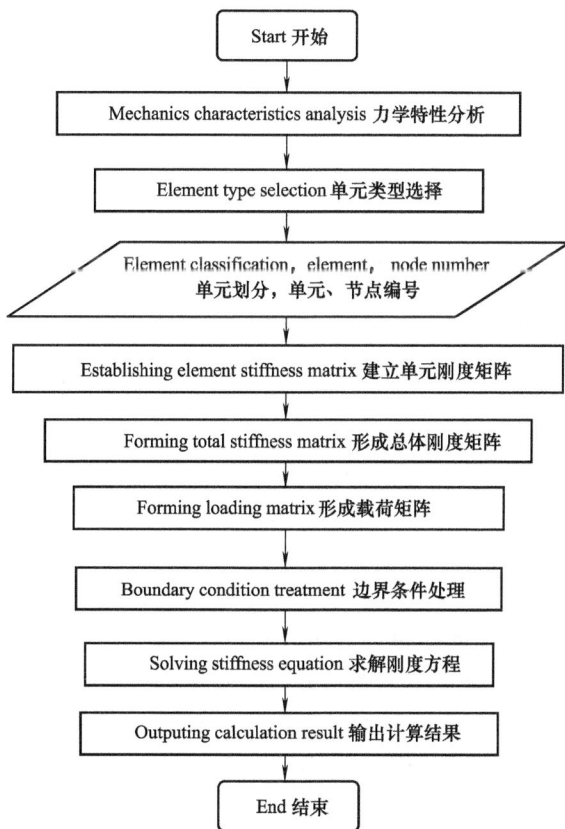

Fig. 2-5　Basic solution process of finite element 有限元基本求解过程

2.3.4　CAE 技术的应用

图 2-6 所示为汽车车身有限元分析模型。建立车身有限元模型，就是根据所研究问题的具体情况，选择合适的有限元单元，对车身进行数学离散化，给这个模型赋予材料属性，进

relation of various physical quantities, the stiffness equation of the elastic body (linear algebraic equations) is established. During solving the stiffness equation, you can find the desired parameters.

2. Finite Element Analysis Procedure

The finite element solution procedure is shown in Fig. 2-5.

2.3.4 Applications of CAE Technology

Fig. 2-6 shows the finite element analysis model of automobile body. According to specific circumstances of the problem, selecting the appropriate finite element unit, carrying out mathematical discretization of the body, giving material properties to the model, simulating boundary conditions, debugging the model, building a finite element model of body is a process with acceptable accuracy simulation model of the body structure. Specific steps based on ANSYS Workbench are as follows:

1) 3D model of automobile body is established based on 3D modeling software (UG NX, Creo, CATIA, etc.).

2) 3D model of automobile body is seamlessly connected to Workbench environment.

3) Adding material properties to the model.

4) Selecting the appropriate element type for the model and setting the appropriate element size.

5) Applying boundary conditions to the model.

6) Solving and viewing finite element analysis results.

2.4 Computer-Aided Process Planning

2.4.1 Introduction of CAPP

1. Basic Concepts of CAPP

The process planning design is an important technical preparation work in the mechanical product manufacturing process. It is also a very strong decision-making process with great experiences and influencing factors. The workload is large and it is easy to make mistakes. If people use the computer to complete the process design work, it can not only greatly reduce the workload, more importantly, but also facilitate knowledge accumulation, data management and system integration, so a method with computer-aided process planning is produced. Computer-aided process planning is designed by automatically inputting geometrical information and process information of machined parts in the computer, and the computer automatically encodes the program until the final output of the optimized part process is programmed.

2. The Basic Composition of the CAPP System

The basic component of CAPP system is related to its development environment, product object and size. In general, the CAPP system consists of four basic components, as shown in Fig. 2-7.

(1) Product design information input   For process design, product design information refers to the structural shape and technical requirements of parts. There are many ways to express the shape of parts and technical requirements, such as commonly used engineering drawing and part model in

行边界条件的模拟，以及模型的调试，最后得到一个具有可接受精度的车身结构仿真模型的过程。ANSYS Workbench 中的具体步骤如下：

1）基于三维造型软件（UG NX、Creo、CATIA 等）建立车身的三维模型。

2）把车身的三维模型无缝链接到 Workbench 环境中来。

3）给模型添加材料属性。

4）给模型选择合适的单元类型，设置适当的单元大小。

5）给模型施加边界条件。

6）求解，查看有限元分析结果。

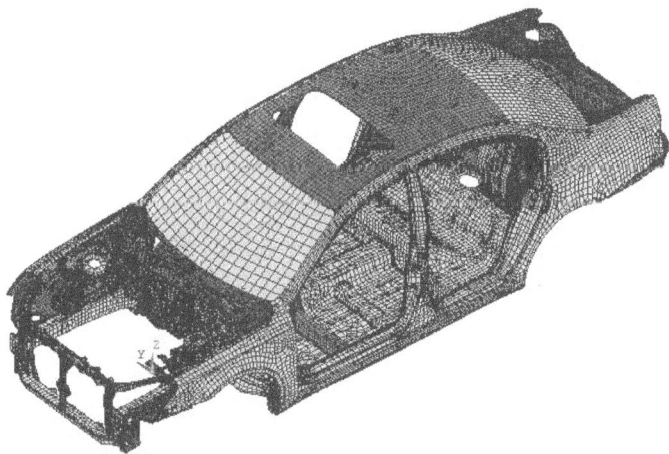

Fig. 2-6 Finite element analysis model of automobile body 汽车车身有限元分析模型

2.4 计算机辅助工艺规程设计

2.4.1 计算机辅助工艺规程设计的概述

1. CAPP 的基本概念

工艺规程设计是机械产品制造过程中一项重要的技术准备工作，又是一项经验性很强、影响因素很多的决策过程，工作量大、容易出错。如果借助计算机来完成工艺设计工作，不仅可以大大减轻工作量，更重要的是便于知识积累、数据管理和系统集成，由此产生了计算机辅助编制工艺规程的方法。计算机辅助工艺规程设计（CAPP）是通过向计算机输入被加工零件的几何信息和工艺加工信息等，由计算机自动进行编码、编程直至最后输出经过优化的零件工艺规程的过程。

2. CAPP 系统的基本组成

CAPP 系统的基本组成与其开发环境、产品对象及规模大小有关。总体上看，CAPP 系统包括 4 个基本组成部分，如图 2-7 所示。

（1）产品设计信息输入 对于工艺过程设计而言，产品设计信息是指零件的结构形状和技术要求。表示零件结构形状和技术要求的方法有多种，如常用的工程图和 CAD 系统中的零件模型。对于 CAPP 系统，必须将这些有关的信息转换成系统所能"读"懂的信息。

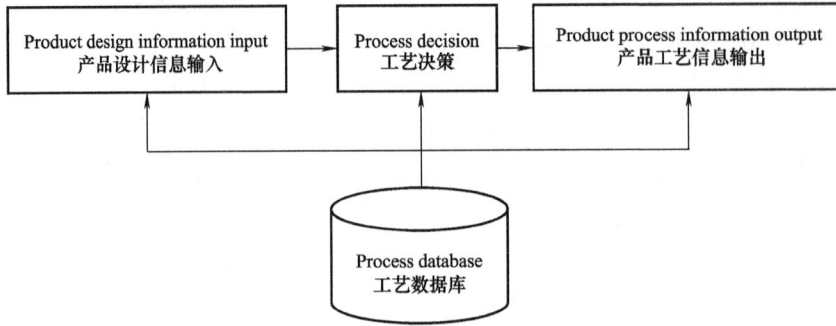

Fig. 2-7 Basic component of CAPP system CAPP 系统的基本组成

CAD system. For CAPP system, the relevant information must be converted into the information which a system can understand.

(2) Process decision   Process decision is to determine the process of the product according to product design information, using process experience and specific production environment. The basic process decision methods used in CAPP system are variant and generative methods. However, a practical CAPP system often uses the variant method synthetically, so as to produce the so-called semi generative method.

(3) Product process information output   Product process information contains process Procedure sheet, process sheet, operation sheet, process diagrams and other types of document output, and can use editing tools to generate process files modified by process documents required. In the CAD/CAPP/CAM integrated system, CAPP system needs to provide process parameter file of CAM numerical control programming.

(4) Process database   Process database is a support tool of CAPP system, which contains all process data needed by process design (such as machining allowance, cutting parameter, machine tools, processing equipment and material, working hour, cost accounting and other information) and process rules (such as the choice of processing methods, process arrangement, etc.).

Contents and steps of CAPP are shown in Tab. 2-1.

3. Functions of CAPP

CAPP has the following functions:

1) The process designer are free from a large number of cumbersome, repetitive manual labor, so that they can focus on the development of new products, process equipment design and new technology research and other creative work.

2) It can overcome the limitations of traditional process design, and greatly improve the efficiency and quality of process design.

3) It can effectively accumulate and inherit the experience of process designers to solve the problem of insufficient experience.

4) It can improve levels of normalization and standardization of enterprise process design and continue to develop towards optimization and intelligent direction.

5) It can achieve the integration of CAD / CAM.

（2）工艺决策　工艺决策是指根据产品设计信息，利用工艺经验和具体的生产环境条件，确定产品的工艺过程。CAPP 系统所采用的基本工艺决策方法有派生型方法和创成型方法。但是一个实用的 CAPP 系统往往会综合使用派生型方法，从而产生所谓的半创成型方法。

（3）产品工艺信息输出　产品工艺过程信息常常以工艺过程卡、工艺卡、工序卡、工序简图等各类文档输出，并可利用编辑工具对生成的工艺文件进行修改，得到所需的工艺文件。在 CAD/CAPP/CAM 集成系统中，CAPP 系统需要提供 CAM 数控编程所需的工艺参数文件。

（4）工艺数据库　工艺数据库是 CAPP 系统的支撑工具，包含了工艺设计所需要的所有工艺数据（如加工余量、切削用量、机床、工艺装备及材料、工时、成本核算等多方面的信息）和工艺规则（如加工方法选择、工序顺序安排等）。

CAPP 的内容与步骤见表 2-1。

Tab. 2-1　The contents and steps of CAPP　CAPP 的内容与步骤

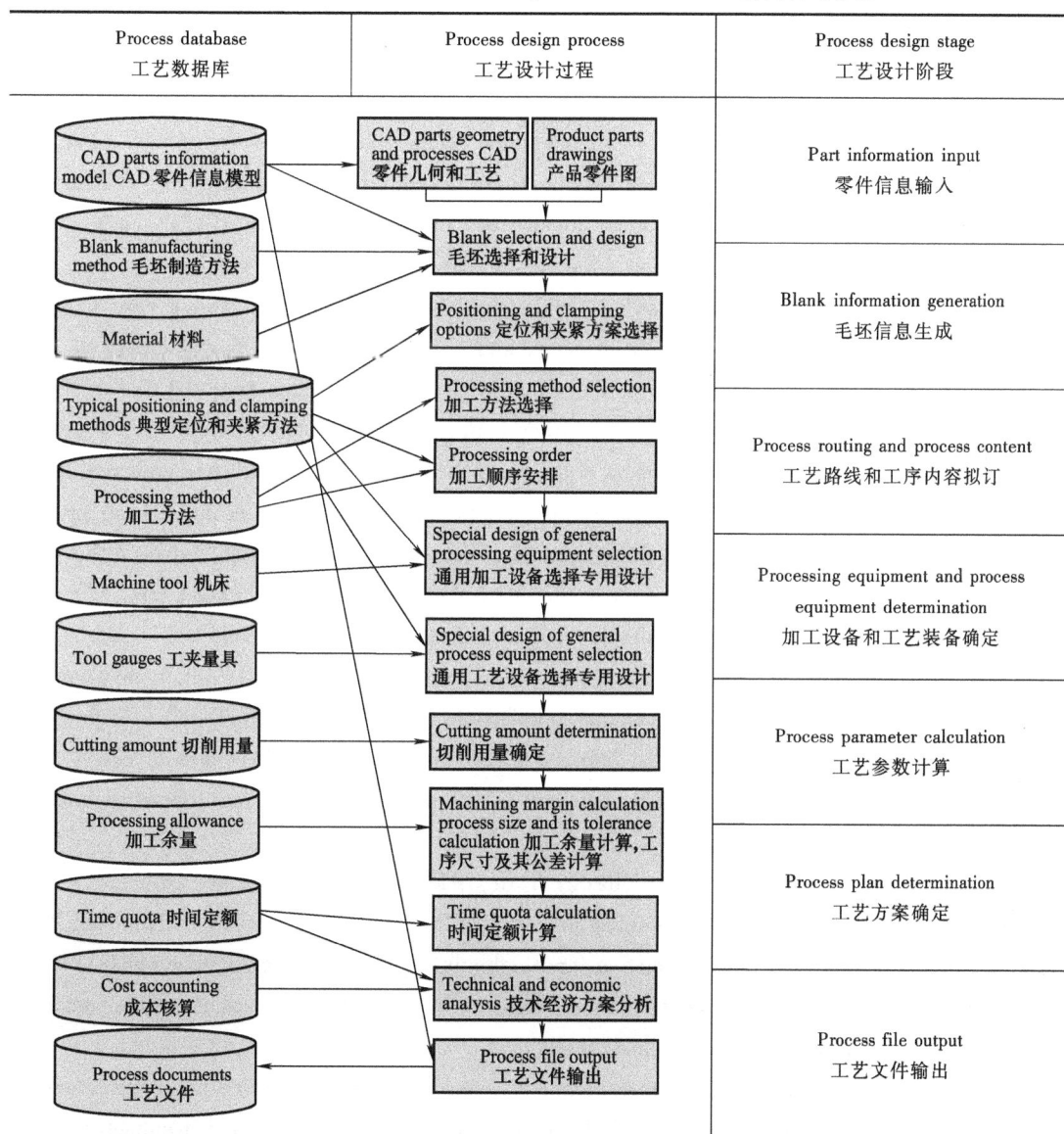

Process database 工艺数据库	Process design process 工艺设计过程	Process design stage 工艺设计阶段

2. 4. 2　Description and Input of Part Information in CAPP System

Part information includes management information, structure shape, size, processing precision, material and heat treatment and so on, which is the object and basis of process design. CAPP system requires part information of a comprehensive and accurate description, and requires the structure of a logical hierarchy and easy to be processed by the computer. Therefore, how to describe and express process feature information of parts, to be easy to retrieve similar parts in the system, or to generate new process rules, it is the first problem to be solved in the development of CAPP system. At present, basic methods of part information description and input can be summarized into the following two categories:

1. Interactive Input Method Based on Engineering Drawings

(1) GT code description method　GT code description method is a method which describes main design and manufacture features of parts by use of part classification coding in group technology. These features are often the basis for making certain process design decisions. GT code description method is characterized by simple input operation, convenience for computer processing, but the description of part information is more rough. There is less input information, and it is suitable for determining process routing. This method is usually used in general variant CAPP system.

(2) Part feature description method　Part feature description method will input various characteristics of parts into the computer by use of human-computer interaction and a certain order, which constitutes mathematical expression of parts, and corresponds to the processing method.

Parts shown in Fig. 2-8 are made up of five features: cylindrical chamfering (I), cylindrical surface (II), cylindrical surface (III), cylindrical surface (IV) and countersinking (V) . In order to input and identify the axial dimension of the part, the end of the part will be numbered. In this regulation, numbering sequence is that the first is the external surface and then the internal surface; the external surface is from right to left, the internal surface is from left to right. Based on this regulation, part shown in Fig. 2-8 can be input in the matrix format shown in Tab. 2-2 (WD, WY, HW represents respectively cylindrical chamfering, cylindrical surface and countersinking) .

Above two methods are developed as an interactive input module in CAPP system developed independently and have no information exchange relationship with other systems. They can't provide a way for automatic acquisition for original information of existing parts in CAD system. This will not only result in duplication work, but also make the CAPP system based on these input methods be unable to integrate with CAD system.

2. Get Product Part Information Directly from CAD Model

It is an ideal method to realize the integration of CAPP with CAD directly by CAD model. It can save human-computer interactive information input, greatly improve running efficiency of CAPP system, reduce artificial information conversion and input error, be helpful for ensuring data consistency. There are three kinds of commonly used methods.

(1) Feature extraction and pattern recognition　Feature provides a common basis for mutual understanding of product information among CAD, CAPP and CAM in an integrated manufacturing sys-

3. CAPP 的作用

CAPP 有如下作用：

1）可将工艺设计人员从大量烦琐、重复性的手工劳动中解脱出来，使其能集中精力进行新产品的开发、工艺装备的设计和新工艺的研究等创造性工作。

2）能克服传统工艺设计的局限性，大大提高工艺设计的效率和质量。

3）能有效地积累和继承工艺设计人员的经验，解决工艺设计人员经验不足的问题。

4）提高企业工艺设计的规范化和标准化水平并不断向最优化和智能化的方向发展。

5）有助于实现 CAD/CAM 的集成。

2.4.2 CAPP 系统零件信息描述与输入

零件信息包括管理信息、结构形状、尺寸、加工精度、材料及热处理等方面的信息，是工艺设计的对象和依据。CAPP 系统要求全面而正确地描述零件信息，而且要求形成逻辑层次分明和易于被计算机处理的结构。因此，如何描述和表达零件的工艺特征信息，便于检索系统中已有的相似零件的工艺规程或生成新的工艺规程，是开发 CAPP 系统必须首先解决的问题。目前，零件信息描述与输入的基本方法可归纳为以下两大类：

1. 基于工程图纸的变互输入方法

（1）GT 代码描述法　GT 代码描述法是利用成组技术（Group Technology，GT）中的零件分类编码对零件的一些主要设计制造特征进行描述的一种方法。这些特征通常是做出某些工艺设计决策的依据。GT 代码描述法的特点是输入操作简单，便于计算机处理，但对零件信息的描述较粗糙，输入的信息量少，适用于只需要确定工艺路线的场合。一般的派生式 CAPP 系统大多采用这种方法。

（2）零件特征描述法　零件特征描述法是将组成零件的各种特征采用人机交互的形式按一定顺序逐个地输入到计算机中，构成零件的数学表达，并能与加工方法等相对应。

图 2-8 所示零件由 5 个特征组成：外圆倒角（Ⅰ）、外圆柱面（Ⅱ）、外圆柱面（Ⅲ）、外圆柱面（Ⅳ）和划窝（Ⅴ）。为了将零件的轴向尺寸输入，并能加以识别，需要将零件的端面编号。在此规定，编号顺序为先外表面后内表面，外表面由右到左，内表面由左到右。按此规定，可将图 2-8 所示的零件以表

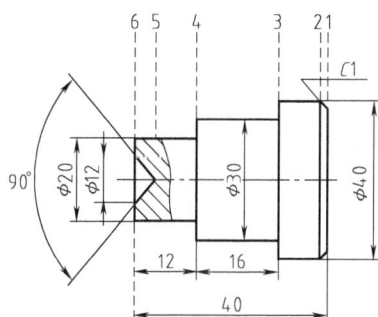

Fig. 2-8　Features of cylindrical parts
圆柱零件的特征

2-2 所列的矩阵格式输入（WD、WY、HW 分别表示外圆倒角面、外圆柱面、划窝表面）。

Tab. 2-2　Features of cylindrical parts 圆柱零件的特征

No. 序号	Surface element 表面元素	Axial dimension 轴向尺寸	Radial dimension 径向尺寸	Start-stop face 起止端面	Remark 备注
Ⅰ	WD	1.0	40.0	2-1	0
Ⅱ	WY	14.0	40.0	2-3	0
Ⅲ	WY	16.0	30.0	3-4	0
Ⅳ	WY	12.0	20.0	4-6	0
Ⅴ	HW	6.0	12.0	6-5	-90

tem. When CAD system uses traditional solid modeling methods, such as constructive solid geometry (CSG) and boundary representation (B-rep), for isolating feature information needed for computer aided process planning design from CAD geometry model, it can carry out feature recognition based on CSG and B-rep model. For example, when B-rep model is used to identify machining features, attributed adjacency graph (AAG) can be used to express features such as grooves, cavities, etc.

The disadvantage of feature recognition and extraction method is that it is necessary to open internal data structure of 3D CAD software, so as to create attributed adjacency graph from given B-rep data, feature information is obtained. But this is difficult to meet requirements of many commercial CAD software. In addition, this method is more suitable for single feature identification. It is difficult to identify combination feature, feature cross and irregular features in complex parts. It is restricted in practice.

(2) Feature expression based on data standard or custom data format　With the emergence of standard data such as initial graphics exchange standard (IGES), some CAPP systems such as TOM system developed by University of Tokyo, Japan, THCAPP system developed by Tsinghua University, which achieve product information transfer by use of IGES file format generated by CAD software. Although IGES is data specification developed with geometric data transformation, which is mainly used for information transmission in CAD systems, but the standard contains feature entity definition and process information, which makes feature information extraction from standard data possible.

(3) Feature modeling system based on product model data exchange standard　Product model data exchange standard initiated and studied by International Organization for Standardization is achieved through international cooperation. Product model data includes not only geometric information such as curve, surface, solid, shape feature, but also a lot of non geometric information such as tolerance, material, and surface roughness and so on. Product model data exchange standard is independent on all kinds of computer aided systems, which provides the system with a neutral expression for data sharing and exchanging. Because of these main characteristics, product model data exchange standard is the basis of establishing unified CAD/CAPP/CAM data model. CAD system based on product model data exchange standard feature modeling will be a most thorough method to solve the integration of CAPP and CAD.

2.4.3　Variant CAPP System

1. Basic Principle

Most variant CAPP systems are based on the principle of group technology, that is, parts will be divided into several parts groups according to their manufacturing features, a standard process planning for each parts group is designed, standard process planning will be stored in a computer database. When programming process planning of a new parts, parts which will be programmed by process planning firstly carry out classification coding. When parts coding is input into the computer, you can retrieve standard process planning for corresponding parts groups. When technical requirements of parts to be programmed are different from standard process planning, standard process planning needs to be edited and modified. Due to simple structure and easy implementation of variant CAPP system, it is still widely used. Figure 2-9 visually shows the work flow of variant CAPP system.

上述两种方法都是作为独立开发的 CAPP 系统中的一个交互式输入模块存在的，与其他系统之间没有信息交换关系，而且它们均不能提供自动获取 CAD 系统中已有零件原始信息的途径。这不仅会造成重复工作，也使基于这些输入方法的 CAPP 系统无法实现与计算机辅助设计系统的集成。

2. 直接由 CAD 模型获取产品零件信息

直接由 CAD 模型获取产品零件信息是实现 CAPP 与 CAD 集成的理想方法。它可以省去人机交互信息输入工作，大大提高 CAPP 系统的运行效率，减少人工信息转换和输入的差错，有助于保证数据的一致性，通常采用的方法有以下 3 种：

（1）特征提取与模式识别　特征提供了集成制造系统中 CAD、CAPP、CAM 等之间相互理解产品信息的共同基础。当 CAD 系统采用传统的实体造型方法，如结构化实体模型（Constructive Solid Geometry，CSG）和边界表示法（Boundary Representation，B-rep）时，为从 CAD 几何模型中分离出计算机辅助工艺规程设计所需要的特征信息，可以在 CSG 和 B-rep 模型基础上进行特征识别。例如，基于 B-rep 模型识别加工特征时，可用属性邻接图（Attributed Adjacency Graph，AAG）来表达由曲面围成的槽、腔等特征。

特征识别与提取方法的缺点是，必须要求三维 CAD 软件开放其内部数据结构，以便由所给出的 B-rep 数据创建属性邻接图，从而得到特征信息。但这一点难以满足许多商品化 CAD 软件的要求。此外，这种方法较适合于识别单一特征。对于组合特征、特征交叉以及复杂零件上的不规则特征等则很难识别，在实用上受到一定限制。

（2）基于数据标准或自定义数据格式的特征表达　随着数据标准如初始图形交换标准（Initial Graphics Exchange Specification，IGES）等的出现，有些 CAPP 系统，如日本东京大学研制的 TOM 系统、清华大学开发的 THCAPP 系统等，可以利用 CAD 软件生成的 IGES 文件格式实现产品信息的传递。虽然 IGES 是因几何数据变换的需要而发展起来的数据规范，主要用于在 CAD 系统之间进行信息传递，但在该标准中已经有了关于特征及工艺信息的实体定义，这就使得从该标准数据中获取特征信息成为可能。

（3）基于产品模型数据交换标准的特征造型系统　产品模型数据交换标准是由国际标准化组织发起和研究，并通过国际性合作来实现的。产品模型数据中不仅包括像曲线、曲面、实体、形状特征等在内的几何信息，还包括许多非几何信息，如公差、材料、表面粗糙度等。产品模型数据交换标准独立于各种计算机辅助系统，它为各系统提供了产品数据及其共享、交换的一种中性表达。由于这些主要特点，产品模型数据交换标准成为建立 CAD/CAPP/CAM 统一数据模型的基础。基于产品模型数据交换标准特征造型的 CAD 系统将成为解决 CAPP 与 CAD 集成的最彻底的方法。

2.4.3　派生型 CAPP 系统

1. 基本原理

大多数派生型 CAPP 系统都建立在成组技术原理的基础上，即将零件按其制造特征分为若干零件组，为每一零件组设计一个标准工艺规程，将标准工艺规程存入计算机数据库中。当编制一个新零件的工艺规程时，先对要编工艺规程的零件进行分类编码。当把零件编码输入计算机后，便可检索出相应零件组的标准工艺规程。当待编工艺规程的零件技术要求与标准工艺规程不同时，需要对标准工艺规程进行编辑修改。由于派生型 CAPP 系统结构简单，

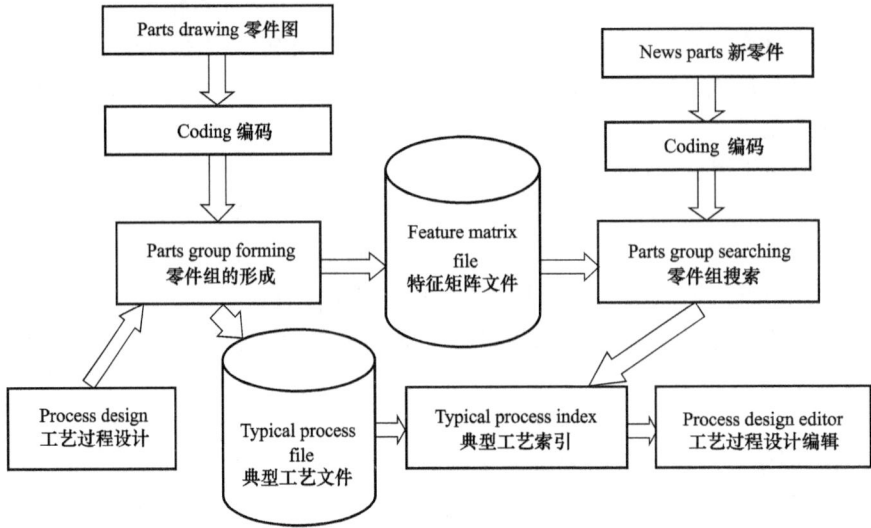

Fig. 2-9　Work flow chart of variant CAPP system 派生型 CAPP 系统的工作流程图

2. System Development and Design

The development and design steps of variant CAPP system are as follows：

1）To encode parts and components, establish feature matrix of parts. Taking the parts shown in Fig. 2-10a as an example, firstly you can select the appropriate encoding system, for example, you can encode parts by use of OPVTZ system, parts code is 04100 3072, two-dimensional array is 1. 0, 2. 4, 3. 1, 4. 0, 5. 0, 6. 3, 7. 0, 8. 7, 9. 2. Among them, first dimension array represents code bit, second dimension array represents code value, and the corresponding feature matrix is shown in Fig. 2-10b. Similarly, the code and feature matrix of other parts can be obtained.

2）To classify parts into groups, establish feature matrix of parts groups. Parts with process similarity and structure similarity can be used with similar processing methods. It is easy to design a typical process, while parts with size similarity can even use the same type of machine tools and process equipment for processing.

3）To design master samples and program standard process planning. The so-called master sample, also known as composite parts, refers to parts that can compound all features in the group together. It can be an actual existence parts in the group, but more is a reasonable combination of all features in group parts to form imaginary parts. Process planning programmed by master sample is a standard process planning. This method is more intuitive, but only applies to rotary parts for the structure that is not complicated.

4）Feature matrix of each parts group and corresponding standard process planning are input into the computer, and relevant procedures for retrieval and modification are compiled, the system is established.

3. Process Information Coding

In order to facilitate identification, storage and call of the computer, it is necessary to convert process information expressed in standard process planning into code form. When designing the sys-

易于实现,故当前应用仍较广。图 2-9 形象地表示出了派生型 CAPP 系统的工作流程。

2. 系统的开发与设计

派生型 CAPP 系统的开发与设计步骤如下:

1)对零部件进行编码,建立零件特征矩阵。以图 2-10a 所示零件为例,首先选择合适的编码系统,如用 OPVTZ 系统对某零件进行编码,得到该零件的代码 04100 3072,用二维数组表示即为 1.0、2.4、3.1、4.0、5.0、6.3、7.0、8.7、9.2。其中,第一维数组表示码位,第二维数组表示码值,相应的特征矩阵如图 2-10b 所示。同理可得其他零件的代码和特征矩阵。

Code value 码值 \ Code bit 码位	1	2	3	4	5	6	7	8	9
0	×			×	×		×		
1			×						
2									×
3						×			
4		×							
5									
6									
7								×	
8									
9									

Workpiece 工件	Code 代码
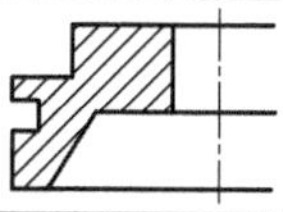	04100 3072

a)   b)

Fig. 2-10　Parts code and feature matrix 零件代码和特征矩阵

a) Parts and code 零件及代码　b) Feature matrix 特征矩阵

2)对零件分类成组,建立零件组的特征矩阵。对工艺相似和结构相似的零件可以采用相类似的加工方法,易于设计出有针对性的典型工艺,而对尺寸相似的零件可以选用同类型甚至同规格的机床和工艺装备进行加工。

3)设计主样件,编制标准工艺规程。所谓主样件,也称复合零件,是指能将零件组内所有型面特征复合在一起的零件。它可以是该组内实际存在的某个零件,但更多的是将组内零件的所有特征进行合理组合而形成的假想零件。根据主样件编制的工艺规程即为标准工艺规程。这种方法比较直观形象,但仅适用于结构不太复杂的回转体零件。

4)将各零件组的特征矩阵和相应的标准工艺规程输入计算机,并编制用于检索和修改的有关程序,建立系统。

3. 工艺信息的代码化

为了便于计算机的识别、储存和调用,必须将标准工艺规程所表达的若干工艺信息转化为代码的形式,在设计系统时首先建立这些代码与各种名称之间的对应关系。

4. 系统的使用

派生型 CAPP 系统的使用步骤如下:

1)输入表头信息,如产品型号、产品名称、零件件号、零件名称、材料牌号等。

2)用各种方法对零件有关信息进行描述和输入。

3)计算机检索和判断该零件属于哪个零件组,调出该零件组的标准工艺。

tem, the corresponding relationship between these codes and various names is firstly established.

4. Usage of the System

The steps to use variant CAPP system are as follows:

1) Input header information, such as product model, product name, parts number, parts name, material grade, etc.

2) Describe and input relevant information of parts in various ways.

3) The computer retrieves and judges which parts group the parts belongs to, and calls out standard process of parts group.

4) According to the need, standard process is modified and edited; a new process standard is generated.

5) Storage and output process standard.

2.4.4　Generative CAPP System

Standard process specification is not stored in a CAPP system in advance. The process planning of new parts is not generated by retrieval methods, but is automatically generated by the system which relies on the knowledge, rules, logical reasoning, and decision algorithms stored in the system without manual intervention according to the information of input parts, by stored knowledge, rules, logical reasoning and decision making algorithm. Therefore, this system is called a generative CAPP system.

Input information of generative CAPP system is design information of parts, output information is process planning of parts. With the support of database and process knowledge database, the system needs to establish a series of logic decision models and calculation programs for process decision. The data stored in the system database is mainly about processing ability of various processing methods, application range of various machine tools, cutting quantity and so on. The process knowledge database in the system is stored in the design process to follow process principle and knowledge. The output result after process decision can be a selected processing methods, or it can also be the processing route of parts, including a detailed operation contents and operation chart.

In theory, generative CAPP system is a complete, advanced system that has all the information needed by process design. In the software system, it contains all decision logics, so it is very convenient to use. However, due to many factors for affecting process design, the development of generative system with a fully automated process planning still has many technical difficulties. The design of many CAPP systems uses a comprehensive method with main variant method and auxiliary generative method; CAPP systems developed in China mostly belong to this type.

2.5　Modular Design

2.5.1　Introduction

After 1980s, with the increasingly intensive market competition and the increasingly rapid product changing speed, traditional product developing and manufacturing modes are unable to adapt to the new

4）根据需要，对标准工艺进行适当修改和编辑，生成新的工艺规程。

5）存储和输出工艺规程。

2.4.4　创成型 CAPP 系统

在一个 CAPP 系统中没有预先存入标准工艺规程，新零件的工艺规程不是依靠检索方法生成的，而是由系统根据输入零件的信息，依靠存储的知识、规则、逻辑推理和决策算法，在无人工干预的情况下自动产生的。故这种系统称为创成型 CAPP 系统。

创成型 CAPP 系统的输入信息是零件的设计信息，输出信息是零件的工艺规程。该系统需要在数据库和工艺知识库的支持下，经过建立在系统内部的一系列逻辑决策模型及计算程序进行工艺过程决策。系统数据库中存储的主要是有关各种加工方法的加工能力、各种机床的适用范围、切削用量等的基本数据。系统的工艺知识库中存储的是设计工艺过程中要遵循的工艺原则和知识。工艺过程决策后的输出结果可以是选择好的加工方法，也可以是零件的加工工艺路线，包括详细的工序设计内容，还可包括工序图。

理论上，创成型 CAPP 系统是一个完备的、高级的系统，拥有工艺设计所需要的全部信息。在其软件系统中包含着全部决策逻辑，因此使用极为方便。但是，由于影响工艺过程设计的因素很多，开发完全自动生成工艺规程的创成型系统还存在着许多技术上的困难。目前许多 CAPP 系统的设计都采用以派生型为主、创成法为辅的综合法，我国自行开发的 CAPP 系统大多为这种类型。

2.5　模块化设计

2.5.1　概述

20 世纪 80 年代后，随着市场竞争的日益激烈和产品更新速度的日益加快，传统的产品开发制造模式无法适应新的要求。模块化设计以其自身的优势得到广泛的应用与发展。各工业发达国家竞相开发系列模块化的机电工业产品，大量模块化产品进入实用化阶段。

模块化设计是通过对一定范围内不同功能或者功能相同而不同性能和规格的产品进行功能分析，从而划分并设计出一系列功能模块，通过模块的选择和组合构成不同的产品，以满足市场的不同需求的设计方法。

模块化设计可以缩短产品开发周期，快速响应市场需求的变化，有利于提高设计的标准化程度，实现优化设计，并且有利于产品维修、升级和再利用。采用模块化设计可以以大批量生产的成本实现多品种、小批量个性化生产。对于机械产品而言，模块化设计是实现大规模定制的前提。

2.5.2　机械模块化设计的分类

在基型产品的基础上进行变型产品的扩展可形成各种系列产品。按机械模块化设计的产品在系列产品中所覆盖的形式和程度，常把机械模块化设计分为以下几种：

1. 横系列模块化设计

横系列模块化设计是指在某一基型品种的基础上，通过变换、增加或减少某些可互换的

requirements. Modular design has found wide application and development with its own advantages. All industrial developed countries developed competitively a series of mechanical and electrical industrial products with modularity. A large number of modular products are put into the practical stage.

Modular design is a kind of product design method to meet the different needs of the market. Its design procedures are as follows: The first is to divide and design a series of functional modules through the functional analysis on the products either with different functions in a certain range or with the same functions but different performance and size; the second is to compose the different products through the selection and combination of modules.

Modular design is helpful to shortening the product developing cycle, quickly responding to the changes in market demand, improving design standards, realizing optimal design, and helpful to the product maintenance, product upgrading and product recycling. Personalized production with multi-variety and small batch production can be achieved in the cost of mass production by means of modular design. For the mechanical products, modular design is the precondition of mass customization.

2.5.2 Classification of Mechanical Modularization Design

The expansion of variant products performed on the basis of base product can obtain different series of products. According to the form and degree covered by the products developed by mechanical modularization design in series products, mechanical modularization design can be generally classified as follows:

1. Horizontal Series Modular Design

Horizontal series modular design means that the variant product is generated by changing, adding or reducing some interchangeable specific modules on the basis of a basic-type product. Its feature is that it does not change the primary parameters of the base product. The changed parts are mainly embodied in certain functions, structure, layout, control system or manipulate way and other aspects. Because this design method does not change the vast majority of basic modules in the base product, the product design is easy and has a very high degree of generalization. In the process of mechanical modularization design, the key is to take necessary measures in the product structure, which, for example, are to set apart adequate place, to design rational interface, to machine the location surface and location hole connection in advance for connection and so on, so as to add and replace various module parts conveniently when the product is modified.

N-038 series automatic machine with high efficiency was designed in light of horizontal series modular design mode. There were 6 groups of modules (i. e. base and bed module, toolrest module, distribution shaft module, headstock and spindle module, threading module, electrical equipment module) including more than 40 modular units in total in the machine tool. Hundreds of different combinations can be obtained by making different selection from more than 40 modular units in 6-group modules, where, there were 8 practical machine tool plans.

2. Longitudinal Series Modular Design

Longitudinal series modular design means that the variant product is designed based on the base product with different specifications in the same type products. Its feature is that it changes the pri-

特定模块而形成变型产品。它的特点是不改变基型产品的主参数，而改变部分主要体现在某些功能、结构、布局、控制系统或操纵方式等方面。由于这种设计方法不会改变基型产品中绝大多数的基本模块，所以在设计上较易实现，且通用化程度很高。在横系列模块化设计过程中，关键是要在结构上采取必要的措施，如留出足够位置、设计合理接口、预先加工出连接的定位面、定位孔等，以便在进行产品变型时可顺利增加和更换各种模块部件。

N-038 系列高效自动机床是按横系列模块化方式设计出来的。该机床采用 6 组模块（底座床身模块、刀架模块、分配轴模块、主轴箱与主轴模块、车螺纹模块、电器模块），共 40 多个模块单元。通过对这 6 组 40 多个模块的不同选择，可以有上百种不同的组合，其中实用的机床方案有 8 种。

2. 纵系列模块化设计

纵系列模块化设计是指在同一类型中对不同规格的基型产品进行变型产品的设计。其特点是要改变基型产品主参数。由于主参数不同的产品，其动力参数往往也不同，这就导致产品的结构及尺寸等发生变化。如果产品的动力参数发生变化，仍试图以同一种尺寸规格的模块去满足不同主参数产品的要求，则势必导致模块的强度或刚度产生欠缺或冗余。因而纵系列模块化设计比横系列模块化设计的难度大得多。

通常在纵系列模块化设计时，对那些与主参数无关且受力不大的模块，诸如控制模块、某些进给系统中的模块等，可以在整个纵系列内采用同一种规格和尺寸。而对于与主参数尤其是与动力参数有关的一些模块，如主运动中的变速器、主轴箱和基础件等，可采用"分段相同"的原则来设计，即在某一段中采用同一种规格尺寸的模块来满足本段内的主参数变化。

3. 跨系列模块化设计

跨系列模块化设计是指针对总体结构相差不大的产品所进行的变型设计。它的特点是在基本相同的基础件结构上选用不同模块而构成跨系列产品；或者在基础件结构不同的跨系列产品上对具有同一功能的单元，选用相同的模块。例如，龙门铣床、龙门刨床和龙门磨床，它们的立柱、横梁、床身的结构形式相差不大。如果能够合理选择上述基础件的形状和尺寸，并兼顾到上述各类机床的刚度要求，那么就可在同一基础件上更换各种铣削头、磨削头、刨削头，从而形成跨系列机床产品。

某组合机床研究所研制的 1HYT 系列液压滑台，按工作行程共有 250mm、300mm、400mm、500mm、630mm 和 800mm 6 个规格。它们全部可以与 1HJT 系列机械滑台、NC—1HJT 系列交流伺服数控机械滑台实现跨系列通用。它们可以通用的基础模块主要有台体模块和滑座模块。

4. 全系列模块化设计

全系列模块化设计与跨系列模块化设计有很大区别。跨系列模块化设计只是横系列兼顾部分纵系列或是纵系列兼顾部分横系列的模块化设计，而全系列模块化设计是指某类产品的全部横、纵系列范围内的模块化设计。只有当跨系列模块化设计覆盖了整个该类产品时，才称为全系列模块化设计。可见，跨系列模块化设计是全系列模块化设计的特例。所以全系列模块化设计的复杂程度和难度都很大，所需的模块种类也更多。

2.5.3　模块化设计过程

根据市场和用户需求，用模块化设计理论与方法开发和研制系列产品。机械产品模块化

mary parameters of the base product. As the products with different primary parameters often have different power parameters, this would cause the change both in structural form and dimension of the products. If the power parameter of product is changed, trying to use the module with the same size to meet the needs of the products with different primary parameters would lead to the deficiency or redundance in module strength or rigidity. Therefore, the longitudinal series modular design is much more difficult than the horizontal series modular design.

Usually, in the longitudinal series modular design, such modules having nothing to do with primary parameter and bearing small force as control module and the modules in some feed systems etc. can use the same specification and size in the whole longitudinal series. But such modules having to do with primary parameter especially with power parameter as the gearbox of primary motion, headstock and base part can be designed by means of "the same in a segment" principle, i. e. using the module with the same specification and size in a segment to meet the change of primary parameter in this section.

3. Cross Series Modular Design

Cross series modular design refers to the variant design performed for the products with a little difference in overall structure. Its feature is either to constitute the cross series product by choosing different modules on the basically same foundation structure, or to choose the same modules for the units with same function on the cross series products with different foundation structure. Take the gantry type milling machine, planer and plano-grinder for example, their columns, cross rails and beds have very a little difference in structure forms. If we are able to choose rationally the shape and dimension of above base parts and take account of the rigidity of each kind of machine tool at the same time, we can replace various milling heads, grinding heads and planer heads, thus forming a cross series of machine tool products.

1HYT series hydraulic sliding table developed by a research institute of aggregate machine tool have 6 specifications in light of working stroke: 250 mm, 300 mm, 400 mm, 500 mm, 630 mm and 800 mm. All of them can be applied for 1HJT-serie mechanical sliding table and NC-1HJT-serie AC servo NC mechanical sliding table. This is a kind of cross-series application.

4. Full Series Modular Design

There are big differences between full series modular design and cross series modular design. Essentially, the cross series modular design is either the horizontal series combining with part of longitudinal series or the longitudinal series combining with part of cross series. The full series modular design refers to the modular design of a kind of product in the range of all horizontal series and longitudinal series. Only when the cross series modular design covers the entire class of products, the modular design can be named as the full series modular design. It can be seen that the cross series modular design is the special case of the full series modular design. Therefore, there are a great complexity and difficulty in the full series modular design which requires more module types.

2.5.3 Modular Design Process

Based on the demands of market and users, series of products are developed by means of modular

设计过程如图 2-11 所示。

（1）用户需求分析 必须准确把握同类产品的市场需求，以及同类产品中基型和各种变型需求的比例，同时进行可行性分析。对于那些需求量很小，而研制费用较高的产品，不宜采用模块化设计。

（2）确定产品系列型谱 合理确定产品的主参数范围和产品系列型谱（即产品种类和规格）是模块化设计的关键一步。产品种类和规格过多，设计难度大。反之，则对市场应变能力减弱。

（3）确定参数范围及主参数 产品的参数（如尺寸参数、运动参数和动力参数等）须合理确定，过大过宽会造成多余，过小过窄则不能满足要求。参数数值大小和数值在参数范围内的分布也很重要，最大、最小值应根据使用要求而定。主参数是表示产品主要性能、规格大小的参数。

（4）模块划分与设计 模块划分的合理性直接影响到模块化产品的性能、外观、模块的通用化程度和产品成本。在设计模块时，应尽量采用标准化结构，保证模块便于制造、组装、维修和更换，还需考虑各主要模块寿命相当。

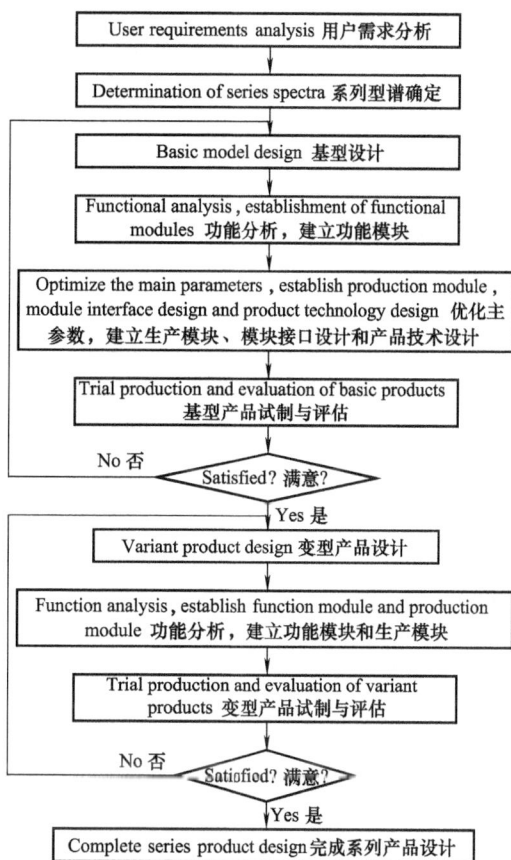

Fig. 2-11 Modular design process of mechanical products
机械产品模块化设计过程

2.5.4 模块化设计的关键技术

1. 模块标准化

模块标准化主要是指模块结构的标准化，特别是模块接口的标准化。为了保证不同功能模块的组合和相同功能模块的互换，模块应具有可组合性和可互换性两个基本特征，而这两个特征主要体现在接口上。

2. 模块的划分

模块化设计的原则是力求以尽可能少的模块组成尽可能多的产品，并在满足要求的基础上使产品精度高、性能稳定、结构简单、成本低廉，且模块结构以及模块间的联系应尽可能简单。因此，合理划分模块是保证模块化设计获得成功的关键。模块划分应遵循以下原则：

1）尽量减少产品包含的模块总数，并最大限度地简化模块自身的结构，以避免模块组合时产生混乱。

2）以有限的模块数获得尽可能多的实用组合方案，以满足用户的多方面需求。

3）划分的模块应具有功能独立性和结构完整性。

4）充分注意模块间的接口要素，以便于模块的组合和分离。

design theory and method. The modular design process of mechanical products is shown in Fig. 2-11.

(1) User requirements analysis (URA)   The market requirements for the same kind of product and the proportion of demand for the base product and various variant products must be known accurately, the feasibility analysis is carried out at the same time. Those products with very small demand amount and high developing cost are not suitable for modular design.

(2) To determine the product series type spectrum   To determine rationally the primary parameter range and the product series type spectrum (i. e. product types and specifications) is a key step in the modular design. Too many product types and specifications would increase the difficulty in design. Conversely, it would weaken market response ability.

(3) To determine the parameter range and the primary parameter   The parameters of product, such as dimensional parameter, kinematic parameter, dynamic parameter etc. , should be determined rationally. Overlarge and over wide would cause surplus, over small and over narrow would not meet the requirement. The magnitude of parameter and distribution of the parameter in the parameter range are also very important. The maximum and the minimum should be determined according to the operation requirements. Primary parameter is the one to indicate the main performance and specification size.

(4) Module division and design   The rationality of module division has a direct influence on the performance and appearance of modular product, degree of module generalization and product cost. In the design of module, standard structure should be used as far as possible so as to ensure that the module is easy to manufacture, assemble, repair and replace. Also, it is necessary to consider that the main modules should have an equivalent life.

2. 5. 4　Key Technology in Modular Design

1. Modular Standardization

Modular standardization refers mainly to the standardization of module structure, especially the standardization of module interface. In order to ensure the combination of modules with different functions and the interchange of modules with identical functions, the module should have two basic features: combinability and interchangeability. These two features are mainly embodied in the interface.

2. Module Division

The principle of modular design is striving to use fewer modules to constitute more products as far as possible, and to make the products have high precision, stable performance, simple structure, low cost on the basis of meeting demands, and the module structure and the connection between modules should be simple as far as possible. Therefore, rational module division is the key to ensure the success of the modular design. Module division should follow the following principles:

1) To minimize the total number of modules included in the product, and to simplify furthest the structure of the module itself so as to avoid confusion when modules are combined.

2) Using the finite module numbers to acquire the practical combination as many as possible so as to meet various needs of users.

3) Divisiory modules should be of functional independence and structural integrity.

5）要考虑对产品精度和刚度的影响，还要考虑经济性。

3. 模块的接口设计

模块化产品多数模块之间的连接属于刚性连接。刚性连接又有固定连接和活动连接之分。模块接口设计直接影响模块化产品的性能（精度、刚度、经济性等）和模块组合的方便性与快捷性。通常，接口设计应考虑的问题有：①应具有易于装配的互换性接口，便于模块的连接与分离；②模块之间定位要准确，连接要可靠；③接口要具有一定的连接刚度和适当的物理性能（如导热性、阻尼特性等）；④制造方便。

2.5.5　机床模块化设计举例

现以某数控成形磨齿机为例，介绍计算机辅助机床模块化设计的方法与过程。

1. 模块划分

数控成形磨齿机采用分级模块化原理，在对机床进行功能分析的基础上，按照不同的功能层次，考虑到机床的结构、功能和各部件之间的连接关系，进行模块的划分。

（1）机床功能分析　该机床的主要功能是实现硬齿面内、外齿轮的成形磨削。在进行模块划分时，必须从数控成形磨齿机的总功能考虑，把总功能分解成相对独立的分功能。除了考虑通用模块外，还要考虑对基型模块的改型和专用模块的设计，以扩大机床的工艺范围。

（2）机床模块划分与接口设计　按照前述的模块划分原则，把该数控成型磨齿机划分成5个基本功能模块，如图2-12所示，并遵循面向装配设计的原则设计了易于装配的互换性接口，以保证模块组合的快速准确。

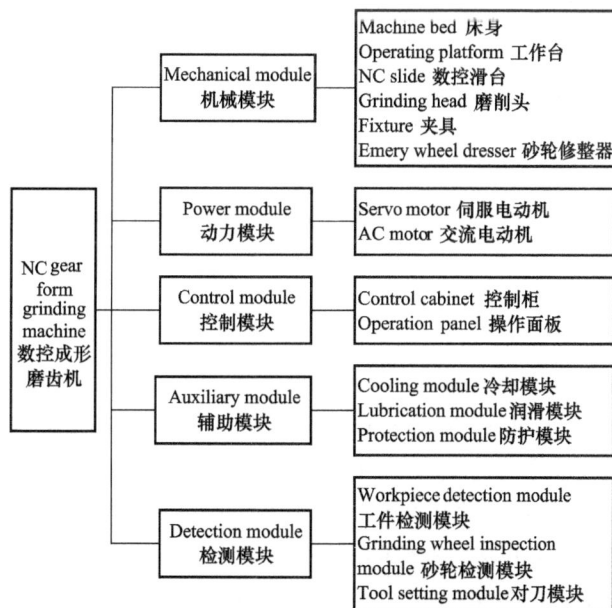

Fig. 2-12　Function module of NC gear form grinding machine 数控成形磨齿机功能模块

2. 模块图形库的建立

参数化建模采用尺寸驱动技术，以约束造型为核心。图2-13所示为数控成形磨齿机通用件设计模块的系统结构。

4）Paying full attention to the interface elements between modules in order to facilitate the combination and separation of modules.

5）To consider the influences on the product precision and rigidity, and to consider the economy also.

3. Module Interface Design

The connections among the most modules in modular products belong to the rigid connection. The rigid connection is divided into stationary connection and movable connection. Module interface design has a direct effect on the performance (precision, rigidity, economy etc.) of modular product and the convenience and rapidity of module combination. Generally, the problems to be considered in module interface design are as follows: ①The interface should have good interchangeability for the convenience of connection and separation of modules; ②the location between modules should be correct, and the connection is reliable; ③the interface should have a certain coupling stiffness and proper physical property (such as thermal conductivity, damping characteristic and so on); ④easy to manufacture.

2.5.5 Example of Machine Tool Modular Design

Take an NC gear form grinding machine for example, the method and process of computer aided machine tool modular design (CAMTMD) are introduced.

1. Module Division

Hierarchical modular principle is used for the NC gear form grinding machine. On the basis of analyzing the functions of machine tool, considering the structure and function of machine tool and the connection between components, different modules are divided based on different functional levels.

（1）Analysis of machine tool function   The main functions of this machine tool are to complete the form grinding of both internal gears and external gears with hard tooth flank. When dividing the machine tool modules, the general function of the gear form grinding machine must be considered, and then the general function is decomposed into relatively independent sub-functions. In addition to considering the general modules, the modification of the basic modules and special-purpose module design should also be considered in order to expand the processing range of the machine tool.

（2）Division of machine tool module   According to the preceding module division principle, the NC gear form grinding machine is divided into five basic function modules as shown in Fig. 2-12. The interchangeable interfaces easy to assemble are designed following the principle of design for assembly (DFA) so as to ensure rapid and accurate modular combination.

2. Establishment of Module Graphics Library

Parametric modeling uses dimension driven technology and takes the constraint modeling as the core. Fig. 2-13 shows the systematic structure of universal part design module of the NC gear form grinding machine.

（1）Establishment of the universal part module graphics library   The graphics library of the universal part modules includes parametric 3D models of various interchangeable parts. It is the basis of modular CAD system for machine tools. The system uses mainly parametric modeling method to es-

（1）通用件模块图形库的建立　通用件模块图形库包括各种通用件的参数化三维模型，是机床模块化 CAD 系统的基础。本系统主要利用 UG 的参数化建模方法直接建立工件模型，用表达式对模型实现全参数控制。有了利用表达式建立的三维模型，当需要得到拓扑结构相同而几何尺寸不同的工件时，只需打开表达式编辑器，修改相应表达式的值，并输出为另一个工件文件即可。

（2）数据库接口程序　选用 Microsoft Access 建立数据库，并使用其中的 ADO. net 技术进行数据库的连接和访问。在通用件设计模块中，所有通用件的参数都存放在数据库中。

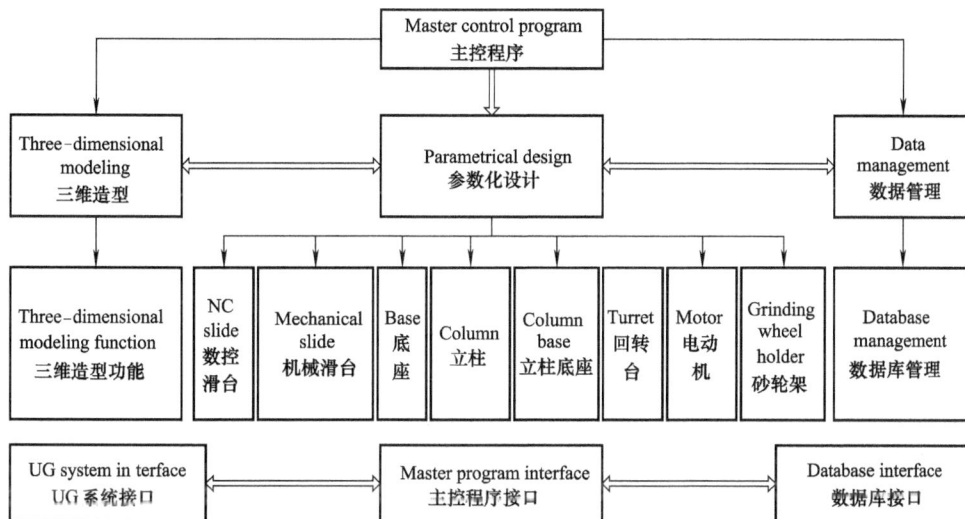

Fig. 2-13　Systematic structure of universal parts design module 通用件设计模块系统结构

设计数据库时，每一个通用件建立一个数据表，数据表的名字为该通用件的代号，该通用件的型号设为主键。这个主键值将作为数据库查询函数的查询条件。数据库访问程序在后台执行。

（3）主控程序　通用件模块工作流程如图 2-14 所示。

3. 模块内部装配结构的控制

WAVE 技术是建立各零件间参数关系的技术。UGWAVE 通过零件间建模来实现部件间的相关。即在同一装配体中，不同零部件或同一零部件内部在建模时共享连接几何体。数控成形磨齿机设计模块对应的各级装配模型，不仅包括部件结构，还包括部件之间的装配关系和装配约束。在数控成形磨齿机的计算机辅助模块化设计系统中，使用 WAVE 技术将模块按设计规则组成一个控制结构，在其中定义所需的几何信息和参数，使机床各级模块装配模型中的子模块之间的几何特征都是整体相关的。图 2-15 所示为利用本系统生成的数控成形磨齿机结构简图。用户只需输入主要参数即可得到所需的三维模型，在此基础上对产品进行细化设计。

由图 2-12 可知，成形磨齿机机械模块下设 6 个子模块，其中床身是基础件。基础件模块化在保证产品的规格和性能的前提下具有变化的可能性。数控滑台、回转工作台是系列通用部件，砂轮修整器和磨削头也是完整的部件。以结构相对独立的部件作为模块单元是最常

tablish the workpiece model directly, and uses the expression formula to control all parameters of model. With the 3D models established by means of the expression formula, when it is required to obtain the workpiece with the same topological structure and different geometric dimensions, simply open the expression editor, modify the value of the corresponding expression, and output another workpiece file.

(2) Database interface program

Microsoft Access is used to establish the database, and the ADO. net technology is used to connect and access the database. In the universal part module, all the parameters of universal parts are stored in the database. In the design of database, every universal part has a data sheet, and the name of the data sheet is the code of the universal part, the model of the universal part is set as major key. The value of the major key is taken as the query criteria of database query function. Database access procedures are performed in the background.

(3) Master control program

The route chart of the general parts module is shown in Fig. 2-14.

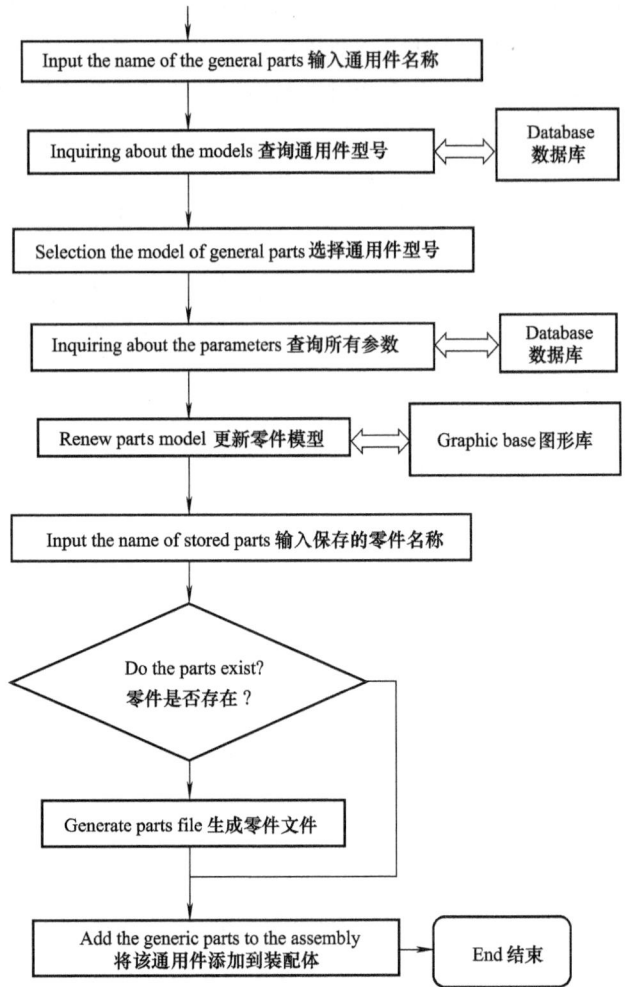

Fig. 2-14　Route chart of the general parts module
通用件模块工作流程图

3. Control of Assembly Structure Inside of Module

Wave technology is the technology to establish the parameters relationship between the parts. UGWAVE can achieve the correlation between assemblies. That is, in the same assembly, different components or the inside of the same component shares the connection geometry. The assembly model at all levels corresponding to the design module of the NC gear form grinding machine includes not only the assembly structure, but the assembly relation and assembly constraint between assemblies. In the computer-aided modular design (CAMD) system of NC gear form grinding machine, the module is designed into a control structure following the design rule by means of WAVE technology, and the required geometric information and parameters are defined in the control structure so as to make all geometric features between sub-modules in module assembly model at different levels relat-

见的模块类型，这样便于模块的设计、制造与互换。另外，只要将该机床上的磨削头更换为铣削头，还可实现内、外齿轮的成形铣削，从而实现数控成形磨齿机的横系列、纵系列和跨系列模块化设计。

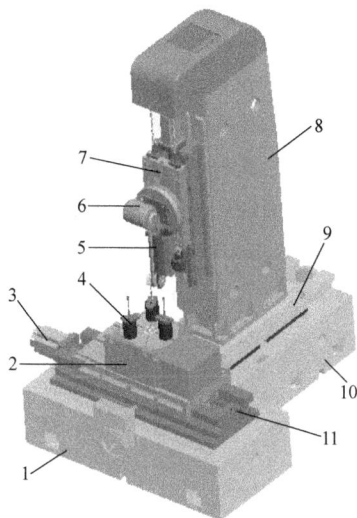

Fig. 2-15　NC gear form grinding machine 数控成形磨齿机

1、10—Base 底座　2—Rotary table 回转工作台　3—Servo motor 伺服电动机　4—Fixture 夹具
5—Grinding headstock 磨削头架　6—AC motor 交流电动机　7、11—NC slide 数控滑台　8—Colum 立柱　9—Sliding seat 滑座

2.6　逆向工程

2.6.1　逆向工程概述

第二次世界大战后，日本为了恢复和振兴本国经济，在科研上确立了"吸收性战略"，这给日本国民经济注入了新的活力，推动了日本经济的高速发展。到 20 世纪七八十年代，日本成为世界上仅次于美国的第二大经济强国。日本所采用的措施正是今天得到普遍重视的逆向工程技术。

逆向工程作为一种技术手段，可使产品研制周期缩短 40%以上。20 世纪 90 年代后，数字化浪潮推动社会飞速发展，世界范围内工业领域的竞争日趋激烈，企业必须不断地开发新产品、缩短产品开发周期、提高产品质量以增强企业在市场的竞争力，从而大大推动了逆向工程技术的研究与发展。

一般来讲，有两种截然不同的新产品开发模式：一种是正向工程，另一种是逆向工程。

1. 正向工程

传统的产品开发往往从市场需求出发，历经产品的概念设计、总体以及零部件的设计、制定工艺规程并设计夹具、完成加工和装配，最后对产品进行检验和性能测试。这种产品开发模式被称为正向工程，如图 2-16 所示。这种开发模式的前提是已经完成了产品的工程图设计或其 CAD 模型。

ed in the whole. Fig. 2-15 shows the structural diagram of gear form grinding machine generated by means of the CAMD system. The user can obtain the required 3D model by inputting the main parameters into the system only, and then design the product in detail.

It can be seen from Fig. 2-12 that the mechanical module of gear form grinding machine consists of six sub-modules, where the bed is a base parts. The modulization of base parts has the possibility to change in structure under the premise of ensuring product specifications and performances. NC sliding table and rotary table belong to the series of general components, the wheel dressing device and grinding head are also the complete assemblies. Taking the relatively independent assembly in structure as a module cell is the most common types of modules, which is convenient for the design, manufacture and interchange of the modules. In addition, as long as the grinding head of the machine tool is replaced by the milling head, the forming milling of the inner and outer gears can be realized, thus realizing the horizontal series, longitudinal series and cross series modular design of NC gear form grinding machine.

2.6　Reverse Engineering

2.6.1　Introduction to Reverse Engineering

After World War II, in order to restore and revitalize its economy, Japan drafted the national policy with "absorbency strategy". This viewpoint put new energy into national economy of Japan, and propelled the economy forward. By 1970s to 1980s, Japan had become the second powerful nation in economy. The measure taken by Japan is just the reverse engineering technology (RE) which receives great interest today.

As a technical means, reverse engineering can shorten the product developing cycle by 40% above. After 1990s, digital wave pushed the society forward rapidly. As the competition of industrial field is more and more intensive in the world, the enterprises have to develop new products continuously, shorten product developing cycle, improve product quality so as to enhance the competitive power in market, thus propelling reverse engineering forward.

General speaking, there are two different patterns in the development of new product: one is the normal engineering, the other is the reverse engineering.

1. Normal Engineering

Traditional product development often starts from market request, and goes through the conceptual design of product, overall and component design, process planning and fixture design, machining and assembling, up to the inspection and performance test of the product. The product development pattern is called normal engineering, as shown in Fig. 2-16. The premise of this development pattern is that the product engineering drawing or CAD model has been completed.

2. Reverse Engineering

In many cases, the initial information state of the product is not a CAD model, but all forms of physical model or sample. If a product needs to be copied or redesigned, the engineering design ex-

2. 逆向工程

在很多情况下，产品的初始信息状态并不是 CAD 模型，而是各种形式的物理模型或实物样件。如果要进行仿制或再设计，必须以现代设计理论、方法、技术为基础，运用各种专业人员的工程设计经验、知识和创新思维，对已有产品进行剖析和再创造，使之成为新产品。这种产品开发模式称为逆向工程，也称反求工程，如图 2-16 所示。

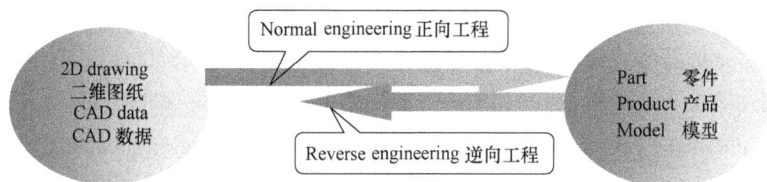

Fig. 2-16　Illustration of normal engineering and reverse engineering 正向工程与逆向工程图例

由图 2-16 可见，逆向工程和正向工程有很大不同。正向工程是一种计算机辅助的产品物化过程，而逆向工程是对已有"物化"产品的再设计。

逆向工程是通过对某种技术进行具体分析以弄清楚它是如何设计或如何工作的总过程。这种探究使每一个人都能参与关于系统和产品运作的建设性学习过程。逆向工程作为一种方法，并不局限于任何特定的目的，但往往是科学方法和技术发展的重要组成部分。这种把东西拆开并揭示其工作方式的过程通常是学习如何开发一项技术或对它进行改进的一种有效方法。

通过逆向工程，研究人员收集为建立一个系统某项技术或组件运作的文档所需的技术数据。在"黑箱"逆向工程中，对系统进行研究时并不审查内部结构，而在"黑箱"逆向工程中，系统内部的运作情况也要检测。

随着 CAD/CAM 技术的成熟和广泛应用，以 CAD/CAM 软件为基础的逆向工程应用越来越广泛。逆向工程的基本过程是：采用合适的测量设备和测量方法对实物模型进行测量，以获取实物模型的特征参数；将所获取的特征数据借助于计算机重构反求对象模型；对重构模型进行必要的创新和改进；进行数控编程并快速地加工出创新的新产品。图 2-17 所示为逆向工程流程图。

2.6.2　逆向工程的基本方法

1. 实物反求法

实物反求的物体往往是比较先进的设备或产品实物。通过对产品的设计原理、结构、材料、精度、制造工艺、包装、使用等方面进行分析研究和进一步再创造，最终研制出与原型产品相近或更佳的新产品。实物反求有两项基本工作，即反求分析与反求设计。实物反求设计是一个认识产品—再现产品—超越原产品的过程。实物反求对象可以是整个产品，也可以是部件、组件或零件。

实物反求的一般步骤为：①识别将要反求的产品或零部件；②观察或拆解关于原型产品是如何工作的信息文档；③把逆向工程产生的技术数据在原型产品的复制或改进中实施；④创造一个新产品，可能的话，并投放市场。

perience, knowledge and innovative thinking of different professionals must be used for analyzing and recreating the existing product based on modern design theory, method and technology, thus making it into a new product. The product development pattern is called reverse engineering, as shown in Fig. 2-16.

It can be seen from Fig. 2-16 that the reverse engineering is very different from the normal engineering. The normal engineering is a computer-aided materialization process of product, whereas the reverse engineering is the redesign of the existed "materialized" product.

Reverse engineering is the general process of analyzing a technology specifically to ascertain how it was designed or how it operates. This kind of inquiry engages individuals in a constructive learning process about the operation of systems and products. Reverse engineering as a method is not confined to any particular purpose, but is often an important part of the scientific method and technological development. The process of taking something apart and revealing the way in which it works is often an effective way to learn how to build a technology or make improvements to it.

Through reverse engineering, a researcher gathers the technical data necessary for the documentation of the operation of a technology or component of a system. In "black box" reverse engineering, systems are observed without examining internal structure, while in "white box" reverse engineering, the inner workings of the system are inspected.

With the maturity and wide application of CAD/CAM technology, the application of reverse engineering based on CAD/CAM software becomes more and more extensive. The basic process of reverse engineering is as follows: using the appropriate measuring equipment and measuring method to measure the physical model so as to gather the characteristic parameters of physical model; using the gathered characteristic data to reconstruct the model of reverse object with the help of computer; performing necessary innovation and improvement on the reconstructed model; preparing NC part program and machining the innovative new product rapidly. Fig. 2-17 illustrates the flow chart of reverse engineering.

2.6.2　Basic Methods of Reverse Engineering

1. Object Reversing

The substance of object reverse is often more advanced equipment or products. Through analyzing the product design principles, structure, materials, precision, manufacturing process, packaging, use etc. and making further recreation, a new product similar to or better than the prototype product is developed. There are two basic tasks in the object reverse, i.e. reverse analysis and reverse design. Object reverse is a process to know a product-reproducing a product-exceeding primary product. The substance of object reverse can be not only the entire product, but also the assembly, subassembly or workpiece.

The general steps of object reverse are as follows: ①identifying the product or component which will be reverse engineered; ②observing or disassembling the information document how the original product operates; ③implementing the technical data generated by reverse engineering in a replica or modified version of the original; ④creating a new product, and putting it into market if possible.

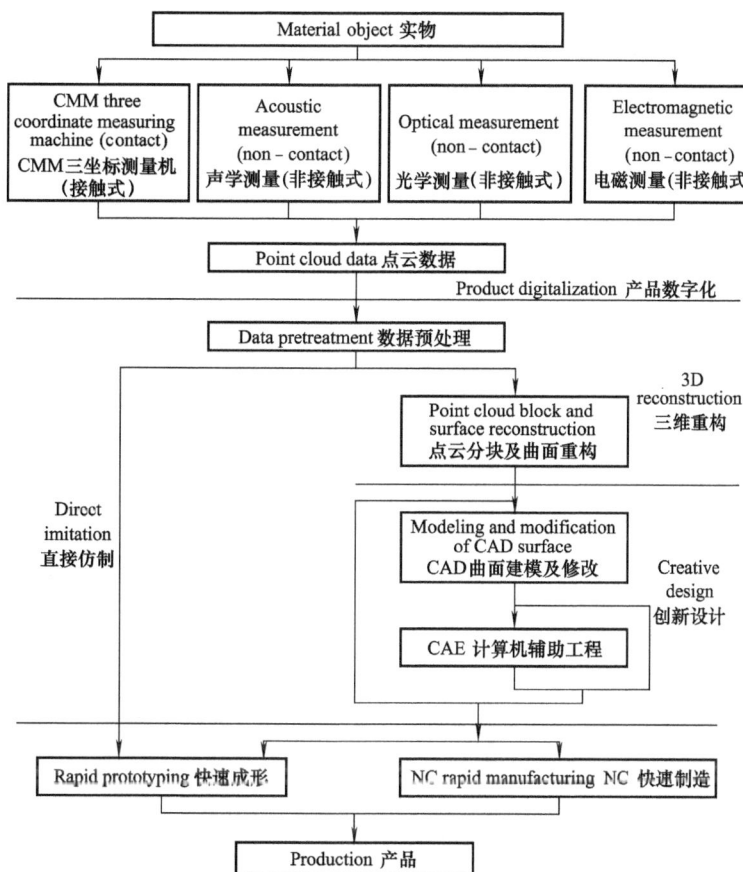

Fig. 2-17 Flow chart of reverse engineering 逆向工程流程图

2. 软件反求法

软件反求法是以产品样本、技术文件、设计书、使用说明书、图样、有关规范和标准、管理规范和质量保证手册等为研究对象的反求工程技术。软件反求的目的是通过对已有技术软件的分析和研究,提高对相关产品的设计和制造能力。

软件反求的一般设计步骤为:①对所反求的产品进行功能分析与结构分析;②分析并验证产品性能参数;③调研国内外同类产品,从中吸取有益成分;④撰写反求设计论证书。

3. 影像反求法

影像反求是以产品照片、广告介绍和影视画面等为参考资料进行分析设计的。影像反求难度最大,目前还未形成成熟的技术。一般要利用透视变换和透视投影,形成不同透视图,从外形、尺寸、比例去琢磨其功能和性能,进而分析其内部可能的结构。

影像反求的主要内容包括方案分析和结构分析。其中方案分析的重点是技术分析和经济分析,而结构分析主要是确定产品结构组成及结构材料。影像反求的一般设计步骤为:①收集参考资料,并对其进行多方面的分析、研究;②产品方案设计;③方案评价;④反求技术设计。

2. Software Reversing

Its research objects are product catalog, technical file, design book, instruction manual, part drawing, related specification and standards, management, management standards, quality ensuring handbook and so on. The purpose of software reversing is to increase the ability to design and manufacture related products by analyzing and studying the existed technical softwares.

The general steps of software reverse are as follows: ① to make function analysis and structure analysis on the product to be reversed; ②to analyze and verify the product performance parameters; ③to investigate the similar products at home and abroad, and to draw the beneficial ingredients; ④to write the argument material of reverse design.

3. Image Reversing

Its reference materials are product pictures, advertisements, video images and so on. This method has the greatest difficulty, and there is no mature technology. Generally, such technologies as the perspective collineation and perspective projection should be used to form different perspective drawings. The functions and performances of a product are studied from contour, dimension and scale and other aspects, further analyzing its possible structure inside.

Main contents of image reverse include scheme analysis and structure analysis. The emphases of scheme analysis are technical analysis and economic analysis, and the structural analysis is mainly to determine the elements of product structure and structural materials. The general steps of image reverse are as follows: ① to collect reference materials and make analysis and research in many aspects; ② to design product schemes; ③to evaluate the schemes; ④ to perform the reverse technology design.

2.6.3 Basic Procedure of Reverse Engineering

The basic procedure of reverse engineering is shown in Fig. 2-18. It can be seen from Fig. 2-18 that the reverse engineering is divided into three periods.

1. Analysis Period

Main tasks in the analysis period are to learn thoroughly the functional principle, structure shape, material property and machining process of the reversed object, to know its key functions and key technologies, to make the evaluation for the design features and deficiencies. This is very important for the reverse design to be smooth and successful or not. The technical indexes of sample parts and the topological relationship among the geometrical elements can be clear through analyzing the relative information of the reversed object.

2. Redesign Period

The specific tasks in this period are as follows: ①to determine the measuring method of the parts according to the analysis result; ② to correct the measured data because of the inevitable measuring errors in the measuring process; ③ to reengineer the geometrical model of reversed object by means of CAD system; ④ to redesign the product model by analyzing further the function of reversed object.

3. Manufacture Period

Main tasks in this period are to draw up the manufacturing process for the innovative product and to complete the manufacture of new product, as well as to test the structure and functions of

2.6.3　逆向工程的基本步骤

逆向工程的基本步骤如图 2-18 所示。由图 2-18 看出，逆向工程分为 3 个设计阶段：

1. 分析阶段

该阶段的主要任务是对反求对象的功能原理、结构形状、材料性能、加工工艺等方面进行全面深入的了解，明确其关键功能及关键技术，对设计特点和不足之处做出评估。这对反求设计能否顺利进行及成功与否至关重要。通过对反求对象相关信息的分析，可以明确样本零件的技术指标以及其中几何元素之间的拓扑关系。

2. 再设计阶段

该阶段的具体任务是：①根据分析结果，确定零件的测量方法、手段、顺序和精度等；②由于在测量过程中难免会有测量误差，需要对测得的数据进行修正；③利用 CAD 系统重构反求对象的几何模型；④进一步分析反求对象的功能，对产品模型进行再设计。

3. 制造阶段

该阶段的主要任务是拟订反求新产品的制造工艺，完成新产品的制造，并对制造后的产品进行结构和功能检测。如果不满足设计要求，可以返回分析阶段或再设计阶段重新修改。

气道是发动机的咽喉，其结构形状的优劣直接影响发动机的功率、油耗、排放和噪声等主要技术指标。发动机气道的形状复杂、分型面特殊、制造难度大，气道设计师通常无法给出准确、详细、能直接指导生产的气道工程图。

运用逆向工程实现气道形状的快速造型和优化有助于大大缩短气道研发周期和提高功效。

2.6.4　逆向工程的关键技术

1. 反求对象的分析

如何根据所提供的反求样本信息，获取反求对象的功能、原理、材料性能、加工工艺等，这是反求工程成败的关键。其具体内容包括：

（1）功能及原理分析　其目的是获得工作原理图，它基于产品原型，但比产品原型好。这也是逆向工程技术的精髓所在。

（2）反求对象的材料分析　它包括材料成分分析、结构和性能试验。

（3）反求对象的加工、装配工艺分析　通常，分析过程是根据反求对象的技术要求，如尺寸精度、几何形状误差和表面质量等去查找设计基准，由后向前一步一步地确定加工工序并制定工艺规程。

（4）精度分析　精度分析的过程是：①明确产品的精度指标；②综合考虑理论和原理误差；③尽可能分析所有误差源，确定总的精度指标；④根据实际生产条件调整和修正精度分配。

（5）反求对象造型的分析　产品造型设计集产品设计和艺术设计于一身。它把工业美学、产品造型原理、人机工程原理应用于产品设计，以提高产品的外观质量、舒适性和便捷性。

2. 逆向工程中的测量技术

目前，实物反求中关于产品几何参数的一些测量方法列于表 2-1 中。

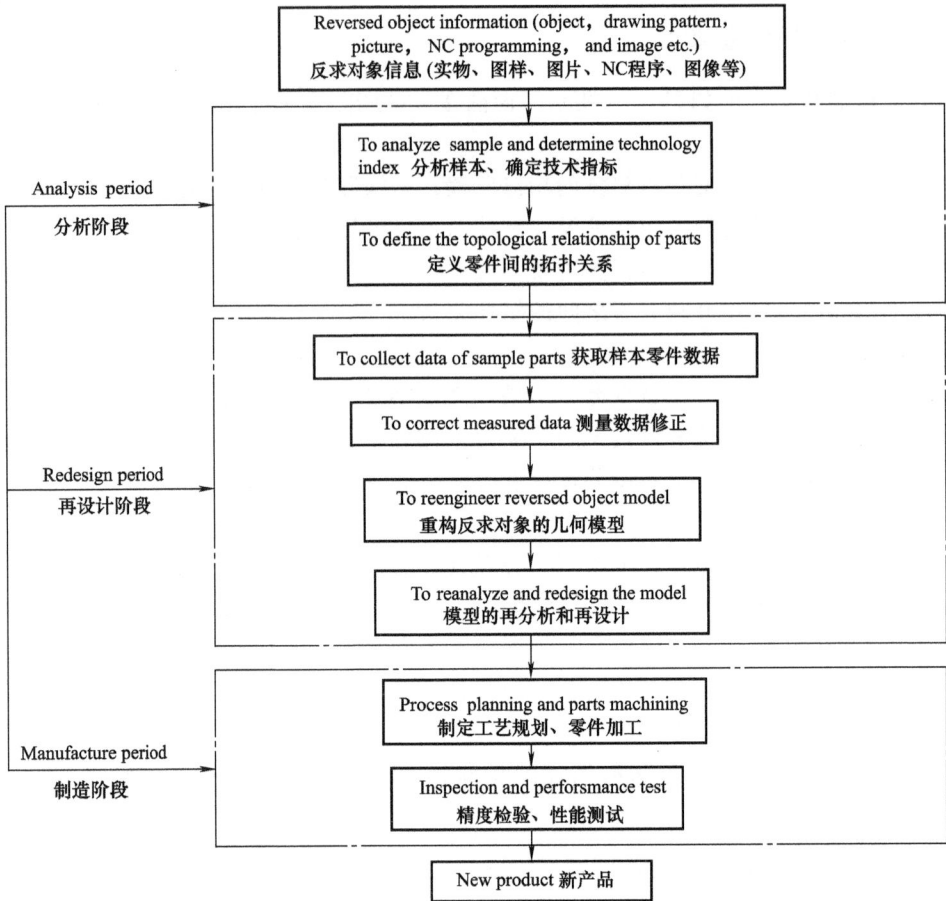

Fig. 2-18　Basic procedure of RE 逆向工程的基本步骤

manufactured product. If the product cannot meet the design requirements, it can return to the analysis period or the redesign period for remodification.

The port of engine is the fauces (vital passage) of the engine; its configuration has direct influence on the power, oil consumption, emission, noise and other main technical indexes. The port of engine has complex shape, special mould joint and is difficult to machine. The designer of port generally has no way to present an accurate, detailed engineering drawing which can be used to guide production directly.

Applying reverse engineering for realizing the rapid modeling and optimizing of port shape is helpful to shortening the port developing cycle greatly and increasing working efficiency.

2.6.4　Key Technology of Reverse Engineering

1. Analysis of Reversed Object

How to obtain the functions, principle, material property, machining process etc. based on the information of the sample to be reversed is the key to carry out reverse engineering successfully. The specific contents include:

(1) Analysis of function and principle　Its purpose is to obtain the working diagram, which is

Tab. 2-1　Basic measuring methods of geometric parameters of products **产品几何参数的基本测量方法**

Coordinate data collecting 坐标数据采集					
Nondestructive measurement 非破坏性测量					Destructive measurement 破坏性测量
Contact type 接触式		Non-contact type 非接触式			
Mechanical hand 机械手	CMM 三坐标测量机	Optical measuring machine 光学测量机	Sonic measuring machine 声学测量机	Magnetic measuring machine 电磁测量机	Automatic layer-cut scanning 自动断层扫描法

　　按测量方法是否会造成被测表面损毁，分为破坏性测量和非破坏性测量。按测头是否与被测表面接触，分为接触式测量和非接触式测量。显然，破坏性测量属于接触式测量。

　　（1）接触式测量方法　接触式测量是通过传感测量头与样件的接触而记录样件表面的坐标位置。它可以细分为点接触式和连续式数据采集方法。三坐标测量机（Coordinate Measuring Machine，CMM）是一种大型精密的三坐标测量仪器，如图2-19所示。其主要优点是测量精度高，对被测物的材质和色泽无特殊要求，适应性强。对于没有复杂内部型腔、特征几何尺寸多、只有少量特征曲面的零件，CMM是一种有效的三维数字化测量手段。但一般接触式测头测量效率低，而且对一些软质表面无法进行测量。三坐标测量机价格昂贵，对工作环境要求高，测量过程尚需人工干预。这些因素限制了它在快速反求领域的应用。

Fig. 2-19　Outside view of CMM
三坐标测量机外观图

　　（2）非接触式测量方法　这种方法有光学测量、超声波测量、电磁测量等。它们是分别基于光学、声学、磁学等原理，将一定的物理模拟量通过适当的算法转化为样件表面的坐标点。其中，基于光学原理的坐标测量设备在反求工程中应用最广。

3. 模型重构技术

　　模型重构是指根据收集的样件几何数据在计算机内重构样件模型。坐标测量技术的发展使得精密测量成为可能。样检测量数据是海量的，常常有数万甚至数百万个点。海量数据为数据处理和模型重构带来一定难度。

　　由测量机测得的所有数据之间没有任何联系。按照数据重构后形成表面的表示形式，模型重构可以分成两种类型：一种是由众多小三角面片构成的网格曲面模型；另一种是分片连续的样条曲面模型。前者应用更广。样件模型重构的基本过程如下：

　　1）数据预处理。它包括过滤、筛选、去噪、平滑、编辑等内容。

　　2）网格模型的生成。

　　3）网格模型的后处理。

　　由于在三角形网格模型中有大量的小三角面片，有必要在精度范围内对模型进行简化。此外，出于多方面的原因，在三角网格曲面上还存在一些瑕疵，如孔洞、缝隙和重叠等，这些瑕疵必须修补。

based on original product but better than it. This is also the quintessence of RE technology.

(2) Analysis of the material of object　It includes the analysis of material composition, structure and performance test etc.

(3) Analysis of the machining and assembly process of reversed object　In general, the analysis procedure is as follows: according to the technical requirement of reversed object, for example, dimension accuracy, geometric tolerance, and surface quality etc., to find out the design datum; to determine the machining operations step by step from the back to the front and to work out the process plan.

(4) Accuracy analysis　It follows the following procedure: ①making clear the accuracy index of product; ②comprehensive considering the theoretical and principle errors; ③analyzing all error sources as far as possible, and determining the general accuracy index; ④adjusting and correcting the accuracy allocation according to the practical production conditions.

2. Measurement Techniques in Reverse Engineering

At present, some main measuring methods of geometric parameters of product in object reversing are listed in Tab. 2-1.

Based on that the surface to be measured would be damaged, the measurement methods are classified into destructive measurement and nondestructive measurement. Obviously, the destructive measurement belongs to the contact type.

(1) Contact type measurement methods　In contact type measurement, the coordinate position of the sample surfaces is recorded through the contact between measurement head of the sensor and the sample piece. It can be divided into two types of data collection methods: point contact type and continuous type. Coordinate measuring machine (CMM) is a kind of large and precise coordinate measuring device, as shown in Fig. 2-19. Its chief advantages are high measurement accuracy, without special demand for the material and colour and lustre of the object to be measured and strong adaptability. For the workpieces without a complex internal cavity and with more characteristic geometry dimensions and a small amount of characteristic surfaces, CMM is an effective 3D digital measurement means. But the ordinary contact type measuring head has low measuring efficiency, and it has no way to measure some soft surfaces. CMM is very expensive and has high demand for work environment, and the measuring process of the CMM requires manual intervention. These factors limit its application in the quick reverse field.

(2) Nondestructive measurement methods　Optical measurement, ultrasonic measurement and electromagnetic measurement are the nondestructive measurement methods. They are based on the optical principle, acoustics principle and magnetic principle respectively, and convert the physical analogue quantity into the sample surface coordinates through the appropriate algorithm. The coordinate measuring facilities based on optics have found most widely application in reverse engineering.

3. Model Reengineering Technology

Model reengineering means reengineering the sample model in computer in term of the collected geometric data of sample. The development of coordinate measuring technology makes the fine measurement possible. Samples measuring data are enormous, often having ten thousands even several mil-

2.6.5　逆向工程的应用

以某发动机的气道作为研究对象，通过获取进气道的表面数据，进行曲面重构，获得曲面模型。采用铸模获得的气道砂芯模型进行数据采集获得气道的表面数据。

发动机气道具有典型的自由曲面特征。气道的曲面形状直接影响发动机的工作性能。传统的设计方法难以设计出合格产品，但采用自由曲面反求技术，则可实现产品快速研发。

1. 气道模型的数据采集与预处理

由于发动机气道是机体内表面，而且形状复杂，如果直接使用三维数字化测量设备进行测量，在后续的模具设计中容易产生比例误差。因此采用铸模获得的气道砂芯模型进行数据采集，以获得气道的表面数据。气道砂芯模型如图 2-20 所示。

采用光学测量设备，多次在不同方向上对测量对象的整体或局部进行大量密集的数据采集，最后通过数据叠加，以获得最完整的数据。在数据预处理过程中，重点是对一些尖锐边和边界附近的测量数据进行规则化处理，对残缺数据进行平滑，对冗余数据进行精简。图 2-21 所示为经过数据处理后气道砂芯模型的完整点云图。

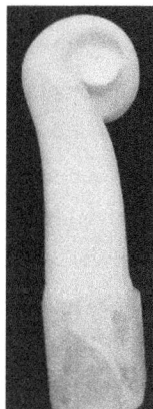

Fig. 2-20　Sand core model of port
气道砂芯模型

图 2-21 所示扫描所获得的点云数据已经能够完整地表达零件的几何信息。但由于离散化的点与数据不容易分辨零件的几何特征，不方便对点云数据进行分块处理，因此还必须对点云进行三角网格化处理。三角网格化处理后的点云数据如图 2-22 所示。

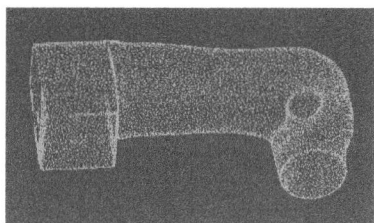

Fig. 2-21　Point cloud of the sand core model of port
气道砂芯模型点云

Fig. 2-22　Point cloud data with triangular mesh processing
三角网格化处理后的点云数据

2. 曲面建模

曲面建模的大致过程是：首先从点云数据中提取"特征线"。特征线是产品的外观流线，或者说是曲面模型的"骨架"，应该首先把它建构好。然后建构出决定产品外观的基础大曲面，并将精度和光顺程度调整到最佳状态。其次再补充基础曲面的过渡曲面，经过适当的倒圆和裁剪后就完成了曲面模型。在曲面模型完成后还应进行一次总的质量检查和评估。最后将曲面缝合加厚后就可以进入产品结构设计阶段。

3. 气道的曲面重构

气道形状不规则，不能由简单的一张曲面构成，而是由多张曲面经过延伸、过渡、裁剪

lions points. Enormous data bring about some difficulties for data processing and model reengineering.

There is not any connection among the original data measured by measuring machine. In the light of the expressing form of surface formed after data reengineering, model reengineering can be classified into two types: One is the net-type contoured surface model composed of many small triangular elements; the other is the continuous in piece splint surface model. The former is more widely used. Basic procedure of sample model reengineering is as follows:

1) Pre-processing of data. It includes filtering, selecting, noise-removing, smooth and editing etc.

2) Generating the net-type model.

3) Post-processing of the net-type model.

Because there are a large number of small triangular pieces in the triangular-net model, it is necessary to simplify the model within the limits of accuracy. Besides, because of many causes/reasons, there are some defects, such as holes, gaps and overlapping etc. on the triangular-net surface. These defects must be mended.

2.6.5 Applications of Reversal Engineering

Taking an engine port as research object, the internal surface data of the port are acquired firstly; then the surface of the port is reconstructed so as to obtain the surface model. In order to acquire the internal surface data of the port easily, here the sand-core model obtained by means of casting mould is used for collecting the data.

The engine port has the characteristic of free form surface. The surface shape of the port has a direct influence on the working performance of the engine. It is difficult to design the qualified products by means of traditional design methods, but the application of the free surface reverse technology can realize the quick development of product.

1. Data Collection and Pre-processing of Port Model

Because the engine port means the inside of the case and has complex shape, if the 3D digital measurement equipment is directly used to measure it, the subsequent mold design would produce scale error. Therefore, the port surface data are obtained through collecting the data of the sand core model of port which is obtained by using the casting mold. The sand core model of port is shown in Fig. 2-20.

The optical measuring equipment is used to collect a large number of intensive data many times on the whole or part of object to be measured from different directions. Finally, the most complete data can be acquired through data superposition. In the process of data preprocessing, the emphases are to carry out the regularization processing for the data measured at some sharp edges and near the boundary, to smooth the incomplete data, and to simplify the redundant data. Fig. 2-21 shows the complete point cloud of the sand core model of port after data processing.

After the preprocessing of the complete point cloud, triangular mesh processing must be done. The point cloud data acquired by scanning the Fig. 2-21 has already been able to express the geometric information of the parts completely. But because the discrete points and data are not easy to distinguish the geometric feature of the parts and are not convenient for the block processing of the

等混合而成。把曲面大体分成 3 块：导向部分、螺旋气道部分和气道与气缸的连接部分，如图 2-23 所示。构造出曲面后，必须对各曲面片进行拼接。曲面拼接要求两参数曲面切平面连续。最后，通过拟合误差分析和曲面修正，完成气道自由曲面整体模型的重构，如图 2-24 所示。

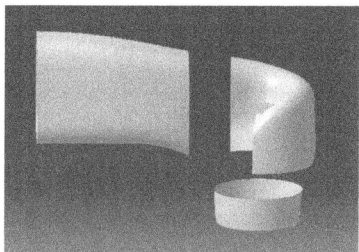

Fig. 2-23　Surface constructed by dividing
分块构建的曲面

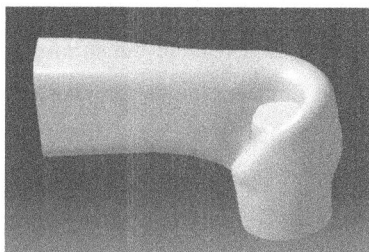

Fig. 2-24　Reengineered surface model
重构的曲面模型

复习题与习题

2-1　试分析现代设计技术的内涵与特点。

2-2　描述先进工程设计技术的体系结构。为什么说 CAD 技术是现代设计技术的主体？

2-3　计算机辅助设计技术包括哪些主要内容？

2-4　简述三维设计技术的发展历程。

2-5　CAE 的全称是什么？CAE 有哪些作用？

2-6　CAE 技术的主要内容有哪些？常用的 CAE 软件有哪些？

2-7　CAPP 系统由哪几部分组成？CAPP 的作用有哪些？

2-8　CAPP 系统的零件信息描述和输入方法有哪些？

2-9　分别简述派生性和创成型 CAPP 系统的基本原理。二者的主要区别是什么？

2-10　什么叫模块化设计？共分为哪几类？

2-11　简述模块化设计的过程。

2-12　模块化设计的关键技术是什么？

2-13　逆向工程的真正含义是什么？简述运用逆向工程技术从事产品开发的基本步骤。

2-14　逆向工程的关键技术是什么？

2-15　什么是模型重构技术？叙述模型重构的基本方法和步骤。

point cloud data, the triangular mesh processing of the point cloud data must be done. The point cloud data after triangular mesh processing are shown in Fig. 2-22.

2. Surface Modeling

On the whole, the surface modeling procedure is as follows. Firstly, the "typical line" is extracted from the point cloud data. The "typical line" is the appearance streamline of a product, or can be named as the "skeleton" of the surface model; hence, it should be built well firstly. Then, the large basis surface to determine the product appearance is constructed, and its accuracy and smoothness are adjusted to the best state. Next, the transition surfaces of the basis surface are further supplemented, and the surface model can be finished through appropriate rounding and trimming. After the surface model is finished, the total quality inspection and evaluation should be done one time. Lastly, the structure design of product can start after the surfaces are sewed up and thickened.

3. Surface Reconstruction of the Port

As the port has an irregular shape, it can't be composed of a simple surface, but is composed of multiple surfaces by stretching, transiting and trimming. The surface is roughly divided into three pieces: guiding part, helical port and the part connecting the port to the cylinder, as shown in Fig. 2-23. After the surface is constructed, the surface patches have to be jointed. The surface jointing requires that the tangential plane of two parameter surfaces should be continuous. Lastly, the reconstruction of the whole model of the free surface of port is accomplished by the fitting error analysis and surface modification, as shown in Fig. 2-24.

Review Questions and Problems

2-1　Try to analyze the connotation and characteristics of modern design technology.

2-2　Describe the architecture of advanced design technology. Why is the CAD technology the main part of modern design technology?

2-3　Does the CAD technology include what are the main contents in CAD technology?

2-4　Briefly describe the development process of 3D design technology.

2-5　What's the full name of CAE? What are the functions of CAE?

2-6　What are the main contents of CAE technology? What are the commonly used CAE software?

2-7　What are the components of a CAPP system? What are the functions of CAPP?

2-8　What are the descriptions and input methods for part information in CAPP systems?

2-9　Briefly state the basic principle of variant CAPP system and generative CAPP system respectively. What are the main differences between them?

2-10　What is modular design? What are the categories?

2-11　Briefly describe the process of modular design.

2-12　What are the key technologies of modular design?

2-13　What is the real meaning of reverse engineering? Briefly state the basic steps of using reverse engineering technology to develop product.

2-14　What are the key technologies of reverse engineering?

2-15　What is model reengineering technology? Narrate the basic methods and procedures to reconstruct model.

Chapter 3   Advanced Manufacturing Process

第3章　先进制造工艺

3.1　Introduction

Advanced manufacturing process refers to the technologies which are directly related to material handling process, and the high quality, high efficiency, low consumption, cleaning and flexibility are required in the manufacturing process. The development of advanced manufacturing process is shown in the following respects.

(1) Continuously increasing machining accuracy　With the progress and development of manufacturing process technology, machining accuracy has been continuously improved. For example, machining accuracy of the boring machine for machining the first steam engine cylinder is only 1 mm in the 18th century. Now the measuring accuracy of the electronic probe for measuring VLSI (very large scale integration circuit) has reached 0.25 nm.

(2) Rapidly increasing cutting speed　With the development of cutting tool material, the cutting speed has improved more than 100 times in nearly a century. For example, before 20th century, the cutting speed of cutting tool materials mainly used in carbon tool steel was around 10 m/min. Now, the cutting speed with ceramic cutting tool, polycrystalline diamond (PCD) cutting tool and polycrystalline cubic boron nitride (PCBN) cutting tool can achieve thousands of meters per minute.

(3) The application of new engineering materials promotes further development of manufacturing process　With the emergence of new materials, such as superhard materials, superplastic materials, composite materials, engineering ceramics, etc. On the one hand, it is required to further improve the cutting performance of tool materials and the machining equipment so that it can be competent for cutting the new materials. On the other hand, it makes people to seek new manufacturing processes in order to adapt the machining of new engineering materials more effectively. Thus, a series of unconventional machining methods have emerged.

(4) Continuous development of near net shape technology　With the enhancement of people's conservation and protection consciousness of human survival resources, it asks the forming accurate of the blanks to develop towards less allowance or no allowance, and makes the formed blank near to or reach the final shape and size of the part. As a result, it leads to the emergence of the near net shape technology, such as precision casting, precision plastic forming, precision connection, etc. Thus, the near net shape technologies, such as precision casting, precision plastic forming and precise connection and so on are presented.

(5) Surface engineering technology is getting more and more attention　Surface engineering technology is a kind of technology which can change the shape morphology, chemical composition and microstructure of a part by using physical, chemical or mechanical processes to obtain different performance requirements from the substrate material. It plays a more and more important role in saving raw materials, improving the performance of new products, prolonging the service life of products, decorating environment, and beautifying life, etc.

Based on the characteristics of the processed materials, advanced manufacturing process can be divided into the following four technologies:

1) Precision and ultra-precision machining technology. It is a technical measure which makes the size and the surface performance of the workpiece meeting product requirements during the re-

3.1 概述

先进制造工艺研究的是与物料处理过程直接相关的各项技术，要求实现加工过程的优质、高效、低耗、清洁和灵活。先进制造工艺的发展表现在以下几个方面：

（1）加工精度不断提高 随着制造工艺技术的进步和发展，机械加工精度得到不断提高。例如，18 世纪，用于加工第一台蒸汽机汽缸的镗床的加工精度仅为 1 mm，现在测量超大规模集成电路所用的电子探针，其测量精度已达 0.25 nm。

（2）切削速度迅速提高 随着刀具材料的发展，在近一个世纪内，切削速度提高了 100多倍。例如，20 世纪以前，以碳素工具钢为主的刀具材料的切削速度为 10 m/min 左右；而现在使用陶瓷刀具、聚晶金刚石（PCD）刀具和聚晶立方氮化硼（PCBN）刀具，切削速度可达每分钟上千米。

（3）新型工程材料的应用推动了制造工艺的进一步发展 超硬材料、超塑性材料、复合材料、工程陶瓷等新型材料的出现，一方面要求进一步改善刀具材料的切削性能、改进机械加工设备，使之能够胜任新材料的切削加工，另一方面迫使人们寻求新型的制造工艺，以便更有效地适应新型工程材料的加工，因而出现了一系列特种加工方法。

（4）近净成形技术不断发展 随着人们对人类生存资源的节约和保护意识的增强，要求零件毛坯成形精度向少无余量方向发展，使成形的毛坯接近或达到零件的最终形状和尺寸，因而出现了诸如精密铸造成形、精密塑性成形、精密连接等近净成形技术。

（5）表面工程技术日益受到重视 表面工程技术是通过运用物理、化学或机械工艺来改变零件表面的形态、化学成分和组织结构，以获取与基体材料不同的性能要求的一种技术。它在节约原材料、提高新产品性能、延长产品使用寿命、装饰环境、美化生活等方面发挥着越来越突出的作用。

基于所处理物料的特征，先进制造工艺可划分为以下 4 种技术：

1）精密、超精密加工技术。它是指对工件表面材料进行去除，使工件的尺寸、表面性能达到产品要求而采取的技术措施。

2）精密成形制造技术。它是指工件成形后只需少量加工或无须加工就可用作零件的成形技术。

3）特种加工技术。它是指利用电、磁、声、光、化学等能力及其组合施加在工件的被加工部位上，从而达到材料去除、变形、改变性能等目的的非传统加工技术。

4）表面工程技术。它是指采用物理、化学、金属学、高分子化学、电学、光学和机械学等学科的技术及其组合，提高产品表面耐磨、耐蚀、耐热、耐辐射、抗疲劳等性能的各项技术。

3.2 超精密加工技术

精密、超精密加工技术是指加工精度达到某一量级的加工技术的总称。超精密加工技术旨在提高零件的几何精度，以保证机器部件配合的可靠性、运动副运动的精确性、长寿命和低运行费用等。

moval of the workpiece surface material.

2) Precision forming manufacturing technology. It is a forming technology that makes formed workpieces becoming parts with little or no machining.

3) Nontraditional machining technology. It is a nontraditional machining technology that can achieve material removal, deformation and change performance by using electricity, magnetism, sound, light, chemistry and other combinations to put on the processed parts of the workpiece.

4) Surface engineering technology. It contains various techniques that can improve wear resistance, corrosion resistance, heat-resistant, radiation resistance, anti-fatigue and other properties of the product surface by using physics, chemistry, metallography, polymer chemistry, electricity, optics, mechanics and other subjects.

3.2 Ultra-Precision Machining Technology

Precision and ultra-precision machining technology is a general term of machining technologies in which machining accuracy has reached a certain level. Ultra-precision technology aims to improve the geometrical precision of parts to ensure the reliability of machine parts, the accuracy, long life and low operating costs of kinematic pairs, etc.

According to the mechanism and characteristics of the machining method, the classification of ultra precision machining is shown in Fig. 3-1.

3.2.1 Key Technologies of Ultra-Precision Machining

1. Machine Tool Technology of Ultra-Precision Machining

(1) Precision spindle   The spindle of ultra-precision machine tool directly supports the movement of workpiece or cutter during the machining process. So, the rotation accuracy of the spindle directly affects the machining precision of the workpiece. Now, the spindle with highest rotation accuracy used in ultra-precision machining machine tool is aerostatic bearing spindle. The roundness of bearing parts and air supply conditions have great influences on the rotation accuracy of aerostatic bearing. Due to the homogenization of the pressure film, the rotation accuracy of bearings can reach the 1/15-1/20 of the roundness of bearing components. So, in order to get 10 nm rotation accuracy, the roundness of the shaft and shaft sleeve is about 0.15-0.20 μm. Now, the rotation accuracy of air bearings spindle is up to 0.05 μm in China, while it is 0.03 μm abroad.

(2) Ultra-precision guideway   The guideway of ultra-precision machining machine tool should have flexible motion, no crawling and good linear precision. Rigid in use should be compatible with use conditions. In addition, it has less calorific value during high speed motion and is easy to maintain. Common guideway form of ultra-precision machining machine tool includes V-V sliding guideway, rolling guideway, hydrostatic guidway and aerostatic guideway. The traditional V-V sliding guideway and rolling guideway have been used in the United States and Germany. Owing to viscous shearing resistance of oil, hydrostatic guide generates bigger heat. Therefore, hydraulic oil must be cooled. In addition, the hydraulic device is large, and the maintenance of the oil circuit is also troublesome. Because the support of aerostatic guide is planar, a larger supporting stiffness can be obtained, and it is almost no heating problem. In the maintenance, it should pay attention to the dust-

根据加工方法的机理和特点，超精密加工分类如图 3-1 所示。

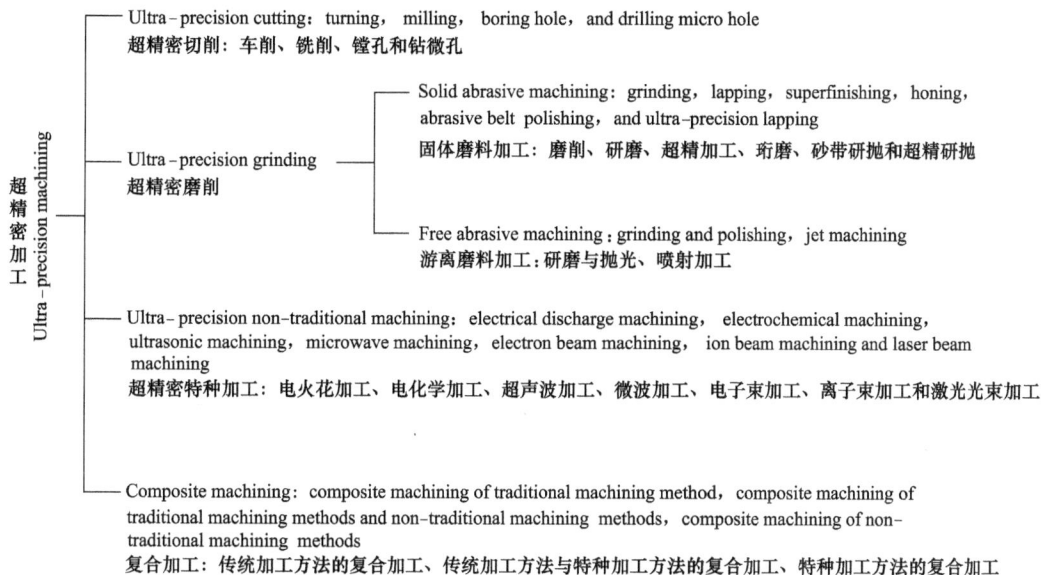

超精密加工 Ultra-precision machining
- Ultra-precision cutting: turning, milling, boring hole, and drilling micro hole
  超精密切削：车削、铣削、镗孔和钻微孔
- Ultra-precision grinding 超精密磨削
  - Solid abrasive machining: grinding, lapping, superfinishing, honing, abrasive belt polishing, and ultra-precision lapping
    固体磨料加工：磨削、研磨、超精加工、珩磨、砂带研抛和超精研抛
  - Free abrasive machining: grinding and polishing, jet machining
    游离磨料加工：研磨与抛光、喷射加工
- Ultra-precision non-traditional machining: electrical discharge machining, electrochemical machining, ultrasonic machining, microwave machining, electron beam machining, ion beam machining and laser beam machining
  超精密特种加工：电火花加工、电化学加工、超声波加工、微波加工、电子束加工、离子束加工和激光光束加工
- Composite machining: composite machining of traditional machining method, composite machining of traditional machining methods and non-traditional machining methods, composite machining of non-traditional machining methods
  复合加工：传统加工方法的复合加工、传统加工方法与特种加工方法的复合加工、特种加工方法的复合加工

Fig. 3-1　Ultra-precision machining methods 超精密加工方法

3.2.1　超精密加工关键技术

1. 超精密加工机床技术

（1）精密主轴　超精密加工机床的主轴在加工过程中直接支持工件或刀具的运动，故主轴的回转精度直接影响到工件的加工精度。现在超精密加工机床中使用的回转精度最高的主轴是空气静压轴承主轴。空气静压轴承的回转精度受轴承部件圆度和供气条件的影响很大，由于压力膜的匀化作用，轴承的回转精度可以达到轴承部件圆度的 1/15~1/20，因此要得到 10 nm 的回转精度，轴和轴套的圆度要达到 0.15 μm~0.20 μm。目前使用的空气轴承主轴回转精度国内可达 0.05 μm，而国外可达 0.03 μm。

（2）超精密导轨　超精密加工机床导轨应动作灵活、无爬行，直线精度好，在使用中应具有与使用条件相适应的刚性，高速运动时发热量少，维修保养容易。超精密加工机床常用的导轨形式有 V-V 形滑动导轨和滚动导轨、液体静压导轨和气体静压导轨。传统的 V-V 形滑动导轨和滚动导轨在美国和德国的应用都取得了良好的效果。液体静压导轨由于油的黏性剪切阻力导致发热量比较大，因此必须对液压油采取冷却措施。另外，液压装置比较大，而且油路的维修保养也较麻烦。气体静压导轨由于支承都是平面，可获得较大的支承刚度，它几乎不存在发热问题，在维修保养方面则要注意导轨面的防尘。

在精度方面，空气导轨是目前最好的导轨。国际上空气导轨的直线度可达 0.1 μm/250 mm~0.2 μm/250 mm，国内可达到 0.1 μm/200 mm，通过补偿技术还可进一步提高导轨的直线度。

（3）传动系统　目前用于精密加工和超精密加工的传动系统主要有：滚珠丝杠传动、静压丝杠传动、摩擦驱动和直线电机驱动。

精密滚珠丝杠是超精密加工机床常采用的驱动方法，超精密加工机床一般采用 C0 级滚

proof of the guideway surface.

In terms of accuracy, air guideway is the best guideway at present. The straightness of international air guideway can reach 0.1 μm/250 mm-0.2 μm/250 mm, the domestic can achieve 0.1 μm/200 mm. The straightness of guideway can be further improved by the compensation technology.

(3) Transmission system   Currently, the transmission system of precision machining and ultra-precision machining mainly includes ball screw drive, hydrostatic screw drive, friction drive and linear motor drive.

Precision ball screw is the common driving method for ultra-precision machining machine tool. C0 grade ball screw is common used in ultra-precision machining machine tool, and the highest positioning accuracy can reach 0.01μm by using closed-loop control. The micro-elastic deformation principle of ball screw can be used to achieve the feed with nano-resolution. But the kinematic precision of guideway can be influenced by the installation error of ball screw, the bending of screw itself, the beating and manufacturing error of the ball, and the pretightening of nut, etc.

The screw and nut of hydrostatic screw do not contact directly, there is a layer of high-pressure membrane. So, there is no crawl and reverse clearance because of the friction, it can keep its accuracy for a long time and the feed resolution can be higher. The feed precision can be improved owing to the homogenization of dielectric film (oil and air). The nano-level positioning resolution can be reached on a longer trip, but its stiffness is relatively small. Liquid hydrostatic screw device is bigger, and there are a lot of auxiliary devices such as oil pump, accumulator, liquid circulation device, cooling device and filtering device, etc. There are still some environmental pollution problems.

Friction driven realizes non-clearance transmission during directly converting rotary motion of servo motor into linear motion by friction. Because of the simple structure, the elastic deformation is greatly reduced and it is a very suitable for ultra-precision machining transmission system. The feed mechanism of Nanoform 600 ultra-precision mirror machining machine tool developed by British Rank Tailor Hobson company uses this device, 1.255 nm resolution and ±0.1μm positioning accuracy can be obtained on the 300 mm trip.

Linear motor is a kind of power device which can directly convert electrical energy into linear mechanical motion. It is suitable for high speed and high precision machining. Usually, high-speed ball screw can work at the speed of 40 m/min and the acceleration of 0.5 g. However, the acceleration of the linear motor can reach 5 g and its velocity and rigidity are more than 30 times and 7 times of ball screw respectively. At present, the positioning precision, resolution and speed of linear motor drive can reach 0.04 μm, 0.01μm and 200 m/s respectively.

(4) Micro-feed device   Micro-feed device with high precision plays an important role in realizing ultra-thin cutting, the machining with high dimension precision and on-line error compensation. In ultra-precision machining, common micro-feed device includes elastic deformation, thermal deformation, fluid film deformation, magnetostrictive deformation and piezoelectric ceramics, etc, where the piezoelectric ceramic material has a better micro-displacement characteristics and controllable properties. Presently, the micro-displacement mechanism based on elastic hinge is the most widely used in piezoelectric ceramic actuator.

2. Ultra-Precision Machining Tool Technology

Natural diamond cutting tool is the most important ultra-precision cutting tool. Because the

珠丝杠，利用闭环控制最高可达 0.01 μm 的定位精度。利用滚珠丝杠的微小弹性变形原理，可实现纳米分辨率的进给。但是，丝杠的安装误差、丝杆本身的弯曲、滚珠的跳动及制造上的误差、螺母的预紧程度等都会给导轨运动精度带来影响。

静压丝杠的丝杠和螺母不直接接触，有一层高压膜相隔，因此没有摩擦引起的爬行和反向间隙，可以长期保持其精度，进给分辨率可更高。由于介质膜（油、空气）有匀化作用，可以提高进给精度，在较长行程上，可以达到纳米级的定位分辨率，但它的刚度比较小；液体静压丝杠装置较大，且必须有油泵、蓄压器、液体循环装置、冷却装置和过滤装置等众多辅助装置，另外还存在环境污染问题。

摩擦驱动传动系统是通过摩擦把伺服电机的回转运动直接转换成直线运动，实现无间隙传动，由于结构上比较简单，因而弹性变形因素大为减少，是一种非常适合超精密加工的传动系统。英国的 Rank Tailor Hobson 公司开发的 Nanoform 600 超精密镜面加工机床的进给机构采用了这种传动系统，在 300 mm 的行程上可获得 1.255 nm 分辨率、±0.1 μm 的定位精度。

滚珠丝杠可在 40 m/min 的速度和 0.5 g 加速度的情况下工作，而直线电机的加速度可达 5 g，其速度和刚度都分别大于滚珠丝杠的 30 倍和 7 倍。目前直线电机传动定位精密度可达到 0.04 μm，分辨率为 0.01 μm，速度可达 200 m/s。

（4）微进给装置 高精度微进给装置对实现超薄切削、高精度尺寸加工和实现在线误差补偿有着十分重要的作用。在超精密加工中，常用的微进给装置有弹性变形式、热变形式、流体膜变形式、磁致伸缩式、压电陶瓷式等多种结构形式。其中，压电陶瓷材料具有较好的微位移特性和可控制性。以压电陶瓷为驱动器的基于弹性铰链支承的微位移机构目前来说是用得最多的。

2. 超精密加工刀具技术

天然金刚石刀具是目前最主要的超精密切削工具，由于它的刃口形状会被直接反映到被加工表面上，因此，金刚石刀具刃磨技术是超精密切削中的一项重要技术。刃磨技术包括晶面的选择、刃口刃磨工艺以及刃磨后刃口半径的测量等。

3. 精密测量技术

目前，在超精密加工领域，尺寸测量技术主要有两种：一种是激光干涉技术，另一种是光栅技术。激光干涉仪分辨率高，最高可达 0.3 nm，一般为 1.25 nm；测量范围大，可达几十米；测量精度高。

近年来超精密加工领域越来越多地选用光栅作为测量工具。从分辨率上看，光栅系统分辨率可达 0.1 nm，测量范围可达 100 mm，精度可达 +0.1 μm。而且光栅对环境的要求相对较低，可以满足超精密加工的使用要求，是很有前途的超精密测量工具。

4. 超精密加工原理

超精密加工的精度要求越来越高，在这种情况下，只是靠改进原来的技术很难提高加工精度，因此，可以从工作原理着手进行研究，以寻求新的解决办法。

近年来，新工艺、新加工方法不断出现。在现代加工中出现了电子束、离子束、激光束、微波加工、超声加工、刻蚀、电火花加工、电化学加工等多地加工方法。电子束、激光束和离子束加工等加工方法虽然使加工效率有相当大的提高，但目前来看都不能满足要求。

发达国家的工艺技术是与新原理、新方法的创新紧密结合的。原始创新不足是制约我国

shape of cutting edge is directly reflected on the machined surface, therefore, the grinding technology of diamond cutting tool is an important technology of ultra-precision cutting. Grinding technology includes the selection of crystal surface, the cutting edge grinding process and the measuring of the edge radius after grinding, etc.

3. Precision Measurement Technology

At present, the dimension measurement technology in ultra-precision machining field mainly has two kinds: one is the laser interference technique; the other is the grating technique. Laser interferometer has a high resolution, the highest resolution is 0. 3 nm and the common resolution is 1. 25 nm. It has a large measuring range up to tens of meters and high measurement accuracy.

In recent years, the grating as measuring tool is used more and more widely in ultra-precision machining. From the resolution, the resolution of the grating system can reach 0. 1 nm. So, its range and accuracy can reach 100 mm and +0. 1 μm respectively. Moreover, the grating has relatively low requirement for environment. So, it can satisfy the use requirement of ultra- precision machining and it is a promising ultra-precision measuring tool.

4. Ultra-Precision Machining Principle

The accuracy of ultra-precision machining is getting higher and higher. In this case, it is very difficult to improve machining accuracy only by improving original technology. Therefore, it can be studied from the working principle in order to find a new solution.

In recent years, new process and new machining methods are appearing constantly. In modern machining, there are many kinds of machining methods, such as electron beam, ion beam, laser beam, microwave machining, ultrasonic machining, etching, electrical discharge machining, and electrochemical machining and so on. The efficiency of machining methods such as electron beam, laser beam and ion beam has been greatly improved, but it cannot meet the requirements at present.

The process technology in developed countries is closely combined with the innovation of the new theory and new method. Lack of original innovation is the main reason for restricting our technology level. Therefore, the basic research of machining technology and development of the innovation process must be strengthened.

5. Ultra-Precision Environmental Control Technology

Ultra-precision machining should be carried out under certain environment so as to achieve the requirements of precision and surface quality. Good work environment is the necessary condition to ensure the quality of ultra-precision machining. The main factors influencing the environment include temperature, humidity, pollution and vibration, etc.

3. 2. 2　Diamond Ultra-Precision Turning

Diamond ultra-precision turning is a precision machining method developed to adapt the machining requirements of precision products, such as computer disks, recorder drums, and all kind of precision optical mirrors, primary mirrors of radio telescopes, camera lenses, plastic lenses, and resin contact lens and so on. It is mainly used in machining of non-ferrous metals and their alloys, such as aluminum, copper and so on. In addition, it is also used in machining of non-metallic materials including optical glass, marble and carbon fiber, etc.

1. Mechanism and Characteristics of Diamond Ultra-Precision Turning

Diamond ultra-precision turning belongs to the micro cutting, and its machining mechanism is

工艺水平提高的主要原因。因此，必须加强工艺技术的基础研究和创新工艺的研发。

5. 超精密环境控制技术

超精密加工要求在一定的环境下进行，这样才能达到在精度和表面质量上的要求。良好的工作环境是保证超精密加工质量的必要条件，影响环境的主要因素有温度、湿度、污染和振动等。

3.2.2 金刚石超精密车削

金刚石超精密车削是为适应计算机用的磁盘、录像机中的磁鼓、各种精密光学镜、射电望远镜主镜面、照相机塑料镜片、树脂隐形眼镜镜片等精密产品的加工而发展起来的一种精密加工方法。它主要用于加工铝、铜等非铁系金属及其合金，以及光学玻璃、大理石和碳素纤维等非金属材料。

1. 金刚石超精密车削机理与特点

金刚石超精密车削属于微量切削，其加工机理与普通切削有较大的差别。超精密车削的加工精度要达到 0.1 μm，表面粗糙度值 Ra 要达到 0.01 μm，刀具必须具有切除亚微米级以下金属层厚度的能力。这时的切削深度可能小于晶粒的大小，切削在晶粒内进行，要求其切削力大于原子、分子间的结合力，刀刃上所承受的切应力可高达 13 000 MPa。

刀尖处应力极大，切削温度极高，一般刀具难以承受，由于金刚石刀具具有极高的硬度和高温强度，耐磨性和导热性能好，加之金刚石本身质地细密，能磨出极其锋利的刃口，因此，可以加工出粗糙度很小的表面。通常，金刚石超精密车削会采用很高的切削速度，故产生的切削热少，工件变形小，因而可获得很高的加工精度。

2. 金刚石超精密车削的关键技术

（1）金刚石刀具及其刃磨　超精密车削刀具应具备的主要条件见表3-1。

Tab. 3-1　Main conditions required for the ultra-precision turning tool 超精密车削刀具应具备的主要条件

Categories 分类	Main conditions 主要条件
Geometry of cutting part in cutting tool 刀具切削部分的几何形状	1. The sharp cutting edge with the minimum edge radius is able to achieve ultra-thin cutting thickness. 具有极小刃口圆弧半径的锋利刀刃口能实现超薄切削厚度 2. The cutting tool geometry does not generate tooth mark, has high strength and very small cutting force. 具有不产生走刀痕迹、强度高、切削力非常小的刀具切削部分几何形状 3. The cutting edge has no defects and can cut a super smooth mirror. 刀刃无缺陷，能得到超光滑的镜面
Physical and chemical properties 物理及化学性能	1. Good anti-adhesion ability, small chemical affinity and low friction coefficient to the workpiece are helpful to obtain excellent machining surface integrity. 与工件材料的抗黏性好，化学亲和力小，摩擦系数低，能得到极好的加工表面完整性 2. Extremely high hardness, wear resistance and elastic modulus so as to ensure a very long tool life and very high dimension durability. 极高的硬度、耐磨性和弹性模量，以保证刀具具有很长的寿命和很高的尺寸耐用度

（2）加工设备　金刚石车床是金刚石车削工艺的关键设备。它应具有高精度、高刚度和高稳定性，还要求抗振性好、热变形小、控制性能好，并具有可靠的微量进给机构和误差

very different from the ordinary cutting. In order to reach the machining error of 0. 1 μm and surface roughness Ra of 0. 01 μm for the ultra-precision turning, the cutting tools must be able to cut the metal layer thickness below sub-micron scale. The cutting depth may be less than the size of grain at this time. Owing to the cutting being carried out inside the grain, it is required that the cutting force be greater than the binding force between the atoms and molecules. As a result, the shear stress borne by the cutting edge can achieve 13, 000 MPa.

It is unbearable for common cutting tool because of enormous stress and cutting temperature on the tip of cutting edge. The diamond tool has the characteristic of sky-high hardness and high temperature strength, good wear resistance and thermal conductive performance. In addition, the sharp edge can be obtained by the grinding method because of fine texture of the diamond itself. Therefore, small surface roughness can be obtained using the diamond tool. Usually, high cutting speed is taken during the process of diamond ultra-precision turning, which can cause less cutting heat and small distortion of the workpiece and obtain high machining accuracy.

2. Key Technologies of Diamond Ultra-Precision Turning

(1) Diamond tool and its grinding   The main required conditions of the ultra-precision turning tools are shown in Tab. 3-1.

(2) Machining equipment   Diamond lathe is the key equipment in diamond turning process. It should have high precision, high rigidity and high stability. In addition, it also needs good vibration resistance, small thermal deformation, good control performance, reliable micro-feed mechanism and error compensation device.

3. Applications of Ultra-Precision Cutting

(1) Mirror turning of magnetic disc substrate   Magnetic disc storage is one of the main peripheral devices of computers. With the rapid development of computer technology, the data storage density of magnetic discs increases continuously, and the floating height of magnetic head decreases rapidly. In order to maintain a gap of less than 0. 3 μm between the floating head and the magnetic disc rotating at 3, 600 r/min so that the head can read/write data rapidly and accurately, the disc surface should have very high precision. For example, if the flying height with 0. 33 μm is required, the surface roughness Ra of the disc must be 0. 015 μm or less. In recent years, the oversea researchers have developed a new running method, i. e. running in skipping way, in order to increase the data storage density of discs by a factor of at least 20. The minimum interval skipped by magnetic head in the world is 3 nm. This means there is no air layer between the head and disc, that is to say, the gap between them approaches to zero. The light-duty micro-miniature magnetic head has been developed in order to alleviate the mechanical damage of discs resulted from direct contact. The subminiaturization of the head requires both the head and the disc should be machined with higher precision. Therefore, high-precision machining of disc substrates is a vital subject in the development of magnetic disc storages.

The blank of magnetic disc substrates is disc-type aluminum alloy with higher surface flatness. Both sides of the blank are turned with diamond cutter before annealing, and then surface polishing is carried out. The magnetic disc substrate can be machined on high-precision lathe by means of diamond cutter sharpened carefully.

(2) Mirror milling of aircraft organic glass   Windows in modern larger passenger planes are made of organic glass. During the planes take off and land on the ground, windows are frequently

补偿装置。

3. 超精密切削的应用

（1）磁盘基片的镜面车削　磁盘存储器是计算机的主要外部设备之一。随着计算机技术的飞速发展，磁盘单位面积的存储密度也在不断提高，磁头在磁盘上的浮动高度急剧减小。要使磁头与以 3600 r/min 速度旋转的磁盘间稳定地保持 0.3 μm 以下的间隙，磁头快速、准确无误地存取信息，这就要求磁盘表面具有很高的精度。例如，如果要求浮动高度为 0.33 μm，则磁盘表面粗糙度值 Ra 就要在 0.015 μm 以下。近年来，国外为了把磁盘记录密度再提高 20 倍以上，正在开发新的磁头走行方式，即磁头跳跃走行方式。目前世界上磁头跳跃的最小间隔为 3 nm，使磁头与磁盘之间没有空气层，间隙近于零。为了减轻直接接触磨损等机械损伤，又研究开发了负荷轻的超小型磁头。由于磁头超小型化，对磁头和盘片平面质量的加工要求更高了。因此，磁盘基片的高精度加工是磁盘存储器开发中的重要课题。

磁盘基片使用平面度好的铝合金圆盘做毛坯，铝板两面用金刚石车刀车削后进行退火处理，再进行表面抛光。磁盘基片可在高精度磁盘车床上采用仔细刃磨过的金刚石车刀加工。

（2）飞机玻璃的镜面铣削　现代大型客机的窗户是使用有机玻璃制成的。飞机起飞和降落时，玻璃屡遭大气中夹带的沙尘碰撞，飞行一定起降次数后，窗户玻璃表面就会变得十分粗糙，直接影响飞行员和乘客的视野，这就需要对玻璃进行重新抛光修复。采用传统抛光方法，修复一块玻璃通常需要 1 h 左右，当玻璃有较深零星刻痕时，加工时间会更长。若采用镜面铣削方法，所需时间不到抛光时间的一半，可以大大缩短飞机的维修时间。此法已被许多大飞机维修中心采用。

镜面铣削切削速度通常为 30 m/s 左右。镜面铣削平面度可达 0.1 μm。粗糙度除取决于机床、刀具因素外，还与工件材料本身特性有关。为了能加工出完美的工件，主轴在换刀后必须进行动平衡，以尽量减少动不平衡对工件表面造成的波纹。

3.2.3　超精密磨削加工

超精密磨削技术是在一般精密磨削基础上发展起来的一种亚微米级加工技术。它的加工精度可达到或高于 0.1 μm，表面粗糙度值 Ra 低于 0.025 μm，并正在向纳米级加工方向发展。镜面磨削一般是指加工表面粗糙度值 Ra 达到 0.01~0.02 μm，使加工后的表面光泽如镜的磨削方法。该方法比较强调表面粗糙度，也属于超精密磨削加工范畴。

3.2.4　超精密研磨与抛光

研磨与抛光都是利用研磨剂使工件与研具之间通过实现相对复杂的轨迹而获得高加工质量的加工方法。近年来，在传统研磨抛光技术的基础上，出现了许多新型的精密和超精密游离磨料加工方法，如弹性发射加工、液中研抛、液体动力抛光、磁性研磨、滚动研磨、喷射加工等，它们以其研磨的高精度、抛光的高效率和低表面粗糙度，成为研抛加工的新方法。

1. 研磨加工机理

研磨加工通常是指在刚性研具（如铸铁、锡、铝等软金属或硬木、塑料等）里注入 1 μm 至十几微米大小的氧化铝和碳化硅等磨料，在一定压力下，通过研具和工件的相对运动，借助磨粒的微切削作用，除去微量的工件材料，以达到高的几何精度和低的表面粗糙度的加工方法。总之，研磨表面是在产生切屑、研具的磨损和磨粒破碎等综合因素作用下形成

suffered from the hitting of gritty dust carried by atmosphere. After a certain number times of take-off and landing, the surfaces of the windows would become very coarse. This influences on the view of passengers and pilots. Therefore, the windows should be renovated by polishing. The time needed to renovate a piece of glass is about an hour if traditional polishing process is adopted. The repairing time would be even longer if there are deeper nicks scattered on the surfaces of these glasses. However, if mirror milling is used instead of traditional polishing, the machining time would be reduced by at least 50%, thus shortening the maintenance time of aircraft window greatly. Mirror milling has been widely used in many maintenance centers of larger aircrafts.

Generally, the cutting speed of mirror milling is about 30 m/s, and the flatness obtained can reach up to 0.1 μm. The surface roughness of machined glass depends on machine tool, cutting tool as well as characteristics of workpiece materials. In order to obtain perfect machining effects, after a new cutting tool is mounted on the spindle, dynamic balance must be performed for the spindle to reduce the ripples on workpiece surface resulted from dynamic unbalance of the spindle.

3.2.3　Ultra-Precision Grinding

Ultra-precision grinding technology is a sub-micron machining technology developed on the basis of the general precision grinding. Its machining precision and surface roughness Ra can reach 0.1 μm and less than 0.025 μm respectively. What is more, it is developing towards the direction of nanometer machining. Generally, mirror grinding refers to the machining method by which the surface roughness Ra can reach 0.01-0.02 μm and the gloss of machined surface looks like a mirror. This method emphasizes more on surface roughness and also belongs to the category of ultra-precision grinding.

3.2.4　Ultra-Precision Lapping and Polishing

Lapping and polishing are the machining methods with high machining quality which is obtained by using the abrading agent to generate the relative complex path between the workpiece and the tool. In recent years, based on traditional polishing technology, there are a lot of precision and ultra-precision machining methods of a new type of free abrasive, such as elastic emission machining, polishing liquid, hydrodynamic polishing, magnetic abrasive jet machining, rolling grinding, etc. They have become new methods of lapping and polishing because of high precision, high efficiency of polishing and low surface roughness.

1. Lapping Mechanism

Grinding is a machining method to achieve high precision and low surface roughness. Usually, the alumina and silicon carbide abrasive with the size ranging from 1 μm to a dozen of micrometer are put into a rigid lapping tool such as iron, tin, aluminum and other soft metal or plastic hardwood, etc. Then, with the aid of the micro cutting action of the abrasive grains, a small amount of workpiece material can be removed by means of the relative movement of the tool and the workpiece under a certain pressure. In a word, the grinding surface is formed due to the influence of some combined factors such as the chip, the wear of the grinding tool and the breakage of the abrasive grains, etc. The lapping model is shown in Fig. 3-2.

2. Polishing Mechanism

During the process of polishing, the same as grinding, the abraser paste is smeared on the pol-

的。图 3-2 所示为研磨加工模型。

2. 抛光加工机理

抛光和研磨一样,是将研磨剂擦抹在抛光器上对工件进行抛光加工。抛光和研磨的不同之处在于抛光用的抛光器一般是软质的,其塑性流动作用和微切削作用较弱,加工效果主要是降低加工表面的粗糙度,一般不能提高工件形状精度和尺寸精度,而研磨用的研具一般是硬质的。图 3-3 所示为抛光加工模型。

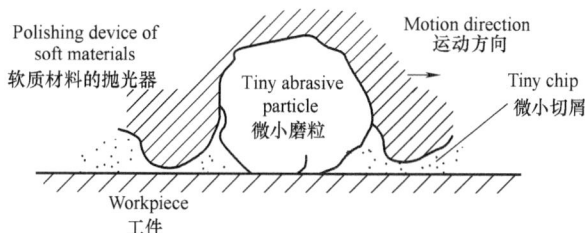

Fig. 3-3　Polishing model 抛光加工模型

3. 化学机械抛光

化学机械抛光(Chemical Mechanical Polishing, CMP)是将化学作用和机械磨削作用综合而形成的加工技术。图 3-4 所示为硅晶片的化学机械抛光设备示意图。

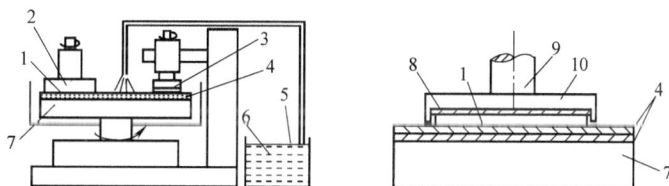

Fig. 3-4　Schematic diagram of chemical mechanical polishing equipment for silicon wafer
硅晶片的化学机械抛光设备示意图

1—Silicon wafer 硅晶片　2—Wafer holder 晶片夹持器　3—Polishing pad dresser 抛光垫修整器

4—Polishing pad 抛光垫　5— SiO_2 particle SiO_2 颗粒　6—Polishing liquid 抛光液

7—Polishing disk 抛光盘　8—Notacoria 背膜　9—Spindle 转轴　10—Chuck 卡盘

4. 采用新原理的超精密研磨抛光

非接触抛光是一种研磨抛光新技术,是指在抛光中工件与抛光盘互补接触,依靠抛光剂冲击工件表面,以获得具有完美结晶性和精确形状的加工表面的抛光方法,其去除量仅为几个到几十个原子级。非接触抛光主要用于功能晶体材料抛光和光学零件的抛光,主要包括弹性发射加工(Elastic Emission Machining, EEM)(见图 3-5)、动压浮动抛光(Hydrodynamic-Type Polishing, HDP)(见图 3-6)、液上漂浮抛光(Hydroplane Polishing)(见图 3-7)及水合抛光(Hydration Polishing)(见图 3-8)。

3.2.5　超精密加工技术的发展趋势

目前,超精密加工技术的发展趋势可总结为以下几点:

1)超精密加工从亚微米级向纳米级发展。超精密加工技术是以高精度为目标的技术,它具有对单项技术的极限、常规技术的突破、新技术的综合 3 个方面永无止境的追求的

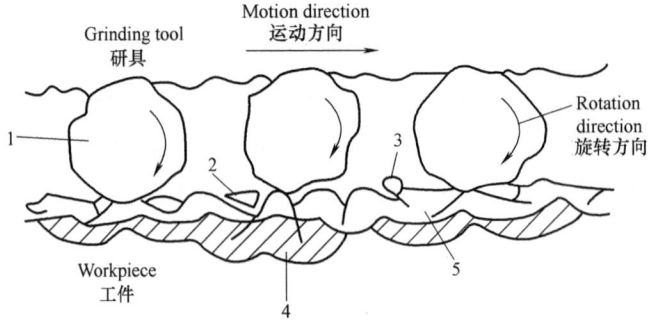

Fig. 3-2　Lapping model 研磨加工模型

1—Abrasive 磨料　2、3—Chip 切屑　4—New machining metamorphic layer 新生的加工变质层

5—Original machining metamorphic layer 原有的加工变质层

ishing machine for polishing the workpiece. There are some differences between polishing and grinding. Usually, the polisher is soft, and it has weak plastic flow and micro cutting effect. In addition, its main machining effect can generally reduce the roughness of the machined surface rather than improving the accuracy of workpiece shape and size. However, the lapping tool is usually hard. The polishing machining model is shown in Fig. 3-3.

3. Chemical Mechanical Polishing

Chemical mechanical polishing is a kind of comprehensive machining technology which combines chemical action and mechanical grinding action. The schematic diagram of chemical mechanical polishing equipment of silicon wafer is shown in Fig. 3-4.

4. Ultra-Precision Grinding and Polishing with New Principle

Contactless polishing is a new technology of grinding and polishing and the contact of the workpiece and the disc has complementarity during the process of the polishing. It is a polishing method, which can obtain a machined surface with perfect crystallinity and accurate shape by using the polishing agent to impact of workpiece surface. Its removal amount is only several to tens of atomic level. Non-contact polishing is mainly used for the polishing of functional crystal materials and polishing of optical parts, mainly including elastic emission machining (see Fig. 3-5), hydrodynamic-type polishing (see Fig. 3-6), hydroplane polishing (see Fig. 3-7) and hydration polishing (see Fig. 3-8).

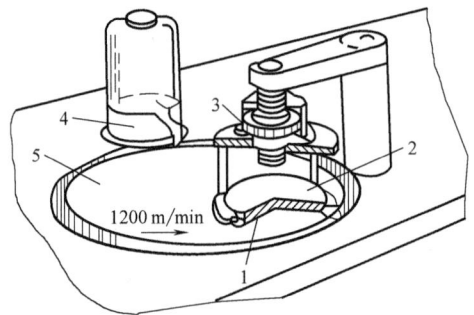

Fig. 3-7　Equipment of hydroplane polishing

液上飘浮动抛光装置

1—Workpiece 工件　2—Crystal panel 水晶平板

3—Adjusting nut 调节螺母　4—Corrosive liquid 腐蚀液

5—Polishing disk 抛光盘

3. 2. 5　The Development Trend of Ultra-Precision Machining Technology

At present, the development trend of ultra-precision machining technology can be summarized as follows:

1) The development of ultra-precision machining is from sub-micron to nanometer. Ultra-precision machining technology aims at high accuracy. It has the characteristics of the limit of single technology,

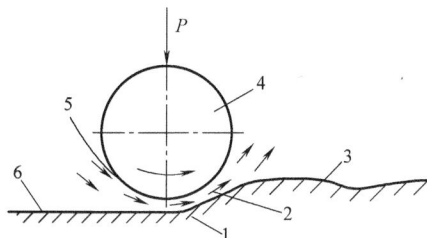

Fig. 3-5　Principle diagram of elastic emission machining 弹性发射加工原理图

1— Workpiece 工件　2—Generated clearance 产生的间隙　3—Unmachined surface 未加工面

4—Resin ball 树脂球　5—Abrading agent 研磨剂　6—Machined surface 已加工面

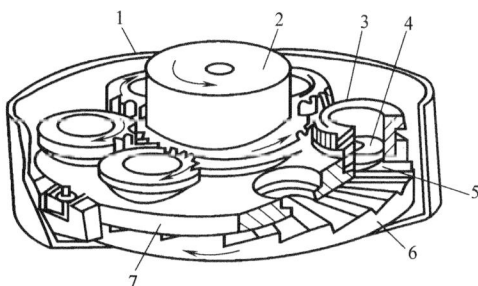

Fig. 3-6　Equipment of hydrodynamic-type polishing 动压浮动抛光装置

1—Container of polishing liquid 抛光液容器　2—Driving gear 驱动齿轮　3—Retaining ring 保持环

4—Workpiece holder 工件夹具　5—Workpiece 工件　6—Polishing disk 抛光盘　7—Load ring plate 载环盘

特点。

2）超精密加工装备思想智能化发展。未来工厂的精密、超精密加工装备将在智能控制理论的指导下，通过在线、在位测量过程建模和优化，取得资源节约、性能优化的效果。

3）研发基于新原理的新一代智能刀具，实现绿色环保、低碳制造。超常制造、智能制造、绿色制造是未来制造业发展的主要方向。在超精密加工环境下，微观尺度效应会导致有别于传统切削的特殊现象。基于新原理的新一代智能刀具，将突破现有的刀具设计理念，通过绿色环保、低碳制造的全新设计和制造技术，实现刀具从被动性加工向主动性和智能化方向发展。

4）超精密测量装置模块化、智能化、集成化及加工检测一体化。

5）加强工艺技术的基础研究和创新工艺研发。特种加工方法和复合加工方法在超精密加工中应用得越来越多，迫切需要进行机理研究；同时，尖端技术和产品的需求日益增长，迫切要求开拓新的加工机理，实现纳米级和亚纳米级加工精度。

3.3　微细和纳米加工技术

3.3.1　微细/纳米加工技术概述

1. 微型机械的提出

人们对许多工业产品的功能集成化和外形小型化的需求，特别是航空航天事业的发展，

breaking through the conventional technology and synthesis of a new technology, and the pursuit of the three aspects is the never-ending.

2）Intelligent development of ultra-precision machining equipment. The precision and ultra-precision machining equipment of future factories will be guided by the theory of intelligent control, and they can achieve the result of resource conservation and performance optimization by the modeling and optimization of online and in the measurement process.

3）Research and development of a new generation of intelligent tools based on

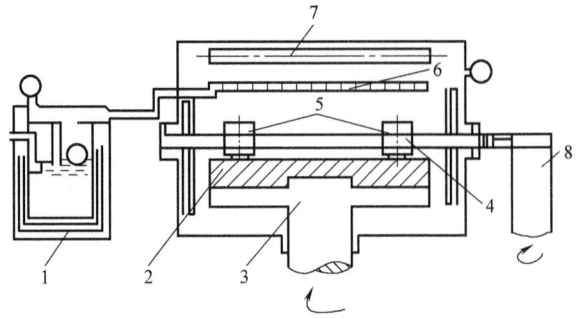

Fig. 3-8 Equipment of hydration polishing
水合抛光装置示意图
1—Water vapor generator 水蒸气发生装置 2—Test specimen 试件
3—Lapping plate 研磨盘 4—Specimen holder 试件夹持器
5—Loader 加载器 6—Spray nozzle of water vapor 喷射水蒸气的喷嘴
7—Heater 加热器 8—Eccentric cam 偏心凸轮

new principles to achieve green environmental protection and low-carbon manufacturing. Superior manufacturing, intelligent manufacturing and green manufacturing will become the main development direction of manufacturing industry in the future. Under the environment of ultra-precision machining, micro scale effect can lead to special phenomenon of different from the traditional cutting. A new generation of intelligent tool based on the new principle will break through the existing concept of tool design. Through a new design and manufacturing technology of green environmental protection and low carbon manufacturing, the cutting tools will be developed towards from passive machining to initiative and intelligent direction.

4）Modularization, intellectualization, integration and integration of machining and testing for ultra-precision measuring devices.

5）Reinforce of the basic research of process technology and development of innovative process technology. The methods of special machining and the composite machining are increasingly increased, and there is urgent need for studying the mechanism. Simultaneously, with the growing need for sophisticated technique and products, it has urgent need for exploiting new machining mechanism in order to realize the machining accuracy of nanometer scale and sub-nanometer scale.

3.3 Micro/Nano Fabrication Technology

3.3.1 Introduction to Micro/Nano Fabrication Technology

1. Emergence of Micromachine

Due to the requirements for functional integration and configural miniaturization of many industrial products, especially the development of the aerospace industry, the request on the miniaturization of many devices is presented, resulting in increasingly microminiaturization of the size of components. Japan first introduced the concept of "micromachine", and then the United States proposed the concept of micro electro-mechanical systems（MEMS）, while Europe also proposed the concept

对许多设备提出了微型化的要求，从而使零部件的尺寸日趋微小化。日本最先提出了微型机械（Micromachine）的概念，接着美国提出了微型机电系统（Micro Electro-Mechanical Systems，MEMS）的概念，而欧洲也提出了微型系统（Micro-Systems）的概念。从广义上说，MEMS是指集微型机构、微型传感器、微型执行器、信号处理系统、电子控制电路，以及外围接口、通信电路和电源等于一体的微型机电一体化产品。微机械按尺寸特征可以分为1～10 mm的微小机械，1μm～1mm的微机械，1nm～1μm的纳米机械。制造微机械的关键技术是微细加工。微机械的发展大体经历了以下几个发展时期：①1959年，著名量子物理学家、诺贝尔物理奖获得者理查德·费曼（Richard Feynman）预言，人类可以用小的机器制作更小的机器，最后将发展到根据人类自己的意愿逐个地排列原子，制造产品；②1960—1962年，世界第一个硅微型压力传感器问世；③1975—1985年是微型机械的酝酿期，这一阶段主要是用制作大规模集成电路的IC技术制作微型传感器；④1986—1989年是微型机械的产生期，主要是用IC技术制作微型机械的零部件。美国是对微机械研制并试制成功最早的国家之一，1988年开始了MEMS的主要项目研究；⑤20世纪90年代以后是其发展期，这一阶段各种超微加工技术相继用于微型机械的制作；⑥1991年，日本启动了一项为期10年耗资250万日元的"微型机械研究计划"；⑦1994年，《美国国防部技术计划》将MEMS列为关键技术项目。

2. MEMS的特点

与常规机械/系统相比，MEMS具有以下一些特点：

（1）微型化　MEMS技术已经达到微米乃至亚微米量级。因此，MEMS器件具有体积小、重量轻、能耗低、精度高、可靠性高、谐振频率高和响应时间短的特点。当器件的结构尺寸缩小到微米/纳米量级时会出现力的尺度效应。

（2）集成化　硅、氧化硅和氮化硅等材料具有良好的机械性能和电气性能，与集成电路（IC）加工工艺完全兼容，易于实现机械与电路的集成，适合于大批量、低成本制造。

（3）多样化　由于系统是由功能不同的单元组合而成的，所以多样性是形成微小型系统的关键。MEMS不仅仅局限于机械，还涉及电子、材料、制造、信息与自动控制、物理、化学和生物等多种学科，并汇集了当今科学技术发展的许多尖端成果。

3. 微细/纳米加工技术的概念与特点

微细加工是指加工尺度为微米级范围的加工方式。微细加工起源于半导体制造工艺，加工方式十分丰富，包含了微细机械加工、各种现代特种加工、高能束加工等方式。

纳米技术（Nano Technology，NT）是在纳米尺度范围（0.1～100nm）内对原子、分子等进行操纵和加工的技术。它是一门多学科交叉的学科，是在现代物理学、化学和先进工程技术相结合的基础上诞生的，是一门与高技术紧密结合的新兴科学技术。纳米级加工包括机械加工、化学腐蚀、能量束加工、扫描隧道加工等多种方法。

微细加工与一般尺度加工有许多不同，主要体现在以下几个方面：

（1）精度表示方法不同　在一般尺度加工中，加工精度是用其加工误差与加工尺寸的比值（即相对精度）来表示的。而在微细加工时，由于加工尺寸很小，精度就必须用尺寸的绝对值来表示，即用去除（或添加）的一块材料（如切屑）的大小来表示，从而引入加工单位的概念，即一次能够去除（或添加）的一块材料的大小。当微细加工0.01 mm尺寸零件时，必须采用微米加工单位进行加工；当微细加工微米尺寸零件时，必须采用亚微米加

of "micro-systems". Broadly speaking, MEMS is a kind of micro-electromechanical integrated product which integrates micro-mechanism, micro-sensor, micro-actuator, signal processing system, electronic control circuit, and peripheral interface, communication circuit and power supply. The size of micro-mechine can be divided into small machine in size of 1-10 mm, micro-machine in 1 μm-1 mm, nano-machine in 1 nm-1 μm. The key technology for manufacturing micro-machines is microfabrication. The development of micro-machine has undergone several periods as follows: ① In 1959, Richard Feynman, a famous quantum physicist and Nobel Prize winner for physics, predicted that humans could make smaller machines with small machines and would eventually develop to arrange atomic and manufacture products based on desire of the mankind; ② from 1960 to 1962, the world's first silicon micro-pressure sensor appeared; ③ it was from 1975 to 1985 that the micro-machine underwent a brewing period, in which IC technology for large-scale integrated circuit was mainly used to produce micro-sensors; ④ from 1986 to 1989 the micro-machine was in generation period, which mainly applying IC technology to produce micromechanical components. The United States is one of the earliest countries to develop and succeed in producing the micro-machine, and the project research on MEMS was started in 1988; ⑤ since the 1990s the micro-machine was developed quickly, in this stage a variety of ultra-micro-processing technology have been used for the production of micro-machine; ⑥ In 1991, Japan launched a "micro-machine research program" with a cost of 2.5 million yen for 10 years; ⑦ In 1994, the U. S. Department of Defense National Defense Technology Program classified MEMS as a key technology project.

2. Characteristics of MEMS

Compared with conventional machinery / systems, MEMS has the following characteristics:

(1) Microminiaturization   MEMS technology has reached the order of microns and even submicrons. Therefore, MEMS devices have the characteristics of small size, light weight, low energy consumption, high precision, high reliability, high resonant frequency and short response time. When the structure size of the device is reduced to the micro/nano scale, the scaling effect of force will appear.

(2) Integration   Silicon, silicon oxide and silicon nitride have good mechanical properties and electrical properties, and can be fully compatible with integrated circuit (IC) processing technology, which making them easy to achieve the integration of machine and circuit, and be suitable for high-volume, low-cost manufacturing.

(3) Diversification   Since the system is a combination of different function units, diversity is the key to form the micro-systems. MEMS is not limited to machinery, but also involves electronics, materials, manufacturing, information and automatic control, physics, chemistry and biology and other disciplines, and brings together nowadays scientific and technological development of many most advanced achievements.

3. Concept and Characteristics of Micro/nano Fabrication Technology

Micromachining refers to a processing method that the processing scale reaches micron scale. The micromachining originated in the semiconductor manufacturing process has abundant processing mode which includes the micro mechanical machining, all kinds of modern non-traditional machining, high energy beam machining, etc.

Nanotechnology is a manipulation and processing technology on atomic, molecular in the nano scale range (0.1-100 nm). It is a multidisciplinary technology that comes into being on the basis of

工单位来进行加工。现今的超微细加工已采用纳米加工单位。

（2）加工机理存在很大差异　由于在微细加工中加工单位急剧减小，此时必须考虑晶粒在加工中的作用。

例如，欲把软钢材料毛坯切削成一根直径为 0.1 mm、精度为 0.01 mm 的轴类零件。对于给定的要求，在实际加工中，车刀至多只允许能产生 0.01 mm 切屑的吃刀深度；而且在对上述零件进行最后精车时，吃刀深度要更小。由于软钢是由很多晶粒组成的，晶粒的大小一般为十几微米，这样，直径为 0.1 mm 就意味着在整个直径上所排列的晶粒只有 20 个左右。如果吃刀深度小于晶粒直径，那么，切削就不得不在晶粒内进行，这时就要把晶粒作为一个个的不连续体来进行切削。相比之下，如果是加工较大尺度的零件，由于吃刀深度可以大于晶粒尺寸，切削不必在晶粒中进行，就可以把被加工体看成是连续体。这就导致了加工尺度在亚毫米、加工单位在数微米的加工方法与常规加工方法的微观机理的不同。另外，还可以从切削时刀具所受阻力的大小来分析微细切削加工和常规切削加工的明显差别。实验表明，当吃刀深度在 0.1 mm 以上进行普通车削时，单位面积上的切削阻力为 （196～294）N/mm^2；当吃刀深度在 0.05 mm 左右进行微细铣削加工时，单位面积上的切削阻力约为 980 N/mm^2；当吃刀深度在 1 μm 以下进行精密磨削时，单位面积上的切削阻力将高达 12 740 N/mm^2，接近于软钢的理论剪切强度 $G/2\pi \approx 13\,720$ N/mm^2（G 为剪切弹性模量，约为 8.3×103 kg/mm^2）。因此，当切削单位从数微米缩小到 1 μm 以下时，刀具的尖端要承受很大的应力作用，使得单位面积上会产生很大的热量，导致刀具的尖端局部区域上升到极高的温度。这就是越是采用微小的加工单位进行切削，就越要求采用耐热性好、耐磨性强、高温硬度和高温强度都高的刀具的原因。

（3）加工特征明显不同　一般加工以尺寸、形状、位置精度为特征；微细加工则由于其加工对象的微小型化，目前多以分离或结合原子、分子为特征。

例如，超导隧道结的绝缘层只有 10 Å 左右的厚度。要制备这种超薄层的材料，只能用分子束外延等方法在基底（或衬底、基片等）上通过一个原子层一个原子层（或分子层）地以原子或分子线度（Å 级）为加工单位逐渐淀积，才能获得纳米加工尺度的超薄层。再如，利用离子束溅射刻蚀的微细加工方法，可以把材料一个原子层一个原子层（或分子层）地剥离下来，实现去除加工。这里，加工单位也是原子或分子线度量级，也可以进行纳米尺度的加工。因此，要进行 1 nm 的精度和微细度的加工，就需要用比它小一个数量级的尺寸作为加工单位，即要用 0.1 nm 的加工方法进行加工。这就明确告诉人们必须把原子、分子作为加工单位。扫描隧道显微镜和原子力显微镜的出现，实现了以单个原子作为加工单位的加工。

3.3.2　微细加工技术

微细加工技术是由微电子技术、传统机械加工技术和特种加工技术衍生而成的。按其衍生源的不同，微细加工可分为微细切削加工、微细特种加工和光刻加工。

1. 微细切削加工

这种方法适合所有金属、塑料和工程陶瓷材料，主要采用车削、铣削、钻削等切削方式，刀具一般为金刚石刀（刃口半径为 100 nm）。这种工艺的主要困难在于微型刀具的制

the combination of modern physics, chemistry and advanced engineering techniques. It is a new science and technology that is closely integrated with high technology. Nano-scale processing includes mechanical machining, chemical corrosion, energy beam machining, scanning tunneling machining, and so on.

There are many differences between the micromachining and the common scale machining, which is shown as follows:

(1) The representing method of accuracy is different   In the common scale machining, the machining accuracy is represent by the ratio of machining error and machining dimension (ie, relative accuracy). In the micromachining, due to the small machining dimension the accuracy must be expressed in terms of the absolute value of the dimension, namely which is represented by the size of a piece of material (such as chips) removed or added. In this way the concept of machining unit is introduced that it is the size of a piece of material removed or added at a time. When machining the parts 0. 01mm in size, it must be processed using micron machining unit. When the micro-sized part is to be micro-processed, it must use sub-micron machining unit for processing. Nowadays, nano-machining unit is applied in super micro-machining.

(2) The machining mechanism is different   Due to the machining unit declines sharply, the role of grain in the machining must be considered.

For example, to cut a blank of mild steel material into a shaft part with 0. 1 mm in diameter of and 0. 01 mm in accuracy. For a given request, in actual machining it is only allowed that the lathe cutting tool keeps a cutting depth in 0. 01 mm, and the cutting depth is smaller during last finishing cutting. As the mild steel is composed of a lot of grains, the grains size is generally more than ten microns, so that the diameter of 0. 1mm means that there is only about 20 grain arranged on the entire diameter. If the cutting depth is less than the grain diameter, then the cutting will have to be carried out in the grain, then each grain should be cut as a discontinuity. In contrast, if a large size part is to be machined, the cutting does not have to be carried out in the grain as the cutting depth may be greater than the grain size. In this case, the machined part can be taken as a continuum. This leads the micro mechanism of the machining method whose scale in sub-millimeter and machining unit in a few microns differs with the conventional machining method. In addition, the significant differences between micro cutting and conventional cutting also can be analyzed from the resistance force of tool born during the process of cutting. Experiments show that during the general lathe cutting the cutting depth is above 0. 1 mm, the cutting resistance force per unit area is 196 – 294 N/mm^2; When the cutting depth is about 0. 05 mm for micro milling, the cutting resistance force per unit area is about 980 N/mm^2; When the cutting depth is less than 1 μm for precision grinding, the cutting resistance force per unit area will be as high as 12740 N/mm^2, which is close to the theoretical shear strength of mild steel $G/2\pi \approx 13720$ N/mm^2 (G is the shear elastic modulus $\approx$ 8. 3×10^3 kg / mm^2). Therefore, when the cutting unit is reduced less than 1μm from a few microns, the tool tip must bear a lot of stress, leading to produce a lot of heat in unit area, and resulting in the local area of tool tip heated to a very high temperature. This is the reason that when applying more smaller machining unit for cutting, the tools with good heat resistance, better wear resistance, high temperature hardness and strength are requested to use.

(3) The machining characteristics are significantly different   The common scale machining dimension accuracy, shape accuracy, location accuracy, while micro-machining is characterized cur-

造、安装及加工基准的转换定位。

目前，日本 FANUC 公司已开发出能进行车、铣、磨和电火花加工的多功能微型超精密加工机床，其主要技术指标为：可实现五轴控制，数控系统最小设定单位为 1 nm；采用编码器半闭环控制及激光全息式直线移动的全闭环控制；编码器与电机直联，具有每周 6400 万个脉冲的分辨率，每个脉冲相当于坐标轴移动 0.2 nm；编码器反馈单位为 1/3 nm，跟踪误差在 ±3 nm 以内；采用高精度螺距误差补偿技术，误差补偿值由分辨率为 0.3 nm 的激光干涉仪测出；推力轴承和径向轴承均采用气体静压支承结构，伺服电机转子和定子用空气冷却，发热引起的温升控制在 0.1℃ 以下。

2. 微细特种加工

（1）微细电火花加工　电火花加工是利用工件和工具电极之间的脉冲性火花放电，产生瞬间高温使工件材料局部熔化或汽化，从而达到蚀除材料的目的。微小工具电极的制作是关键技术之一。利用微小圆轴电极，在厚度为 0.2 mm 的不锈钢片上可加工出直径为 40 μm 的微孔，如图 3-9 所示。

当机床系统定位控制分辨率为 0.1 μm 时，最小可实现孔径为 5 μm 的微细加工，表面粗糙度值 Ra 可达 0.1 μm。这种方法的缺点是电极的定位安装较为困难。为此常将切削刀具或电极在加工机床中制作，以避免装夹误差。

（2）复合加工　它是指电火花与激光复合精密微细加工。针对市场上急需的精密电子零件模具与高压喷嘴等使用的超高硬度超微硬质合金及聚晶金刚石烧结体的加工要求，特别是大深径比的深孔加工要求，开发出一种高效率的微细加工系统，它采用了电火花加工与激光加工的复合工艺。首先利用激光在工件上预加工出贯穿的通孔，以便为电火花加工提供良好的排屑条件，然后再进行电火花精加工。

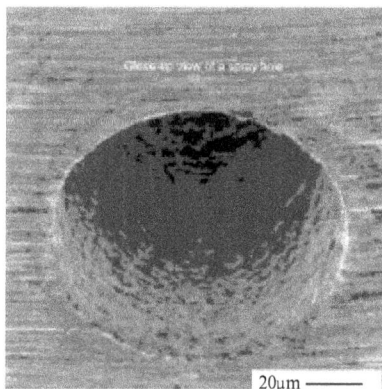

Fig. 3-9　Micro-hole machined by Micro-EDM
微细电火花加工出的微孔

3. 光刻加工

光刻加工是利用光致抗蚀剂（感光胶）的光化学反应特点，在紫外线照射下，将照相制版上的图形精确地印制在有光致抗蚀剂的工件表面，再利用光致抗蚀剂的耐腐蚀特性，对工件表面进行腐蚀，从而获得极为复杂的精细图形。

目前，光刻加工中主要采用的曝光技术有：电子束曝光技术、离子束曝光技术、X 射线曝光技术和紫外准分子激光曝光技术。其中，离子束曝光技术具有最高的分辨率，电子束曝光代表了最成熟的亚微米级曝光技术，紫外准分子激光曝光技术则具有最高的经济性，是近年来发展速度极快且实用性较强的曝光技术，在大批量生产中保持主导地位。

典型的光刻工艺过程为：①氧化，使硅晶片表面形成一层 SiO_2 氧化层；②涂胶，在 SiO_2 氧化层表面涂布一层光致抗蚀剂，即光刻胶，厚度在 1~5 μm；③曝光，在光刻胶层面上加掩模，然后用紫外线等方法曝光；④显影，曝光部分通过显影而被溶解除去；⑤腐蚀，将加工对象浸入氢氟酸腐蚀液，使未被光刻胶覆盖的 SiO_2 部分被腐蚀掉；⑥去胶，

rently the separation or combination of atoms and molecules because of the miniaturization of the machining object.

For example, the insulation layer of superconductor tunnel junction has only a thickness of about 10 Å. To prepare this ultra-thin layer material, the methods such as molecular beam epitaxy only can be applied to obtain the ultra-thin layer in nano-scale, through the gradual deposition of one atomic layer (or molecular layer) with one atomic layer (or molecular layer) with atomic or molecular scale (Å scale) on the substrate (or underlayment, supporting base, etc.). Another example is the use of ion beam sputtering etching, which can peel off the material one atomic layer by one atomic layer (or molecular layer) to achieve removal machining. Here, the machining unit is also atomic or molecular scale, which can achieve nano-scale machining. Therefore, to achieve the machining accuracy or micro scale of 1 nm, it needs a machining unit less than it, namely a machining method for 1 nm scale should be applied. It is suggested that atomic or molecular must be as the machining unit. The appearance of scanning tunneling microscopy and atomic force microscopy has realized the machining of taking a single atom as machining unit.

3. 3. 2　Micro-machining Technology

Micro-machining technology is derived from micro-electronics technology, traditional mechanical machining technology and non-traditional machining technology. According to different derivative sources, the micro-machining can be divided into micro-cutting machining, micro nontraditional machining and lithography machining. Some typical micro-machining methods are to be introduced as follows:

1. Micro-Cutting Machining

This method is suitable for all metal, plastic and engineering ceramic materials, which the cutting methods mainly used include turning, milling, drilling and so on and the cutter is generally diamond cutting tool (radius of cutting edge is 100 nm) . The main difficulty of this process lies in the manufacturing and installation of micro-cutter, the transformation and location of machining reference.

At present, Japan FANUC company has developed a multifunctional ultra-precision machining micro-machine tool that can achieve turning, milling, grinding and EDM. The main technical indicators of this micro-machine tool involve that five-axis control can be achieved, which the minimum setting unit of the numerical control system is 1nm; semi-closed loop control and the whole closed-loop control with laser holographic linear motion are applied to; the encoder is connected directly to the motor and has a resolution of 64 million pulses per week, each pulse corresponds to an axis movement of 0. 2 nm. The encoder feedback unit is 1/3 nm and the tracking error is within ±3 nm; using the high precision pitch error compensation technique, the error compensation value is measured by a laser interferometer with a resolution of 0. 3 nm; both the thrust bearing and the radial bearing adopt the air hydrostatic supporting structure, the rotor and the stator of servo motor are cooled by air, the temperature rising caused by heat is controlled below 0. 1 ℃.

2. Micro Nontraditional Machining

(1) Micro EDM　EDM is the use of the pulse spark discharge between the workpiece and the tool electrode resulting in instantaneous high temperature to make the workpiece material local melting or vaporization, so as to achieve the purpose of the material erosion removing. Fabrication of micro tool electrode is one of the key technologies. Using a micro circular shaft electrode, a micro-pore with diameter of 40 μm can be processed on stainless steel with thickness of 0. 2 mm, as shown in Fig. 3-9.

腐蚀结束后，光致抗蚀剂就完成了它的作用，此时要设法将这层无用的胶膜去除；⑦扩散，即向需要杂质的部分扩散杂质，以完成整个光刻加工过程。图3-10为半导体光刻加工过程示意图。

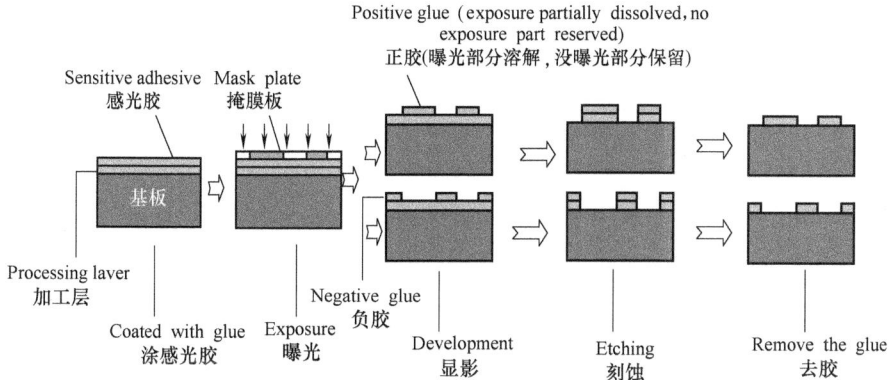

Fig. 3-10 Schematic diagram of semiconductor lithography process 半导体光刻加工过程示意图

3.3.3 纳米加工技术

纳米技术作为一个全新的交叉技术领域，覆盖范围很广，如纳米电子、纳米材料、纳米机械、纳米加工、纳米测量等。纳米加工技术是一门新兴的综合性加工技术，以现代物理学、化学和先进工程技术等为基础，在纳米尺度范围（0.1~100 nm）内对原子、分子等进行操纵和加工。纳米加工技术的发展面临两大途径：①将传统的超精加工技术，如机械加工、电化学加工、离子束蚀刻、激光加工等向极限精度逼近，使其达到纳米的加工能力；②开拓新效应的加工方法，如用扫描隧道显微镜对表面的纳米加工，可操纵试件表面的单个原子，实现单个原子的和分子的搬迁、去除、增添和原子排列重组，并对表面进行刻蚀等。如美国IBM公司用STM将35个原子排出"IBM"字样；中国科学院化学研究所用原子摆出我国的地图；日本用原子拼成了"Peace"一词。

纳米器件之所以得到广泛的关注，是因为它们具有独特的性能、前所未有的功能，并且在与多种形式的能量相互作用时会呈现出奇特现象，进而带来材料的高能量效率、内置式的智能和性能改善等。但纳米器件的制造方法相当复杂，制作成本很高。目前，器件的纳米制造方法有两类：

1）"自上而下"法（减材法）。例如电子束光刻加工、离子束光刻加工等。由于传统的"自上而下"法的微电子工艺受经典物理学理论的限制，依靠这一工艺来减小电子器件尺寸将变得越来越困难，而且这些技术大多只能用于制作形状简单的二维图形，不适用于大批量的纳米制造。

2）"自下而上"法（增材法）。该方法是从单个分子甚至原子开始，一个原子一个原子地进行物质的组装和制备。如扫描探针显微镜显微加工、自装配、直接装配、纳米印刷、模板制造等都属于"自下而上"的纳米加工方法。这个过程没有原材料的去除和浪费。

传统微纳器件的加工是以金属或者无机物的体相材料为原料，通过光刻蚀、化学刻蚀或两种方法结合使用的"自上而下"的方式进行加工，在刻蚀加工前必须先制作"模具"。长

When the positioning resolution of the machine tool is controlled within 0. 1 μm, the micro-machining of a minimum pore with diameter of 5 μm can be achieved, and the surface roughness can reach up to 0. 1 μm. The shortcoming of this method is that the positioning installation of the electrode is much more difficult. For this purpose, the cutting tool or electrode is often made in the machine tool to avoid clamping error.

(2) Composite machining   It refers to a kind of precision micromachining combined EDM and laser beam machining. Aiming at the machining need for the precision electronic part mold and ultrafine cemented carbides and polycrystalline diamond sinters with super high hardness used for high pressure jet nozzle, especially for the machining of deep hole with high-aspect-ratio, to develop a kind of high efficient micro-machining system, which uses the composite process of EDM and laser machining. First of all, a via on the workpiece is pre-processed by using laser, in order to provide a good chip removal conditions for EDM, and then conduct the EDM finishing.

3. Lithography Machining

Lithography machining is to use photochemical reaction characteristics of photoresist (photosensitive adhesive) to print accurately the graphics on the photographic plate onto the workpiece surface with photoresist in the ultraviolet radiation, and then erode the workpiece surface using the corrosion resistance of the photoresist, so that a very complex fine graphic is obtained.

At present, the exposure technology used in lithography machining involves: Electron beam exposure technology, ion beam exposure technology, X-ray exposure technology and ultraviolet excimer laser exposure technology. As to these technologies, the ion beam exposure technology has the highest resolution; the electron beam exposure represents the most mature submicron exposure technology. While the ultraviolet excimer laser exposure technology exhibits the highest economy, which is very practical exposure technology developed extremely fast in recent years and maintains the dominant position in mass production.

The typical photolithography process can be described as follows. ①Oxidation, to form a layer of SiO_2 oxide layer on the silicon wafer surface; ②coating, on the surface of SiO_2 oxide layer a layer of photoresist is coated with the thickness of 1-5 μm; ③exposure, to add mask on the surface of the photoresist layer, and then to be exposure using ultraviolet and other methods; ④development, the exposure part of the development is dissolved to remove; ⑤corrosion, the machined object is immersed in hydrofluoric acid solution to the part of SiO_2 that is not masked by photoresist is eroded away; ⑥removing of photoresist, the role of the photoresist is completed at the end of the erosion, it is time to try to remove this layer of useless glue; ⑦diffusion, to diffuse impurities to where they need so as to finish the whole photolithography process. Fig. 3-10 shows the photolithography process of semiconductor.

3. 3. 3   Nanofabrication Technology

As a new field of cross technology, nanotechnology has a wide range of applications, such as

期以来推动电子领域发展的以曝光技术为代表的"自上而下"方式的加工技术即将面临发展极限。如果使用蛋白质和 DNA（脱氧核糖核酸）等纳米生物材料，将有可能形成运用材料自身具有的"自组装"和相同图案"复制与生长"等特性的"自下而上"方式的元件。图 3-11 所示为采用"自下而上"方法加工出的纳米碳管和量子栅栏。

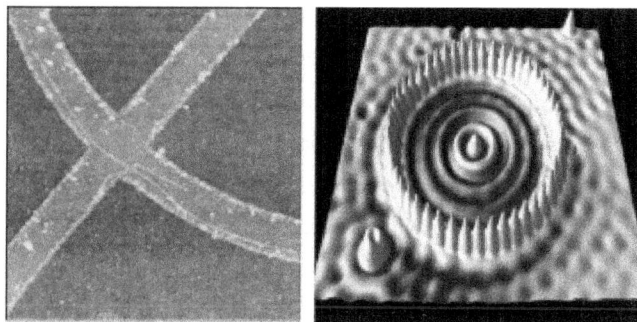

Fig. 3-11　Carbon nanotubes and quantum fence processed by bottom-up method
"自下而上"方法加工出的纳米碳管和量子栅栏

纳米加工技术主要有以下几种：

1. LIGA 加工工艺

LIGA 加工工艺是由德国科学家开发的集光刻、电铸和模铸于一体的复合微细加工新技术。LIGA 是德文 Lithographic（光刻），Galvanoformung（电铸）和 Abformung（注塑成形）3 个词的缩写。LIGA 技术是三维立体微细加工最有前景的加工技术，尤其对于微机电系统的发展有很大的促进作用。

20 世纪 80 年代中期，德国的 W. Ehrfeld 教授等人发明了 LIGA 加工工艺。这种工艺包括 3 个主要步骤：深层同步辐射 X 射线光刻、电铸成形和注塑成形。其最基本和最核心的工艺是深层同步辐射 X 射线光刻，而电铸成形和注塑成形工艺是 LIGA 产品实用化的关键。LIGA 适合用多种金属、非金属材料制造微型机械构件。采用 LIGA 技术已研制成功的产品有微传感器、微电机、微执行器、微机械零件等。

用 LIGA 工艺加工出的微器件侧壁陡峭、表面光滑，可以大批量复制生产、成本低，因此广泛应用于微传感器、微电机、微执行器、微机械零件、集成光学和微光学元件、真空电子元件、微型医疗器械、流体技术微元件、纳米技术元件等的制作。现在已将牺牲层技术融入 LIGA 工艺中，使获得的微型器件中有一部分可以脱离母体而移动或转动；还有学者研究控制光刻时的照射深度，即使用部分透光的掩模，使曝光时同一块光刻胶在不同处曝光深度不同，从而使获得的光刻模型可以有不同的高度，用这种方法可以得到真正的三维立体微型器件。LIGA 技术原理图如图 3-12 所示。

LIGA 技术的特点：X 射线具有良好的平行性、显影分辨力和穿透性能，克服了光刻法制造的零件厚度过薄的不足（最大深度 $40~\mu m$）；原材料的多元性，几何图形的任意性、高深宽比、高精度；X 射线同步辐射源比较昂贵。

2. 扫描隧道显微加工技术

通过扫描隧道显微镜（Scanning Tunneling Microscope，STM）的探针来操纵试件表面的

nano electronics, nano materials, nano machinery, nano machining, nano measurement and so on. Nanofabrication technology is an emerging comprehensive machining technology, based on modern physics, chemistry and advanced engineering technology, which is a manipulation and processing technology on atomic, molecular in the nano-scale range (0.1-100 nm). The development of nano machining technology is facing two major pathways: ①To make the traditional ultra precision machining, such as mechanical machining, electrochemical machining, ion beam etching, laser beam machining and so on, approach to the accuracy of approximation, achieving to the nano-scale machining capacity; ② to exploit machining methods with new effects, such as the nano machining for the surface using the scanning tunneling microscopy, which can manipulate a single atom on the surface, to make a single atom and molecular move, remove, addition and atomic arrangement of re-organization, as well as conduct the surface etching. For instance, the United States IBM company arranged the characters "IBM" with 35 atoms using STM. The Institute of Chemistry of Chinese Academy of Sciences used atoms to put out a map of China. Japan spelt out the word "Peace" with atoms.

The nano device has received extensive attention because of their unique properties, unprecedented function, as well as the strange phenomenon appears in various forms of energy interaction, and then bring the high energy efficiency of materials, and improvement in the intelligent and performance built-in etc. However, the fabrication methods of nano devices are very complex and the fabrication cost is very high. At present, there are two kinds of nano device fabrication methods:

1) "Top-Down" method. Such as the electron beam lithography machining and ion beam lithography machining belong to this method. Due to the limitation of the classical physics theory, it becomes more and more difficult to reduce the size of electronic devices using the traditional "top-down" method. And most of these techniques can only be used to make the shape of a simple two-dimensional.

2) "Bottom-Up" method. This method is to start from a single molecule or even an atom, and to assemble and prepare the material in an atom and an atom. The scanning probe microscopy, self-assembly, direct-assembly, nano printing, template manufacturing, are all regarded as the "bottom-up" nano-machining method. This process does not remove and waste raw materials.

The traditional fabrication of micro/nano devices is to take bulky materials of metal or inorganic matte as the raw material. The machining process of "top-down" is carried out by photolithography, chemical etching and "molds" must be made before etching. The top-down machining technology, which has been driven by exposure technology for long-term development in the electronics field, is facing development limits. If nano-biomaterials such as proteins and DNA (deoxyribonucleic acid) are used, it will be possible to form a bottom-up element that utilizes the self-assembly of the material itself and the same pattern "copy and grow". Fig. 3-11 shows the carbon nanotubes and quantum fence processed using the bottom-up method.

The nanofabrication process mainly includes as follows:

1. LIGA Technology

LIGA machining process, developed by German scientists, is a new composite microfabrication

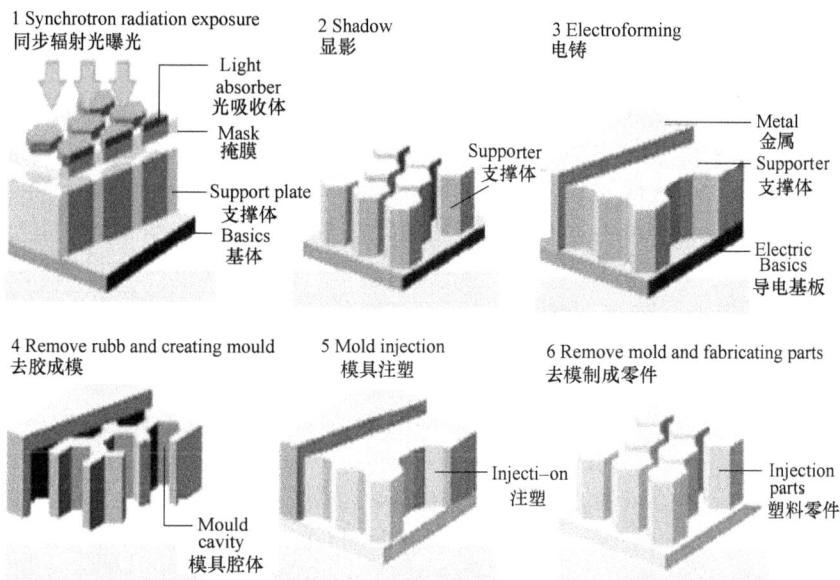

Fig. 3-12　Schematic diagram of LIGA technology LIGA 技术原理图

单个原子，实现单个原子的和分子的搬迁、去除、增添和原子排列重组，实现极限的精加工。目前，原子级加工技术正在研究对大分子中的原子搬迁、增加原子、去除原子和原子排列的重组。

利用 STM 进行单原子操纵的基本原理是：当针尖与表面原子之间距离极小时（<1 nm），会形成隧道效应，即针尖顶部原子和材料表面原子的电子云相互重叠，有的电子云双方共享，从而产生一种与化学键相似的力。同时，表面上其他原子对针尖对准的表面原子也有一定的结合力，因此探针可以带动该表面原子跟随针尖移动而又不脱离试件表面，实现原子的搬迁。当探针针尖对准试件表面某原子时，在针尖和样品之间加上电偏压或脉冲电压，可使该表面原子成为离子而被电场蒸发，从而实现去除原子形成空位；在有脉冲电压存在的条件下，也可以从针尖上发射原子，实现增添原子填补空位。

3. AFM 机械刻蚀加工

原子力显微镜（Atomic Force Microscope，AFM）在接触模式下，通过增加针尖与试件表面之间的作用力会在接触区域产生局部结构变化，即通过针尖与试件表面的机械刻蚀的方法进行纳米加工。

4. AFM 阳极氧化法加工

该工艺为通过扫描探针显微镜（Scanning Probe Microscope，SPM）针尖与样品之间发生的化学反应来形成纳米尺度氧化结构的一种加工方法。针尖为阴极，样品表面为阳极，吸附在样品表面的水分子充当电解液，提供氧化反应所需的 OH^- 离子，如图 3-13 所示。该工艺早期采用 STM，后来多采用 AFM，主要是由于 AFM 法自身采用氧化过程，简单易行，刻蚀处的结构性能稳定。

technology which integrates the process of lithography, electroforming and casting together. LIGA is abbreviation of three German words "Lithographie" "Galvanoformung" and "Abformung". LIGA technology is the most promising machining technology in three-dimensional microfabrication, especially has a great role in promoting the development of micro-electromechanical systems.

In mid 1980s, the German professor W. Ehrfeld et al invented the LIGA machining process. This technology includes three main steps: synchrotron radiation X-ray deep lithography, electroforming and injection moulding. The most basic and key process technology is the X-ray deep lithography, while the critical process for applying the LIGA technology is the electroforming and the injection moulding. LIGA is suitable for making micro-mechanical components with a variety of metal and non-metallic materials. The products developed successfully by using LIGA technology contain micro sensor, micro motor, micro actuator, micro mechanical part, etc.

The micro devices fabricated by LIGA process have steep side-wall and smooth surface, which can be copied in large quantities production with low cost. Therefore, LIGA process is widely used in the production of micro sensor, micro motor, micro actuator, micro-mechanical part, integrated optical and micro-optical part, vacuum electronic element, micro-medical device, micro-fluid component, as well as nanotechnology component. The sacrificial layer technology has now been incorporated into the LIGA process so that a portion of micro-device obtained can be moved or rotated away from the matrix. Some scholars have studied to control the depth of exposure during photolithography, that is, to use the mask partly nonopaque to make the same photoresist at different exposure position have different exposure depth, so that the lithography model obtained has different height and a real three-dimensional micro-device can be fabricated. The schematic diagram of LIGA technology is shown in Fig. 3-12.

The features of LIGA technology include that the X-ray contains good parallelism, development resolution and penetration performance, overcomes some insufficient which the thickness of parts made by lithography is too thin (the maximum depth is 40 microns); the raw materials used for LIGA process is diversity, and the object machined can be random geometric figure, high aspect ratio and high precision; the X-ray synchrotron radiation source is more expensive.

2. Scanning Tunnel Micro Machining Technology

The single atom in the surface of the specimen can be manipulated by the probe of the scanning tunneling microscope (STM), to implement the moving, removal, addition and atomic arrangement of individual atom and molecule, finally achieving the ultimate finish machining. At present, atomic scale machining technology is focus on in moving, addition, removal and arrangement and reorganization of atoms in macromolecules.

The basic principle of single atom manipulated by STM is that when the distance between the tip and the surface atom is very small (less than 1 nm), the tunnel effect will be formed, that is, the electron clouds of the atom at the tip and the atom of the material surface overlap each other, some of electron clouds share with each other, which results in a force similar with the chemical bond. At the same time, the other atoms on the surface also have a certain adhesion on the surface atom pointed by the tip. Hence, the probe can drive the surface atom follow the tip move without away from the specimen surface, to achieve the atomic migration. When the probe tip is aligned with an atom on the specimen surface, an electrical bias or pulse voltage is imposed between the tip and the sample, which can change the surface atom into ion and evaporate by the electric field, to form

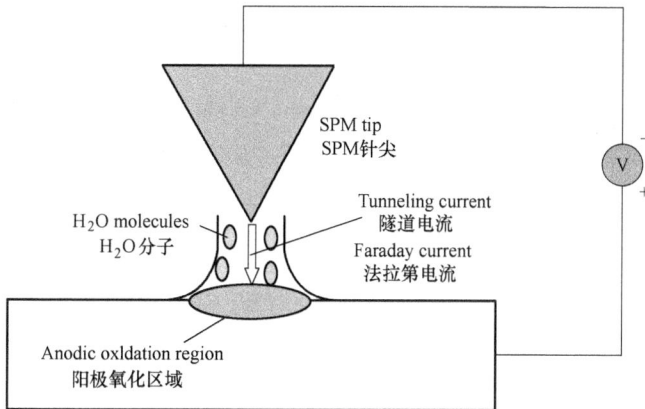

Fig. 3-13　AFM anodic oxidation process　AFM 阳极氧化法加工

3.3.4　微纳加工技术的发展与应用

进入 20 世纪 90 年代以来，MEMS 技术高速发展，在光信息处理、生物医学、机器人、汽车、航空航天、军事及日用消费电器等方面得到了越来越多的应用。

2000 年，微型机械产品就有约为 140 亿美元的销售市场，2005 年 MEMS 市场销售额达到 650 亿美元。当前 MEMS 业界的销售额年增长率为 10%~20%，预计 MEMS 将是今后复杂高新技术发展的推进器和增长点。

在生物医学领域，已开发出对细胞进行操作的微型机械装置，如微操作台、微夹钳、微型剪子和锯子，用于视网膜上切除结疤。用微型机械工具可从被阻塞的动脉内切除脂肪堆积物；通过微型阀门可把限定药量的烈性药丸送入人体或动物体内。

在信息仪器领域，利用扫描隧道显微镜 STM 可将 1M 地址的信息储存在 $1~\mu m^2$ 的芯片上，显著降低存储系统的尺寸与重量、存取等待时间、失效率和成本低，且存储数据量大。通过集成三轴 MEMS 陀螺和加速度计，构成一个结构灵巧价格便宜的惯性测量器件，可取代传统的惯性装置，用于车辆摄像机等装置的稳定控制、姿态调节和个人导航系统。

在航空航天和军事领域正在研制的微型卫星，重量不到 0.1 kg，可由 1 枚运载火箭同时向太空发射多个微型卫星形成覆盖全球的微卫星网络。美国陆军研制的长、宽、高均小于 15 cm、重量不超过 120 g 的微型飞行器能以 30~60 km/h 的速度，连续飞行 20~30 min，而被广泛地用于战场侦察通信和反恐活动中。

当前 MEMS 的主要产品有：①微型机械构件，如微齿轮、微弹簧、微连杆、微轴承、微刀具等；②微执行器，如微阀、微泵、微开关、微电动机等；③微系统，如微型机器人、微型机床、微型惯性仪表等。

目前，各国对 MEMS 的研究非常关注，如美国正在研制的无人驾驶飞机仅有蜻蜓大小，并计划进一步缩小成蚊子机器人，用于收集情报和窃听。医用超微型机器人可进入人的血管，从主动脉管壁上刮去堆积的脂肪，疏通患脑血栓病人阻塞的血管。日本制订了采用机器人外科医生的计划，并正在开发能在人体血管中穿行、用于发现并杀死癌细胞的超微型机器人。

vacancy by removing the atoms. Under the condition of the pulse voltage, the atom can also be emitted from the tip to achieve the addition of atom and fill up vacancy.

3. AFM Mechanical Etching Machining

Under the contact mode of atomic force microscope (AFM) the local structural variation will be produced by increasing the force between the tip and the specimen surface, that is, the nano-machining can be realized through the mechanical etching of the tip and the specimen surface.

4. AFM Anodic Oxidation Machining

The process is a kind of machining method to form nano-scale oxide structure by the chemical reaction between the tip of the scanning probe microscope and the specimen. The anode is tip, and the cathode is the sample surface. The water molecules adsorbed on the sample surface act as the electrolyte, providing the OH^- ions required for the oxidation reaction, as shown in Fig. 3-13. During the early stage of this process STM was used, and AFM was often applied later, mainly due to the oxidation process of AFM method itself is simple and easy to operate, and the etched structure has stable property.

3. 3. 4 Development and Application of Micro/Nano Fabrication Technology

Since the 1990s, MEMS technology has been rapidly developed and has been increasingly applied in many fields such as optical information processing, biomedicine, robotics, automobile, aerospace, military and daily consumer appliances, etc.

In 2000, the micro-mechanical products have shared about 14 billion dollars in sales market. In 2005 MEMS market shared reached 65 billion dollars. The current annual growth rate of the MEMS industry is 10% to 20%. It is expected that MEMS will be a propeller and one growth point of the complex development in future.

In the field of biomedicine, a lot of micro-mechanical devices, such as micro-manipulator, micro-clamp, micro-scissors and saw have been developed for the operation of cells, which can be used for resection of scars on the retina. Fat deposits can be cut from the blocked artery by using the micro-mechanical tools. The strong pills limited dose can be fed into the human body or animal body through the micro-valve.

In the field of information equipment, using the STM can store the information for 1 M address in one chip with a size of one square micron, which reduces significantly the size and weight of storage system, latency and access time, and has low failure rate and cost, and high quantity storage data. Through the integration of three-axis MEMS gyroscope and accelerometer, an inertial measurement device with agile structure and cheap price can be constituted, which can replace the traditional inertial device. It can be used to stability control, attitude regulation and personal navigation system for the vehicle camera and other devices.

In the aerospace and military fields the microsatellites which are being developed with weight less than 0. 1 kg, can form a global micro-satellite network by a number of microsatellites which can be launched by a carrier rocket at the same time. The length, width and height of the micro-aircraft developed by the army of the United States, are all less than 15 cm, the weight of which is less than 120 g. This micro-aircraft can fly at the speed of 30-60 km/h and can continuously flight 20-30 min, and it is widely used in battlefield communication surveillance and anti-terrorism activities.

At present, the main products of MEMS are as follows: ①Micro-mechanical components, such as micro gear, micro spring, micro connecting rod, micro bearing and micro cutting tool and so on;

3.4　高速加工技术

3.4.1　高速加工的概念

高速加工（High-Speed Machining，HSM）或高速切削是指采用超硬材料刀具和磨具，利用高速、高精度、高自动化和高柔性机床，以提高切削速度来达到提高材料切除率、加工质量的先进加工技术。

高速切削最早由 Carl Salomon 于 1931 年提出。在分别对钢、非铁金属及轻合金在不同切削速度下（钢：440 m/min；青铜：1600 m/min；黄铜：2840 m/min；铝：16 500 m/min）进行大量切削试验的基础上，Salomon 指出：在常规切削范围内，切削温度随切削速度的增大而提高，但当切削速度增大到某一数值以上，切削温度会随切削速度的增大而下降。图 3-14 所示为切削速度与切削温度关系图。

Salomon 的研究还表明，对于任何材料存在一个速度范围，在这一范围内，切削温度超过了任何刀具材料的熔点，造成加工无法进行（该范围被称为"死谷"）。因此高速切削也可定义为切削速度大于该极限值下的切削加工。

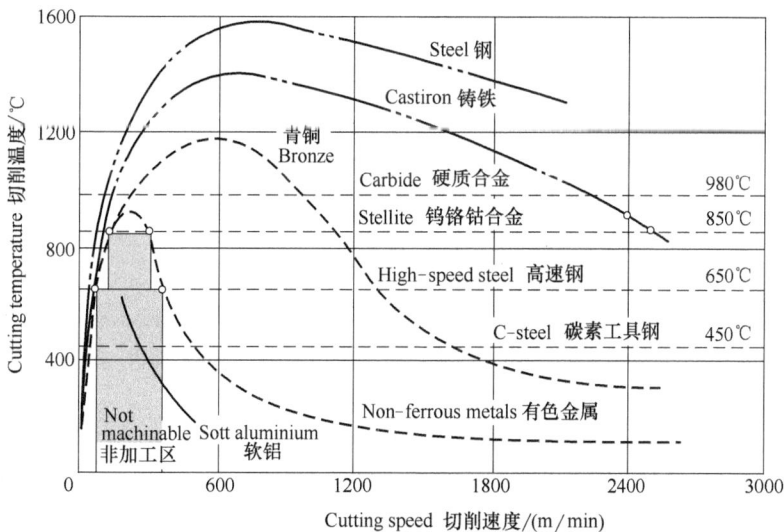

Fig. 3-14　Relationship between cutting speed and cutting temperature 切削速度与切削温度关系图

从加工材料的观点出发，高速切削是一个相对的概念，因为要保证刀具有足够寿命，不同的工件材料所采用的切削速度大不相同。例如，以 1800 m/min 的线速度加工金属铝和以 180 m/min 的线速度加工金属钛，相比之下，前者更容易实现。而 500 m/min 的切削速度对于合金钢加工来讲属于高速切削，但对于铝材加工来讲则属于传统加工范畴。目前沿用的两种高速切削定义分别是由国际生产工程研究协会（CIRP）和 Darmstadt 工业大学的生产工程与机床研究所（PTW）提出的。CIRP 提出：以线速度 500～7000 m/min 进行的切削为高速切削。PTW 则提出：以高于 5～10 倍普通切削速度进行的切削称为高速切削，而超高速切削

②micro-actuators, such as micro valve, micro pump, micro switch and micro motor and so on; ③the micro system, such as micro robot, micro machine tool and miniature inertial instrument and so on.

Currently, all the countries in the world are very concerned with the research on MEMS. For example, the United States is developing unmanned aircraft which has the same size with a dragonfly, and going to further reduce its size into mosquito robot used for information collection and eavesdropping. Medical ultra-micro robot can enter the blood vessels to scrape the accumulation of fat from the aortic wall, opening the blood vessels blocked of patients with cerebral thrombosis. Japan has made a plan to use robotic surgeons and is developing ultra-miniature robots that can walk through human blood vessels and discover and kill cancer cells.

3.4　High-Speed Machining Technology

3.4.1　Concept of High-Speed Machining

High-speed machining (HSM) or high-speed cutting is an advanced machining technology which uses super-hard cutting tools and grinding wheels and utilizes the machine tools with high speed, high precision, high automation and high flexibility to improve cutting speed so as to acquire high stock removal and better machining quality.

High-speed cutting was first proposed by Carl. J. Salomon in 1931. Based on his metal cutting studies on steel, non-ferrous and light metals at cutting speeds of 440 m/min (steel), 1, 600 m/min (bronze), 2, 840 m/min (copper) and up to 16, 500 m/min (aluminum), he assumed that under conventional machining conditions, machining temperatures rise with the increase of cutting speed, while from a certain cutting speed upward machining temperatures start dropping again. Fig. 3-14 shows the schematic diagram of high-speed machining hypothesis put forward by Salomon.

Salomon's fundamental research showed that for any material there is a certain range of cutting speeds where machining cannot be made due to excessively high temperatures because the temperatures exceed melting points of any cutting tool material (called "the death valley"). For this reason, HSM can also be termed as cutting speeds beyond that limit.

From the viewpoint of processing materials, high-speed cutting is a relative term because different part materials should be cut at very different speeds in order to ensure adequate tool life. For example, it is easier to machine aluminum at 1800 m/min than to machine titanium at 180 m/min. And a cutting speed of 500 m/min is considered high-speed machining for cutting alloy steel whereas this speed is considered conventional in cutting aluminum. At present, the common-used two definitions of HSM were given by the International Academy for Production Engineering (CIRP) and the Institute of production Engineering and Machine Tool (PTW) at the Darmstadt University of Technology Darmstadt. The CIRP termed machining processes performed at the speed of 500-7000 m/min as HSM. The PTW defines HSM as being such that conventional cutting speeds are exceeded by a factor of 5 to 10. In ultra-high speed machining, the cutting speeds should be about 10 times higher than those used in conventional processes.

At present, the cutting speed ranges for different machining methods and for different workpiece materials are shown in Tab. 3-2.

则是以比常规切削速度高 10 倍左右的速度进行的切削。目前，不同加工方法和不同工件材料的高速切削速度范围见表 3-2。

Tab. 3-2　High-speed cutting speed ranges for different machining methods and part materials

不同加工方法、不同工件材料的高速切削速度范围

Machining method 加工方法	Cutting speed 切削速度 /(m/min)	Part material 工件材料	Cutting speed 切削速度 /(m/min)
Turning 车削	700~7000	Aluminum alloy 铝合金	2000~7500
Milling 铣削	300~6000	Copper alloy 铜合金	900~5000
Drilling 钻削	200~1100	Steel 钢	600~3000
Broaching 拉削	30~75	Cast iron 铸铁	800~3000
Reaming 铰削	20~500	Heat-resistant alloy 耐热合金	500 以上
Sawing 锯削	50~500	Titanium alloy 钛合金	150~1000
Grinding 磨削	5000~10 000	Fiber-reinforced plastics 纤维增强塑料	2000~9000

3.4.2　高速切削的优缺点

高速切削的切削速度高，切削深度小，进给速度高，因而具有许多传统加工所不具有的优点。

1）多数情况下，高速切削时刀具和工件的温度较低，因而可以延长刀具寿命。另一方面，高速切削的切削深度小，刀刃切削时间短。有研究表明，高速加工中 90%~95% 的热量可被高速流走的切屑带走，工件内部累积的热量很少。因此，高速切削适合于加工易于热变形的零件。

2）高速切削中切削力小，刀具变形一致，变形量也小。这是实现高效、安全生产的必要条件。另外，由于高速切削的切削深度浅，作用在刀具和主轴上的径向力小，对主轴轴承、导轨和滚珠丝杠非常有利。

3）由于高速切削中切削速度和进给速度比传统切削参数高 5~10 倍，因此材料去除率可提高 3~6 倍，加工时间也可相应缩减到原来的 1/3。此外，加工出的工件表面质量高，因此很多情况下可完全或部分消除后续的精加工。

4）高速切削可用于难加工材料的切削，且生产率高，刀具磨损小，零件加工质量高。采用高速切削可以实现硬、脆材料的塑性去除加工，以及韧性材料的高表面完整性加工。

高速切削存在的问题与所要加工工件的材料和几何形状有关。常见的缺点有几个方面：①加工中由于加速度大以及主轴的快速启停加速了导轨、滚珠丝杠和主轴轴承的磨损，需要高性能的机床主轴及控制器，因此机床价格昂贵，且维修成本高；②高速切削刀具会发生过度磨损，因此要求更优的刀具材料和刀具涂层。

3.4.3　高速加工技术的应用

高速切削在生产率和切削效率方面具有很大优势，目前几乎在各个领域均有应用。以下对车削和铣削的运动学原理进行比较。车削加工要求工件旋转。对于大型零件，很难使工件

3.4.2 Advantages and Limitations of HSM

HSM is characterized by very high cutting speed, small depths of cut and high feed rates which give HSM many advantages when compared with conventional machining processes.

1) In HSM, cutting tool and workpiece temperature are kept low which gives a prolonged tool life in many cases. On the other hand, the cuts are shallow and the engagement time for the cutting edge is extremely short. Research shows 90%-95% cutting heat produced in HSM would be taken away by chips flowing at high rates. And heat accumulated in the workpiece is not so much. Therefore, HSM is suitable to machine workpieces that are prone to deform in high temperature.

2) In HSM, the low cutting force gives a small and consistent tool deflection. This is one of the prerequisites for a highly productive and safe process. As the depths of cut are typically shallow in HSM, the radial forces on the tool and spindle are low. This saves spindle bearings, guide-ways and ball screws.

3) Since both cutting speeds and feed rates in HSM are 5-10 times higher than traditional machining, material removal rates can be improved by 3-6 times higher and processing time of a workpiece can be reduced to 1/3 correspondingly. In addition, owing to the high surface quality produced in HSM it is possible in many cases to eliminate subsequent finish machining entirely or in part.

4) HSM provides a better way to machine hard-to-machine materials with high productivity, less tool wear and better machining quality. It can be used to accomplish plastic working of hard and brittle materials. Also, it is suitable for machining ductile materials with high surface integrity.

The problems existed in the application of HSM have to do with the material and geometry of the workpiece to be machined. The common disadvantages are claimed to be the following aspects: ①As the high acceleration and deceleration, rapid start and stop of spindle would speed up the wear of guide ways, ball screws and spindle bearings, the spindles and controllers with high performances are required. This kind of machine tool is quite expensive and has higher maintenance costs. ②Excessive tool wear associated with HSM calls for advanced cutting tool materials and coatings.

3.4.3 Applications of HSM

HSM, with many advantages in productivity and efficiency, currently finds its way into almost every field of machining. Let us compare the kinematics principle of turning and milling procedures. The workpiece rotates in turning. For the large parts, it is quite difficult to rotate it at high speed and to clamp it in safety. Compared with milling operation, turning is generally not suitable for high-speed cutting. Consequently, high-speed turning does not have wide industrial application yet.

Compared to the short chips produced in high-speed milling, high-speed drilling produces long chips which have to be taken out of the hole. Unlike the milling, it is impossible to remove the longer chips from the holes conveniently and quickly in drilling. During drilling, the heat is absorbed by the drill and the wall of the drilled hole, so high-speed drilling is impossible without internal cooling of the solid carbide drill that allows the cutting fluid to go directly to the contact positions between cutting edge and the material machined. Drilling is defined as a high speed procedure if the cutting parameters exceed the conventional ones by a factor of at least 2.

Consequently, the main focus of HSM is on the milling processes. The following considerations are dedicated to milling technology. HSM will only succeed if there is a perfect interaction among

高速旋转，也很难保证工件的安全装夹。与铣削相比，车削一般不太适合于高速切削。因而，高速车削仍然没能得到广泛的应用。

与高速铣削产生的短切屑相比，高速钻削产生的切屑较长。这些切屑需要从加工孔中排出。对于钻削，切屑不能像在铣削中一样可方便、及时地排出。因此，高速钻削需要对硬质合金钻头进行内部冷却，使切削液直接流向刀刃和加工材料的接触部位。高速钻削是切削参数增加到常规钻削加工参数两倍以上的钻削加工。

基于上述几点，高速切削主要应用于铣削加工。以下均专门针对铣削加工而言。高速铣削只有在机床、刀具、工件和刀具装卡、切削液、切削参数（如主轴转速、切削速度和进给量）等方面相互协调的情况下才能顺利进行。

高速切削主要应用于三大领域：①轻合金的加工，可加工汽车零件、计算机用小零件或医疗装置用小零件，这些零件的加工工序较多，需要较高的加工速度；②航空工业，用于加工尺寸较长的薄壁铝件；③模具行业。此外，石墨或铜电极也非常适合采用高速铣削。借助于覆有 TiC（N）复合粉或金刚石涂层的整体硬质合金立铣刀可实现高生产率的石墨材料高速铣削。

3.4.4 高速加工的关键技术

1. 高速切削加工机床

高速切削加工机床是实现高速加工的前提和基本条件。这类机床一般都是数控机床和精密机床。与普通机床的最大区别在于高速切削加工机床要具有很高的主轴转速和加速度，且进给速度和加速度也很高，输出功率很大。例如，高速切削加工机床的转速一般都大于 10 000 r/min，有的高达 100 000 ~ 150 000 r/min；主轴电动机功率为 15 ~ 80 kW；进给速度约为常规机床的 10 倍，一般为 60 ~ 100 m/min；无论是主轴还是移动工作台，速度的提升或降低往往要求在瞬间完成，因此主轴从起动到达到最高转速，或从最高转速降低到零，要在 1 ~ 2 s 内完成，工作台的加速度、减速度由常规机床的 0.1 ~ 0.2 g 提高到 1 ~ 8 g。这就要求高速切削加工机床要具有很好的静态、动态特性，数控系统以及机床的其他功能部件的性能也得随之提高。高速切削加工机床的关键技术包括高速主轴系统、快速进给系统、高性能 CNC 控制系统、先进的机床结构等。

（1）高速主轴系统 高速主轴单元是高速加工机床最关键的部件。目前高速主轴的转速范围为 10 000 ~ 25 000 r/min，加工进给速度在 10 m/min 以上。为适应这种切削加工，高速主轴应具有先进的主轴结构、优良的主轴轴承、良好的润滑和散热等新技术。

1）电主轴。在超高速运转的条件下，传统的齿轮变速和带传动方式已不能适应要求，代之以宽调速交流变频电动机来实现数控机床主轴的变速，从而使机床主传动的机械结构大为简化，形成一种新型的功能部件——主轴单元。在超高速数控机床中，几乎无一例外地采用"电主轴"（motorized spindle），如图 3-15 所示。由于电主轴取代了从主电动机到机床主轴之间的一切中间传动环节，主传动链的长度缩短为零，故把这种新型的传动方式称为"零传动"。

电主轴振动小，由于直接传动，减少了高精密齿轮等关键零件，消除了齿轮的传动误差。同时，集成式主轴也简化了机床外形设计，容易实现高速加工中快速换刀时的主轴定位等。这种电主轴和以前用于内圆磨床的内装式电动机主轴有很大的区别，主要表现在：①有

machine tool, tool, workpiece and tool clamping techniques, cutting fluids, cutting parameters, such as spindle speeds, cutting speeds and feeds.

The main application fields of HSM are three industry sectors. The first category is industry which deals with machining light alloys to produce automotive components, small computer parts or medical devices. This industry needs fast metal removal because the technological process involves many machining operations. The second category is aircraft industry which involves machining of long aluminum parts, often with thin walls. The third industry sector is the die and mould industry. Besides, milling of electrodes in graphite and cooper is another excellent area for HSM. Graphite can be machined in a productive way with Ti (C, N) or diamond coated solid carbide end mills.

3.4.4 The Key Technologies of HSM

1. High-Speed Cutting Machine Tool

High-speed cutting machine tool is the precondition and basic condition of HSM. This type of machine tools are generally CNC machine tools and precision machine tools. The biggest difference from the ordinary machine is that the spindle of high-speed cutting machine has a high RPM and acceleration, and the feed rate and acceleration is also high, the output power is very large. For instance, the RPM of high-speed cutting machines is generally larger than 10, 000 r/min, some are as high as 100, 000-150, 000 r/min; the power of the spindle motor is 15-80 kW; the feed rate is about 10 times of conventional machine tools, i. e. 60-100 m/min in general. Whether it is a spindle or a moving table, it is required that the increase or decrease of speed be completed in moment. The action of the spindle from the start to the highest RPM, or from the highest RPM to zero should be completed within 1-2 s, the acceleration and deceleration of the worktable should be increased from 0. 1-0. 2 g of conventional machine tool to 1-8 g. This requires that high-speed cutting machines should have a good static and dynamic characteristics, and the performances of numerical control system and other functional components of the machine tool has to be improved. The key technologies of high-speed cutting machine includes high-speed spindle system, fast feed system, CNC control system with high performance, advanced machine tool structure, etc.

(1) High-speed spindle system   High-speed spindle unit is the most critical component of high-speed machine tool. At present, the RPM range of high-speed spindle is 10, 000-25, 000 r/min, and the working feed rate is above 10 m/min. In order to adapt to this kind of cutting, high-speed spindle should have advanced spindle structure, excellent spindle bearing, good lubrication and heat dissipation and other new technologies.

1) Motorized spindle. Under the condition of ultra-high speed rotation, the traditional gear transmission and belt transmission cannot meet the requirements. Instead, AC variable frequency motor with wide speed regulation is used to change the RPM of NC machine tool spindle, which simplifies the main drive structure of machine tool, and forms a new type of functional component—the spindle unit. The motorized spindle is used almost without exception in the ultra-high speed NC machine tools, as shown in Fig. 3-15. Because all intermediate transmission links between the main motor and the spindle of machine tool are replaced by the motorized spindle, the length of the primary drive chain is reduced to zero. Therefore, this new type of transmission pattern is called "zero transmission".

The vibration of motorized spindle is small. Due to the direct drive, the high-precision gears and

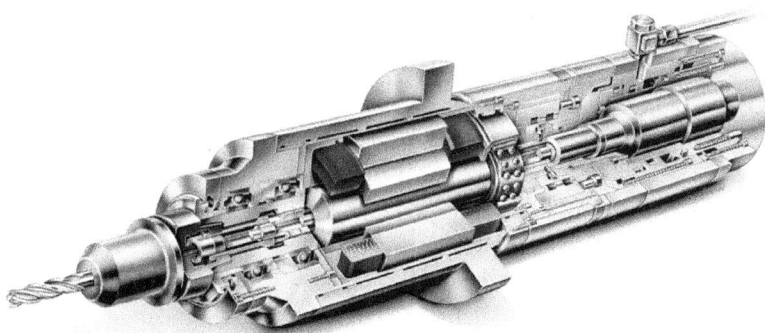

Fig. 3-15　Motorized spindle structure 电主轴结构

很大的驱动功率和转矩；②有较宽的调速范围；③有一系列监控主轴振动、轴承和电动机温升等运行参数的传感器、测试控制和报警系统，以确保主轴超高速运转的可靠性与安全性。

2）静压轴承高速主轴。目前，在高速主轴系统中广泛采用液体静压轴承和空气静压轴承。由于静压轴承为非接触式，具有磨损小、寿命长、回转精度高、阻尼特性好的特点，另外其结构紧凑，动态、静态刚度较高。但静压轴承价格较高，使用维护较为复杂。

液体静压轴承高速主轴的最大特点是运动精度很高，回转误差一般在 0.2 μm 以下。液体静压轴承刚度高、承载能力强，但结构复杂、使用条件苛刻、消耗功率大、温升较高。采用空气静压轴承可以进一步提高主轴的转速和回转精度，其最高转速可达 100 000 r/min，回转误差在 50 nm 以下。气体静压轴承刚度差、承载能力低，主要用于高精度、高转速、轻载荷的场合。

3）磁浮轴承高速主轴。磁浮轴承的工作原理如图 3-16 所示。电磁铁绕组通过电流而对转子产生吸力，与转子重量平衡，转子处于悬浮的平衡位置。转子受到扰动后，偏离其平衡位置。传感器检测出转子的位移，并将位移信号送至控制器。控制器将位移信号转换成控制信号，经功率放大器变换为控制电流，改变吸力方向，使转子重新回到平衡位置。位移传感器通常为非接触式，在一个磁浮轴承内一般安装 5~7 个。磁浮轴承高速主轴结构如图 3-17 所示。

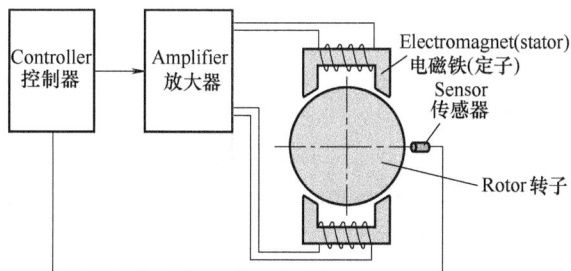

Fig. 3-16　Operation principle of magnetic bearing 磁浮轴承的工作原理

磁浮主轴的优点是高精度、高转速和高刚度；其缺点是机械结构复杂，而且需要一整套的传感器系统和控制电路，所以磁浮主轴的造价较高。另外，主轴部件内除了驱动电机外，还有轴向和径向轴承的线圈，每个线圈都是一个附加的热源。因此，磁浮主轴必须有很好的

other key parts can be saved, and the gear transmission error is eliminated. At the same time, the integrated spindle also simplifies the configuration design of machine tool in some of the key work, such as simplifying the design of machine tools, and makes it easy to position the spindle for rapid changing tool in high-speed machining. The motorized spindle is very different from the spindle with built-in motor used in the internal grinding machine, which is mainly manifested in: ①large drive power and torque; ②wider speed range; ③ there are a series of sensors to monitor the spindle vibration, bearing and motor temperature and other parameters, and test control and alarm system in order to ensure the reliability and safety of the spindle running at a high speed.

2) High-speed spindle with static bearing. At present, the hydrostatic bearing and aerostatic bearing are widely used in high-speed spindle system. Because of non-contact, the static bearing has the characteristics of small wear, long service life, high rotary accuracy and good damping, compact structure and high dynamic and static rigidity. But the price of static bearing is high, the maintenance is more complex.

The chief feature of high-speed spindle with hydrostatic bearing is the high kinematic accuracy, and its rotary error is less than 0.2 μm. The hydrostatic bearing has high rigidity and high bearing capacity, but complex structure, rigorous use conditions, larger power consumption is, and higher temperature rise. Aerostatic bearings can further increase the RPM of spindle and rotary accuracy. Its max. RPM can reach 100, 000 r/min, the rotary error can be less than 50 nm. Aerostatic bearings have poor rigidity and low bearing capacity, and are mainly used in the situation with high precision, high speed and light duty.

3) High-speed spindle with magnetic bearing. The operation principle of magnetic bearing is shown in Fig. 3-16. The electromagnet winding with electric current generates the suction force to the rotor and balances the weight of the rotor, thus keeping the rotor in a suspended equilibrium position. When the rotor is disturbed, it deviates from its equilibrium position. The sensor detects the displacement of the rotor and sends the displacement signal to the controller. The controller converts the displacement signal into a control signal, which is converted to the control current by the power amplifier. The direction of suction force is changed so that the rotor return to the balance position again. Generally, displacement sensors are the non-contact type, and there are 5-7 displacement sensors in a magnetic bearing. The structural sketch of high-speed spindle with magnetic bearings is shown in Fig. 3-17.

The advantages of the magnetic spindle are high precision, high speed and high rigidity; its disadvantages are that the mechanical structure is complex, and a set of sensor system and control circuit are required. Therefore, the cost of the magnetic spindle is higher. In addition to the drive motor, there are coils of axial and radial bearing in the spindle unit, each coil is an additional heat source. Therefore, the magnetic spindle must have a good cooling system.

(2) Feed system of ultra-high speed cutting machine tool The feed system of the ordinary NC machine tool adopts the structure equipped with the ball screw pairs and the rotary servo motor. Because the ball screw has low torsional rigidity, it is easy to generate torsional vibration in the high-speed running. The long transmission chain of the feed system would generate larger integrated transmission error and nonlinear error, thus influencing on the machining accuracy. It is required in the ultra-high speed cutting that the feed rate increases while increasing the speed of the spindle, and the feed movement reaches the high speed and stops instantly. Using the linear motor as the executive component of the feed servo system can drive directly the worktable to move. The length of transmission

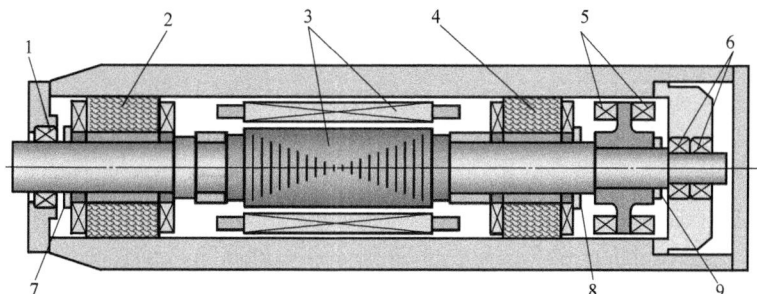

Fig. 3-17　High-speed spindle with magnetic bearing 磁浮轴承高速主轴结构

1—Front auxiliary bearing 前辅助轴承　2—Front radial bearing 前径向轴承　3—Electric spindle 电主轴

4—Rear radial bearing 后径向轴承　5—Double axial thrust bearing 双面轴向推力轴承

6—Rear auxiliary bearing 后辅助轴承　7—Front radial sensor 前径向传感器

8—Rear radial sensor 后径向传感器　9—Axial sensor 轴向传感器

冷却系统。

（2）超高速切削机床的进给系统　普通数控机床的进给系统采用的是滚珠丝杠副加旋转伺服电机的结构。由于丝杠扭转刚度低，所以高速运行时易产生扭振。进给系统的机械传动链较长会产生较大的综合传动误差和非线性误差，从而影响加工精度。超高速切削在提高主轴速度的同时也必须提高进给速度，并且要求进给运动能在瞬时达到高速和瞬时准停等。使用直线电机作为进给伺服系统的执行元件，可直接驱动机床工作台运动。其传动链长度为零，并且不受离心力的影响。这种进给伺服系统的结构简单、重量轻，容易实现很高的进给速度（80~180 m/min）和加速度（2~10 g）；动态性能好、运动精度高，并且无机械磨损。

（3）机床床身结构　为了适应高速加工的要求，机床床身，包括工作台都要具有高的静刚度、动刚度、抗振性、精度保持性，以及优良的抗热变形的能力。图 3-18 所示为第 1 代高速铣龙门结构床身，床身为铸铁。图 3-19 所示为第 2 代高速铣 O 形结构床身，床身为人造大理石。质量大的人造大理石床身具有极高的热稳定性和良好的吸振性能（通常是铸铁的 6 倍），从而可保证极好的零件加工精度。

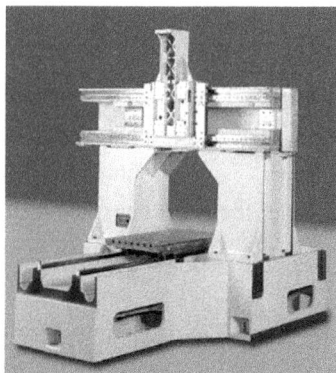

Fig. 3-18　The first generation planer-type bed
of high-speed milling machine
第 1 代高速铣龙门结构床身

Fig. 3-19　The second generation o-type bed
of high-speed milling machine
第 2 代高速铣 O 形结构床身

chain is zero, and will not be influenced by centrifugal force. The feed servo system has the advantages as follows: simple structure, light weight, easy to reach high feed rate (80-180 m/min) and acceleration (2-10 g), good dynamic performance, high precision (0.1-0.01 m), and without mechanical wear.

(3) Bed structure of machine tool   In order to meet the requirements of high-speed machining, the machine tool bed the work table should have high static and dynamic rigidity, vibration resistance, precision retention and excellent resistance to thermal deformation. Fig. 3-18 shows the first 1generation planer-type bed of high-speed milling machine, which is made of cast iron. Fig. 3-19 shows the second generation O-type bed of high-speed milling machine, which is made of artificial marble bed. The heavy artificial marble bed has a high thermal stability and good vibration absorption performance (usually 6 times of cast iron), thus, the excellent part machining accuracy can be ensured.

2. High-Speed Cutting Tool

The metal removal rate in high-speed cutting is very high. The high strain rate of the material being cut makes the chip formation process and the phenomena occurred between cutting tool and the chip be different from the traditional cutting. The heat resistance and abrasion resistance have become the chief contradiction. At the same time, the high RPM of the spindle makes the cutting tool rotated at high speed produce a large centrifugal force, which will not only affect the machining accuracy, but also cause the cutter body to break. Therefore, the high-speed cutting has a strict requirement for the cutter material, geometrical parameter, the cutter body structure as well as the installation of cutting tool and so on.

(1) Requirements for cutting tool materials

1) High strength and high toughness.

2) Extremely high hardness and high wear resistance.

3) High heat resistance and high resistance to thermal shock. The higher the hardness at high temperature of the tool material is, the better the heat resistance is. And the higher the cutting speed is. The thermal shock resistance prevents the cutting tool in intermittent cutting from endurance failure caused by the thermal stress.

(2) High-speed cutting tool material   At present, the cutting tools used in high-speed cutting are ceramic, cubic boron nitride (CBN), diamond and coated blade. Tab. 3-3 lists the properties and applications of cutting tool materials commonly used in high-speed cutting.

(3) Structure and safety of high-speed cutting tool   In high-speed cutting, it is not allowed to clamp the indexable face milling cutter by means of friction force. Instead, the blade with a central hole is used and clamped by screw with suitable pretightening force.

The design of the cutter body should reduce the weight and the diameter but increase the height. Milling cutter structure should avoid using the through slot in order to reduce the sharp angle and stress concentration. The static and dynamic balance of the cutter, chuck, spindle and its combination should be carried out in order to avoid influencing on the machine tool spindle and bearing life, and on the machined surface quality and the tool life because of the imbalance of high-speed cutting tool.

3. High-Speed Cutting Process Technology

High-speed cutting process mainly includes the tool path mode, specialized CAD/CAM programming strategy, optimized high-speed machining parameters, and environmentally conscious cooling method and so on.

2. 高速切削刀具

高速切削的金属切除率很高。被加工材料的高应变率使切屑形成过程及刀具-切屑间发生的各种现象都和传统切削不一样。刀具的耐热性和耐磨性成为主要矛盾。同时，由于主轴转速很高，高速转动的刀具会产生很大的离心力，不仅影响加工精度，还有可能致使刀体破裂。因此，高速切削对切削刀具材料、刀具几何参数、刀体结构乃至刀具的安装等提出了很高的要求。

（1）对刀具材料的要求

1）高强度与高韧性。

2）极高的硬度和高耐磨性。

3）高耐热性和高抗热冲击性。刀具材料的高温硬度越高，耐热性越好，允许的切削速度越高。抗热冲击性使刀具在断续切削受到热应力冲击时，不致产生疲劳破坏。

（2）高速切削刀具材料　目前适用于高速切削的刀具材料主要有陶瓷、立方氮化硼、金刚石和涂层刀片。表3-3列出了高速切削常用刀具材料的性能和用途。

（3）高速切削刀具的结构及其安全性　高速切削时，可转位面铣刀不允许采用摩擦力夹紧方式，而采用带中心孔的刀片，使用合适的预紧力用螺钉夹紧。

Tab. 3-3　Properties and applications of cutting tool materials commonly used in high-speed cutting

高速切削常用刀具材料的性能和用途

Tool materials 刀具材料	Advantages 优点	Disadvantages 缺点	Applications 用途
Ceramic 陶瓷	Good wear resistance, good thermal shock resistance, good chemical stability, good bonding resistance, dry cutting 耐磨性好、抗（热）冲击性能好、化学稳定性好、抗黏结性好、干式切削	Poor toughness, high brittleness, prone to collapse, large affinity to aluminum at high temperature 韧性差、脆性大、容易产生崩刃、和铝的高温亲和力大	Hardened cast iron, hard steel, nickel-based superalloy, stainless steel 淬火铸铁、硬钢、镍基高温合金、不锈钢
CBN 立方氮化硼	High hardness, good thermal stability, low friction coefficient, high thermal conductivity, without BUE 硬度高、热稳定性好、摩擦系数小、热导率高、不产生积屑瘤	Poor strength and toughness, low bending strength, prone to collapse 强度和韧性差、抗弯强度低、易崩刃	Finishing such materials as hardened steel, high-temperature alloy, tool steel, high-speed steel 淬硬钢、高温合金、工具钢、高速钢等高硬度材料的精加工
Diamond 金刚石	High hardness, little friction coefficient, high thermal conductivity, excellent abrasion resistance, quite sharp 硬度极高、摩擦系数很小、热导率高、耐磨性极好、十分锋利	Poor toughness, low bending strength, prone to collapse, not suitable for cutting iron and titanium materials 强度和韧性差、抗弯强度低、易崩刃，不宜切削含铁和钛的材料	Precision / ultra-precision cutting of monocrystalline Al, Si and Ge, aluminum alloy, brass, magnesium alloy 单晶铝、单晶硅、单晶锗、铝合金、黄铜、镁合金的精密/超精密切削
Coating 涂层	High surface hardness, good wear resistance 表面硬度高、耐磨性好	Poor heat resistance and abrasion resistance 耐热性和耐磨性较差	High-hard aluminum alloy, titanium alloy 高硬铝合金、钛合金

刀体的设计应减轻质量，减小直径，增加高度。铣刀结构应尽量避免采用贯通式刀槽，

In principle, the layered contour-parallel cutting is most often used in high-speed cutting. Directly vertical feed down would cause cutter collapse easily and should not be used. As the milling force produced in the slant-path feed is gradually increased, the impact on the tool and the spindle is smaller than that of the vertical feed, and the cutter collapse phenomenon can be reduced obviously. The spiral-path feed method using spiral cut down into the workpiece is most suitable for the high-speed cutting of cavity.

The principle of CAD/CAM programming is to keep a constant tool load as far as possible, and to minimize the variation of feed rate and maximize program processing speed. The main methods are: to reduce the program block and to improve the speed of program processing as far as possible; by adding some arc transition section into program segment to reduce the rapid changes of speed as far as possible; in roughing, instead of removing the material simply, paying attention to uniforming the machining allowance in this operation and subsequent operations so as to reduce the milling load changes as far as possible; to carry out the tangential feed through continuous spiral path and circular path as far as possible so as to ensure constant cutting conditions; making full use of the simulation function in CNC system to verify the correctness of the tool position data, whether the tool interferes with the workpiece, whether the tool collides to the fixture accessories before machining the workpiece in order to ensure product quality and operation safety.

In general, the machining efficiency increases with the increase of milling speed, but the tool wear would increase. Except for the higher feed per tooth, the surface roughness decreases with the increase of cutting speed. For tool life, there is an optimum feed per tooth and an optimum axial depth of cut, and the optimum value range is relatively narrow. The general principle in choosing high-speed milling parameters is the high-cutting speed, medium feed f_z per tooth, smaller axial depth of cut a_p and suitably large radial depth of cut a_e.

Because of the increases of metal removal rate and cutting heat in high-speed cutting, the colling media must have the ability to remove chips from the workpiece quickly, and to reduce cutting heat and lubricate the cutting interface. Conventional coolants and pouring methods are difficult to send the coolant into the cutting area. Instead, it would increase the temperature changes of milling cutter in the engaging and disengaging process, thus causing the cutter thermal fatigue and shortening the tool life. Modern tool materials, such as cemented carbides, coated tools, ceramics, CBN and so on have a higher red hardness and can be used in dry cutting.

On the one hand, using micro quantity oil mist can reduce the friction of cutter-chip-workpiece; on the other hand, the tiny oil mist particles can be gasified rapidly when they contact with the tool surface, thus taking away more heat than the coolant heat transfer doing. At present, this cooling medium has become the preferred one in high-speed cutting.

3.5　Modern Non-traditional Machining Technology

3.5.1　Introduction to Non-traditional Machining

1. Emergence and Development of Non-traditional Machining

Traditional machining has a long history, which plays a great role in the production and materi-

以减少尖角和应力集中。对刀具、夹头、主轴及其组合分别进行静、动平衡，以避免高速回转刀具的动不平衡影响机床主轴和轴承的使用寿命，以及影响工件已加工表面的质量和刀具寿命。

3. 高速切削工艺技术

高速切削工艺主要包括加工走刀方式、专门的 CAD/CAM 编程策略、优化的高速加工参数以及具有环保意识的冷却方式等。

高速切削的加工方式原则上多采用分层环切加工。直接垂直向下进刀极易出现崩刃现象，不宜采用。斜线轨迹进刀方式的铣削力是逐渐加大的，因此对刀具和主轴的冲击比垂直下刀小，可明显减少下刀具崩刃的现象。采用螺旋向下切入的螺旋式轨迹进刀方式最适合于型腔的高速加工。

CAD/CAM 编程原则是尽可能保持恒定的刀具载荷，把进给速率变化降到最低，使程序处理速度最大化。其主要方法有：尽可能减少程序块，提高程序处理速度；在程序段中加入一些圆弧过渡段，尽可能减少速度的急剧变化；粗加工不是简单的去除材料，要注意保证本工序和后续工序加工余量均匀，尽可能减少铣削负荷的变化；尽量采用连续的螺旋和圆弧轨迹进行切向进刀，以保证恒定的切削条件；充分利用数控系统提供的仿真功能，在零件加工前要验证刀位数据的正确性、刀具各部位是否与零件发生干涉、刀具与夹具附件是否发生碰撞等，以确保产品质量和操作安全。

通常，加工效率会随着铣削速度的提高而提高，但刀具磨损加剧。除较高的每齿进给量外，加工表面粗糙度随切削速度提高而降低。对于刀具寿命，每齿进给量和轴向切深均存在最佳值，而且最佳值的范围相对较窄。高速铣削参数一般的选择原则是高的切削速度、中等的每齿进给量 f_z、较小的轴向切深 a_p 和适当大的径向切深 a_e。

在高速切削时，由于金属去除率和切削热的增加，冷却介质必须具备将切屑快速冲离工件、降低切削热和润滑切削界面的能力。常规的切削液及加注方式很难把切削液送到加工区域，反而会加大铣刀刃在切入切出过程的温度变化，从而使刀具热疲劳、降低刀具寿命。现代刀具材料，如硬质合金、涂层刀具、陶瓷、CBN 等具有较高的红硬性，可以用于干切削。

微量油雾冷却一方面可以减小刀具-切屑-工件之间的摩擦，另一方面，细小的油雾粒子在接触到刀具表面时能快速气化，从而比冷却液热传导方式能带走更多的热量。目前，微量油雾已成为高速切削首选的冷却介质。

3.5　现代特种加工技术

3.5.1　特种加工概述

1. 特种加工的产生及发展

传统的机械加工已有很久的历史，它对人类的生产和物质文明起了极大的推动作用。例如，19 世纪 70 年代就发明了蒸汽机，但苦于制造不出高精度的蒸汽机气缸，无法推广应用。直到有人创造和改进了气缸镗床，解决了蒸汽机主要部件的加工工艺，才使蒸汽机获得广泛应用，引起了世界性的第一次工业革命。这一事实充分说明了加工方法对新产品的研制、推广和社会经济等起着多么重大的作用。随着新材料、新结构的不断出现，加工方法将

al civilization of human beings. For example, the steam engine was invented in the 1870 s. Howev-er, the steam engine was not spread and applied, owing to an air cylinder with high precision could not be machined. Until the cylinder boring machine was created and improved, together with the ma-chining process of the main components of the steam engine was developed, the steam engine was widely used, which led to the first Industrial Revolution in the world. This fact fully demonstrates that the machining methods have significant effects on the development of new products, promotion and social economy, etc. With the continuous emergence of new materials, new structures, the ma-chining methods will become more important.

However, from the First Industrial Revolution until to the Second World War, in this long ages of up to 150 years by mechanical machining (including grinding), there was no urgent need for non-traditional machining, and there was no sufficient conditions for the development of non-traditional machining. During that period, people's thinking had been limited in the traditional use of mechani-cal energy and cutting force to remove excess metal, to achieve the machining requirements.

Until 1943, Lin Ke of couples in the Soviet Union studied the phenomena and causes of the spark discharge corrosion damage of the switch contacts. They found that the instantaneous high temperature of electric spark could make metal melt, gasification locally and be eroded away, so created and invented the EDM (electro-discharge machining) method. This method was be used to machine small holes in quenching steel with copper wire, and can be used to machine metal materials of any hardness with soft tools. It was the first time that get rid of the traditional cutting machining and directly use electricity and heat to remove the metal, obtained the effect of "overcoming hardness with softness".

After the Second World War, particularly since 1950s, with the need of the production devel-opment and scientific experiment many industrial departments especially the defense industrial de-partments required advanced science and technology products to promote the development of high precision, high speed, high temperature, high pressure, high power, small size, and so on. The used materials are more difficult to be processed, the parts shape becomes more complex, and the surface precision, roughness and some special is required increasingly high, Therefore, new require-ments for the machinery manufacturing department are put forward as follows:

1) To solve machining problems of all difficult-to-cut materials, such as cemented carbide, iron alloy, heat-resistant steel, stainless steel, hardened steel, diamond, sapphire, quartz, as well as germanium, silicon, which are high hardness, high strength and high toughness, high brittle met-als and non-metallic materials.

2) To solve machining problems of all kinds of special complex surface, such as the three-di-mensional forming surface of jet turbine blades, integral turbine, engine casing, forging mould and injection mold, the shape hole of special section of all stamp die and cold drawing die, the barrel ri-fling, nozzle, grid, spinneret holes, and the narrow slit.

3) To solve machining problems of all kinds of ultra precision, finishing, or the parts with spe-cial requirements, such as aerospace or aviation gyroscope and servo valve with high surface quality and precision, as well as low rigidity parts including slender shaft, thin wall parts and elastic ele-ment, etc.

In order to solve the above-mentioned process problems, it is rather difficult or even impossible to implement only relying on the traditional cutting machining. Therefore, people have been exploring many new machining methods, and the non-traditional machining (NTM) emerges and is developed

变得更加重要。

但是自第一次工业革命以来，一直到第二次世界大战以前，在这段长达150多年都靠机械切削加工（包括磨削加工）的漫长年代里，并没有产生特种加工的迫切要求，也没有发展特种加工的充分条件，人们的思想一直还局限在传统的用机械能量和切削力来除去多余的金属，以达到加工要求。

直到1943年，苏联的拉扎林柯夫妇研究开关触点遭受火花放电腐蚀损坏的现象和原因，发现电火花的瞬时高温可使局部的金属熔化、气化而被蚀除掉，开创和发明了电火花加工方法，用铜丝在淬火钢上加工出小孔，可用软的工具加工任何硬度的金属材料，首次摆脱了传统的切削加工方法，直接利用电能和热能来去除金属，获得"以柔克刚"的效果。

第二次世界大战后，特别是进入20世纪50年代以来，随着生产发展和科学实验的需要，很多工业部门，尤其是国防工业部门，要求尖端科学技术产品向高精度、高速度、高温、高压、大功率、小型化等方向发展。所使用的材料越来越难加工，零件形状越来越复杂，对零件的表面精度、粗糙度和某些特殊要求也越来越高，因此，对机械制造部门提出了下列新的要求：

1）解决各种难切削材料的加工问题。例如，硬质合金、铁合金、耐热钢、不锈钢、淬火钢、金刚石、宝石、石英以及锗、硅等各种高硬度、高强度、高韧性、高脆性的金属及非金属材料的加工。

2）解决各种特殊复杂表面的加工问题。例如，喷气涡轮机叶片、整体涡轮、发动机机匣和锻压模和注射模的立体成形表面，各种冲模、冷拔模上特殊截面的型孔，炮管内膛线、喷油嘴、栅网、喷丝头上的小孔、窄缝等的加工。

3）解决各种超精、光整或具有特殊要求的零件的加工问题。例如，对表面质量和精度要求很高的航天、航空陀螺仪、伺服阀，以及细长轴、薄壁零件、弹性元件等低刚度零件的加工。

要解决上述一系列工艺问题，仅仅依靠传统的切削加工方法很难实现，甚至根本无法实现，于是人们相继探索研究新的加工方法，特种加工就是在这种前提下产生和发展起来的。

2. 特种加工的特点

切削加工的本质和特点：一是靠刀具材料比工件更硬；二是靠机械能把工件上多余的材料切除。一般情况下这是行之有效的方法。但是，在工件材料越来越硬，加工表面越来越复杂的情况下，原来行之有效的方法却会限制生产率并影响加工质量。于是人们开始探索用软的工具加工硬的材料，不仅用机械能而且还采用电能、化学能、光能、声能等能量来进行加工。为区别于现有的金属切削加工，这类新加工方法统称为特种加工（Non-Traditional Machining，NTM 或 Non-Conventioned Machining，NCM）。它们的特点是：①不是主要依靠机械能，而是主要用其他能量，如电能、化学能、光能、声能、热能等去除金属材料；②工具硬度可以低于被加工材料的硬度；③加工过程中工具和工件之间不存在显著的机械切削力。

正因为特种加工工艺具有上述特点，所以就总体而言，特种加工可以加工任何硬度、强度、韧性、脆性的金属或非金属材料，且专长于加工复杂、微细表面和低刚度零件。而且，有些方法还可用于进行超精加工、镜面光整加工和纳米级（原子级）加工。

3. 特种加工的分类

依据加工能量的来源及作用形式列举出常用的特种加工方法，见表3-4。

under this situation.

2. Characteristics of NTM

The nature and characteristics of the cutting machining are as follows: ①The cutting tool material is harder than the workpiece; ②mechanical energy is relied on to remove excess material on the workpiece. This is an effective way in general case. However, when the workpiece material is much harder, surface shape becomes more and more complex, the original effective method changes to unfavorable factor which limits the productivity and effects of machining quality. People attempt to use soft tools to machine hard materials, such as using electrical energy, chemical energy, light energy, sound energy for processing besides mechanical energy. Different from the existing metal cutting machining, these new machining methods are collectively referred to as the NTM or Non-Conventional Machining (NCM). The features of this kind of machining methods include that: ①Not rely mainly on mechanical energy, but mainly on other energy, such as electricity energy, chemical energy, light energy, sound energy and heat energy to remove metal materials. ②The hardness of tool can be lower than that of the material to be machined. ③There is not significant mechanical cutting force between the tool and the workpiece in the machining process.

Because the NTM technology has the above-mentioned characteristics, in terms of overall, NTM can process metal or non-metallic materials with any hardness, strength, toughness, brittle, specially can machine the complex, micro and tiny surface, and low rigidity parts. Moreover, some methods can also be used for ultra finishing, mirror finishing and nano-scale (atomic level) machining.

3. Classification of NTM

According to the source of machining energy and action form, the NTM methods commonly used are listed in Tab. 3-4.

3.5.2 Laser Beam Machining

1. Technology and Characteristics of Laser Beam Machining (LBM)

Laser technology is an emerging science developed in the early 1960s. In material machining, a new kind of machining method has formed, called the laser beam machining, which is a machining technology of using the photothermal effect of laser beam on materials. Laser beam machining technology is a comprehensive technology involving many disciplines of optical, mechanical, electrical, materials and testing etc.

Because the laser has the characteristics of high brightness, high directivity, high monochromatic and high coherence, the laser beam machining has the following characteristics:

1) After focusing, the power density of laser beam machining can be as high as 10^8-10^{10} W/cm^2, light energy is transformed into heat energy, so that almost any material can be melted and vaporized. For example, heat-resistant alloy, ceramics, quartz, diamond and other hard brittle materials can be machined.

2) The size of laser spot can be focused to micron scale, and the output power can be adjusted, so it can be used for precision micro machining.

3) The tool used for machining is a laser beam, which is non-contact machining, so there is no obvious mechanical force and no tool wear. The machining speed is fast, the heat effect zone is small, and the machining automation is easy to realize. The machining can be conducted at normal temperature and atmospheric pressure in the air, but also be conducted through the transparent ob-

Tab. 3-4　Commonly used non-traditional machining methods 常用的特种加工方法

Machining method 加工方法		Main form of energy 主要能量形式	Action form 作用形式
Electrical discharge machining 电火花加工	Electrical discharge forming 电火花成形加工	Electric energy, heat energy 电能、热能	Melting, gasification 熔化、气化
	Wire electrical discharge machining 电火花线切割加工	Electric energy, heat energy 电能、热能	Melting, gasification 熔化、气化
Electrochemical machining 电化学加工	Electrochemical machining 电解加工	Electrochemical energy 电化学能	Ion transfer 离子转移
	Electroforming machining 电铸加工	Electrochemical energy 电化学能	Ion transfer 离子转移
	Plating machining 涂镀加工	Electrochemical energy 电化学能	Ion transfer 离子转移
High energy beam machining 高能束加工	Laser beam machining 激光束加工	Light energy, heat energy 光能、热能	Melting, gasification 熔化、气化
	Electron beam machining 电子束加工	Electric energy, heat energy 电能、热能	Melting, gasification 熔化、气化
	Ion beam machining 离子束加工	Electric energy, mechanical energy 光能、热能	Abscission 切蚀
	Plasma arc machining 等离子弧加工	Electric energy, heat energy 电能、热能	Melting, gasification 熔化、气化
Material abscission machining 物料切蚀加工	Ultrasonic machining 超声加工	Sound energy, mechanical energy 声能、机械能	Abscission 切蚀
	Abrasive flow machining 磨料流加工	Mechanical energy 机械能	Abscission 切蚀
	Liquid jet machining 液体喷射加工	Mechanical energy 机械能	Abscission 切蚀
Chemical machining 化学加工	Chemical milling processing 化学铣切加工	Chemical energy 化学能	Corrosion 腐蚀
	Photographic plate processing 照相制版加工	Chemical energy, light energy 化学能、光能	Corrosion 腐蚀
	Lithography processing 光刻加工	Light energy, chemical energy 光能、化学能	Photochemical, corrosion 光化学、腐蚀
	Photoelectric forming electroplating 光电成形电镀	Light energy, chemical energy 光能、化学能	Photochemical, corrosion 光化学、腐蚀
	Etching processing 刻蚀加工	Chemical energy 化学能	Corrosion 腐蚀
	Bond 粘接	Chemical energy 化学能	Chemical bond 化学键
	Explosion processing 爆炸加工	Chemical energy, mechanical energy 化学能、机械能	Explosion 爆炸
Forming machining 成形加工	Powder metallurgy 粉末冶金	Heat energy, mechanical energy 热能、机械能	Hot-embossing 热压成形
	Superplastic forming 超塑成形	Mechanical energy 机械能	Superplastic forming 超塑性
	Rapid forming 快速成形	Heat energy, mechanical energy 热能、机械能	Hot melt forming 热熔化成形
Combined machining 复合加工	Electrochemical-arc machining 电化学电弧加工	Electrochemistry energy 电化学能	Melting, gasification corrosion 熔化、气化腐蚀
	Electrochemical- discharge-grinding machining 电解电火花机械磨削	Electric energy, heat energy 电能、热能	Ion transfer, melting, cutting 离子转移、熔化、切削
	Electrochemical corrosion process 电化学腐蚀加工	Electrochemistry energy, heat energy 电化学能、热能	Melting, gasification corrosion 熔化、气化腐蚀
	Ultrasonic assisted electrical discharge machining 超声放电加工	Sound energy, heat energy, electric energy 声能、热能、电能	Melting, abscission 熔化、切蚀
	Combined electrochemical machining 复合电解加工	Electrochemistry energy, mechanical energy 电化学能、机械能	Abscission 切蚀
	Combined cutting machining 复合切削加工	Mechanical energy, sound energy, magnetic energy 机械能、声能、磁能	Cutting 切削

jects, such as welding machining for vacuum tube inside.

4) Compared with the electron beam machining, laser machining device is quite simple without complicated vacuum system.

5) The laser processing is a kind of heat machining of instantaneous, partial melting, gasification, with many influence factors. Therefore, during the micro machining, the accuracy, especially the repeat accuracy and the surface roughness is not easy to guarantee. It must be repeated to test and find the reasonable parameters, in order to achieve a certain machining requirements. Due to the reflection of light, the machining of the surface gloss or transparent material must be carried out before the color or coarsen treatment, so that more light energy is absorbed into heat energy for machining.

The disadvantage of laser beam machining is that its equipment is still expensive at present.

2. Laser Beam Machining Technology and Its Application in Industry

(1) The principle of laser drilling   Laser cutting and drilling is the result of a series of thermal physical phenomena generated by the surface of the workpiece, which is irradiated by high energy laser beam on the surface of the workpiece. It is related to the characteristics of the laser beam and the thermophysical properties of the material. Laser drilling is non-contact machining, and has no mechanical pressure on the workpiece, its thermal effect is minimal, so it is more advantageous to the machining of precision parts. The energy and the trajectory of the laser beam are easy to realize precise control, so it can finish the precision machining of complex parts. The laser can make micro pores in almost any material, has been presently used to machine the fuel nozzle of the rocket engine and the diesel engine, holes on chemical fiber spinneret, holes of gem bearing in watch and instrument, and diamond wire drawing die, etc.

(2) Laser cutting technology   Laser cutting is that a laser beam of high power density is used to scan the material surface, which in a very short period of time the material will be heated to thousands and tens of thousands of degrees to make the material melting or gasification, and be blew away from the joint-cutting with high pressure gas, so as to achieve the purpose of cutting materials. The laser cutting has lots of characteristics, including fast speed of cutting, smooth incision, generally without the need for subsequent processing, small heat effect zone, small deformation, thin kerf (0. 1-0. 3 mm), notch without mechanical stress, free cutting burr, high machining accuracy, good repeatability, no damage to the surface of material. Laser cutting is suitable for automatic control, small parts of a variety of precision cutting, and can be used to cut all kinds of materials.

(3) Laser welding technology   Laser welding is that the material is melt to form welded joint based on the high power laser beam as the heat source. Laser welding registers the molten pool cleaning effect, which can clean metal of welded seam, is suitable for the intermetallic welding of same or different materials and different thickness. It is particularly advantageous for metal welding with high melting point, high reflectivity, high thermal conductivity and physical properties. Generally, during the laser welding process there is no need solder and flux, just need "hot melt" of the processing area of the workpiece together. The main characteristics of the laser welding include high power density of laser beam, deep penetration of weld seam, fast speed, high efficiency, narrow weld seam, small heat effect zone, small deformation of the workpiece, homogeneous microstructure of weld seam, small grain size, few porosities and defects, etc. The laser welding can realize precise welding, in which laser power can be controlled and be easy to realize automation, is superior to conventional welding method in mechanical properties, the corrosion resistance and electromagnetic performance.

3.5.2　激光加工

1. 激光加工技术及其特点

激光技术是 20 世纪 60 年代初发展起来的一门新兴科学。在材料加工方面，已形成一种崭新的加工方法——激光加工（Laser Beam Machining，LBM），它是利用激光束对材料的光热效应来进行加工的一门加工技术。激光加工技术是涉及光、机、电、材料及检测等多门学科的一门综合技术。

由于激光具有高亮度、高方向性、高单色性和高相干性的特性，激光加工具有以下优点：

1）聚焦后，激光加工的功率密度可高达 $10^8 \sim 10^{10}$ W/cm^2，光能转化为热能，几乎可以熔化、气化任何材料。例如，耐热合金、陶瓷、石英、金刚石等硬脆材料都能加工。

2）激光光斑大小可以聚焦到微米级，输出功率可以调节，因此可用于精密微细加工。

3）加工所用工具是激光束，是非接触加工，所以没有明显的机械力，没有工具损耗问题。加工速度快、热影响区小，容易实现加工过程自动化。能在常温、常压下于空气中加工，还能通过透明体进行加工，如对真空管内部进行焊接加工等。

4）与电子束加工等相比，激光加工装置比较简单，不要求复杂的抽真空装置。

5）激光加工是一种瞬时、局部熔化、气化的热加工，影响因素很多，因此，精微加工时，精度，尤其是重复精度和表面粗糙度不易保证，必须进行反复试验，寻找合理的参数，才能达到一定的加工要求。由于光的反射作用，对于表面光泽或透明材料的加工，必须预先进行色化或打毛处理，使更多的光能被吸收后转化为热能用于加工。

激光加工的不足之处在于激光加工设备目前还较昂贵。

2. 激光加工技术及其在工业中的应用

（1）激光打孔原理　激光切割打孔是利用高能激光束照射在工件表面，表面材料所产生的一系列热物理现象综合的结果。它与激光束的特性和材料的热物理性质有关。激光打孔加工是非接触式的，对工件本身无机械冲压力，热影响极小，从而对精密配件的加工更具优势。激光束的能量和轨迹易于实现精密控制，因而可完成精密复杂的加工。激光几乎可以在任何材料上打微型小孔，目前已应用于火箭发动机和柴油机的燃料喷嘴加工、化学纤维喷丝板打孔、钟表及仪表中的宝石轴承打孔、金刚石拉丝模加工等方面。

（2）激光切割技术　激光切割是利用高功率密度的激光束扫描材料表面，在极短时间内将材料加热到几千至上万摄氏度，使材料熔化或气化，再用高压气体将熔化或气化物质从切缝中吹走，达到切割材料的目的。激光切割的特点是速度快，切口光滑平整，一般无须后续加工；切割热影响区小，板材变形小，切缝窄（0.1 ~ 0.3 mm）；切口没有机械应力，无剪切毛刺；加工精度高，重复性好，不损伤材料表面。激光切割适于自动控制，宜于对细小部件进行各种精密切割，可以用于切割各种材料。

（3）激光焊接技术　激光焊接是以高功率聚焦的激光束为热源，熔化材料形成焊接接头的。激光焊接具有溶池净化效应，能纯净焊缝金属，适用于相同或不同材质、不同厚度的金属间的焊接，对高熔点、高反射率、高导热率和物理特性相差很大的金属焊接特别有利。激光焊接一般无须焊料和焊剂，只需将工件的加工区域"热熔"在一起就可以。激光焊接的特点主要有：激光束功率密度很高，焊缝熔深大，速度快，效率高；激光焊缝窄，热影响

（4）Laser surface heat treatment technology　Laser surface heat treatment is a method of using high power density laser beam for the surface treatment of metal materials. It can achieve the surface treatment of materials of phase transformation hardening, rapid melting and solidification, alloying, cladding, which produce changes in surface composition, structure, and properties that cannot be achieved by other surface quenching. Because laser has very strong penetration ability, when the metal surface is heated to the critical transition temperature which is only below the melting point, the surface is rapidly austenitizing and then rapidly self-quenching, and then the metal surface is strengthened rapidly, namely laser phase transformation hardening. Among them, the technology of phase transformation hardening and melting and solidification technology have become mature and industrialization. Because the application range and the improvement scope of the base material performance by the alloying and the cladding technology are better than the first two, the development prospect of the alloying and the cladding technology is broad.

3.5.3　Water Jet Machining

1. Emergence of Water Jet Machining Technology

Water jet machining (WJM) is also called liquid jet machining (LJM), which integrates the impact of high pressure and high speed water (or water with additives) on the workpiece and the free grinding effect of abrasive suspending in the water, applying to the process for all kinds of materials, such as cutting, detaching, perforating, crushing and surface material removal. The water jet technology is a kind of method, which uses water as an energy carrier to process materials. The first patent for water jet technology was invented in 1968 by the United States, Dr. Norman Franz, who studied how to use the new method of cutting wood. In 1980s, the United States Dr. Mohamed Hashish and Flow International company introduced the abrasive water jet technology based on water jet technology, which has a wider range of applications, comprising automotive, aerospace, metallurgy, electronics, construction, petrochemical, shipbuilding and other industries.

2. Water Jet Machining System and Its Characteristics

Machining system which aims to cutting, is generally composed of an ultra high pressure jet pump which can generate hundreds MPa (or water through the pump and is pressurized by the supercharger), a liquid accumulator (it makes the pulse flow steady), an artificial sapphire nozzle with 0.1-0.5 mm of diameter, a horizontal workbench, a robot and a comprehensive control device. The movement mode of the cutting device can be divided into the following parts: the workpiece is fixed with nozzle moving, the workpiece and nozzle move mutually. The movement of the cutting device and is controlled by using the water jet NC device and the multi-joint robot. At present, the water jet NC system has been widely used, and the multi-joint robot and gantry type robot has entered the practical use. The complex shape of coordinate surface by using computer controlled machining has been successfully tested several times.

Water jet technology is a kind of cold cutting technologies which are environmentally-friendly, and it meets the requirements of sustainable development manufacturing technology. It has the following characteristics:

1）The water jet can be used for flexible and soft processing due to it is also a high-energy jet beam. In NC cutting, the nozzle position is very important. If the direction of injection is perpendicular to the cross section, uniform incision can be obtained and be independent of the cutting path. The

区很小，工件变形很小，激光焊缝组织均匀，晶粒很小，气孔少，夹杂缺陷少。激光焊接可实现精密焊接，且功率可控，易于实现自动化，在机械性能、抗蚀性能和电磁学性能上优于常规焊接方法。

（4）激光表面热处理技术　激光表面热处理就是利用高功率密度的激光束对金属进行表面处理的方法，可对材料实现相变硬化、快速熔凝、合金化、熔覆等表面处理，产生用其他表面淬火达不到的表面成分、组织、性能的改变。激光的穿透能力极强，当把金属表面加热到仅低于熔点的临界转变温度时，其表面迅速奥氏体化，然后急速自冷淬火，金属表面迅速被强化，即激光相变硬化。其中相变硬化和熔凝处理技术已趋于成熟并产业化，而合金化和熔覆工艺对基体材料的适应范围和性能改善的幅度较前两种好，发展前景广阔。

3.5.3　水射流加工

1. 水射流加工技术的产生

水射流加工（Water Jet Machining，WJM）又称液体喷射加工（Liquid Jet Machining，LJM），是综合了高压高速的水（或带有添加剂的水）对工件的冲击作用和由悬浮于水中的磨料的游离磨削作用来对各类材料施以切割、分离、穿孔、破碎和表层材料去除等加工。水射流技术是一种以水为能量载体，对材料进行加工的方法。首个水射流技术专利于1968年由美国的诺曼·弗朗兹博士在研究如何使用新方法切割木材时获得。在20世纪80年代，美国的穆罕默德·哈希什博士和Flow International公司在水射流技术的基础上推出了磨料水射流技术，使其应用范围更加广泛，涵盖了汽车、航空航天、冶金、电子、建筑、石油化工、船舶等多个行业。

2. 水射流加工系统及加工特点

以切割为加工目的加工系统，一般是由产生数百MPa的超高压射流泵（或水经过水泵后通过增压器增压）、贮液蓄能器（使脉冲的液流平稳）、孔径为0.1~0.5 mm的人造蓝宝石喷嘴、水平工作台、机器人和综合控制装置所组成。切割装置运动的方式可分为：工件固定喷嘴移动式，工件喷嘴相互移动式；采用NC装置及多关节机器人进行控制。目前，水射流NC装置系统已推广使用，多关节机器人及龙门架式机器人也进入实用化，使用计算机控制加工坐标曲面的复杂形状，也已多次试验成功。

水射流技术是一种绿色冷切割技术，符合可持续发展制造技术的要求，具有以下特点：

1）由于水射流也是一种高能束流射线，所以可以进行灵活柔软的加工。在进行NC切割时，喷嘴的位置极为重要，如果喷射方向与断面相垂直，就可以得到与切割路线无关的均匀切口，切割加工可以从任意位置开始，到任意位置结束，这是其他机械加工方法所达不到的。因此，这种方法适用于任何不规则表面形状的加工。

2）与使用热的、机械的加工方法不同，在加工点处温度很低，因此对加工件的热应力极小，不会引起其表层组织的变化，可以在易燃、易爆、有毒的多种危险场所作业，安全可靠。

3）由于使用水作为加工介质，辅助材料的管理十分方便，喷孔直径相当小，水的用量不大。作为工具的射流束是不会变钝的，喷嘴寿命也相当长。

4）在高速流体力学作用下，水射流能够与多种化学、物理作用相复合。若流量大到某种程度，就能在加工的同时顺利地排除切屑，因而具有清洗作用，在粉尘严重时，更能显示

cutting can start from any position and be the end to any location, which is the other machining method that cannot achieve. Therefore, this method can be applied to the machining of any irregular surface shape.

2) Different from the thermal, mechanical processing methods, the temperature at processing point of the water jet is very low. Therefore, during water jet machining the thermal stress of the workpiece is rather small and will not cause the changes of surface structure. Water jet machining can safely and reliably to operate in flammable, explosive, toxic and others dangerous places.

3) Because water is used as the processing medium, the management of the auxiliary material is very convenient. The orifice diameter is quite small, and the water consumption is small. The jet beam as tool is never blunt, nozzle life is quite long.

4) Under the action of high-speed fluid mechanics, water jet can be combined with a variety of chemical and physical effects. If the water flow is increased to a certain extent, it can smoothly remove the chips while machining. So water jet has a cleaning effect, especially shows superiority besides the strong dust case.

5) While the tiny nozzle is used, the force bearing on the part is minimal. So water jet can smoothly cut the soft materials (such as rubber, paper, wood, asbestos, composite materials) and the components easy deformation (such as honeycomb sandwich of aircraft skin, sponge structure).

3. Application and Development Trend of Water Jet Machining

In recent years, as the unique advantages of water jet machining become more and more obvious and the ultra high pressure processing system has been increasingly perfected, the water jet machining technology has entered the ranks of the high technology. Advanced industrial countries in the world are competing to develop this technology.

The water jet machining has wide applications, which is suitable for processing almost any material, besides iron and steel, aluminum, copper and other metal materials, also for processing special rigid brittle, very thin, very soft or flying debris of materials, such as plastic, leather, paper, cloth, wood, asbestos, marble, glass, ceramics and composites, etc.

In automotive industry, the Swedish VOLVO automobile company is the first to use water jet machining. In the early 1990s, water jet cutting has been widely applied in the automotive industry in the United States. It has been utilized to cut asbestos brake pads, rubber carpet, composite plate, glass fiber reinforced plastic, and components like chassis, wagon box, instrument panel, cover, car door, etc. In aviation industry, water jet machining has been widely used in cutting of fiber reinforced composite materials and titanium alloy, as well as used to remove the burr of engine turbine wheel disc, casing, cylinder block, hole edge of flame tube, groove, screw, cross hole and blind hole, which will not cause change of surface structure. In aerospace industry, it is used for cutting of advanced composites, honeycomb sandwich plate, titanium alloy component and printed circuit board, to improve the fatigue life. In mechanical processing industry, water jet machining can be used for a variety of metals (such as titanium, aluminum, molybdenum, copper, iron, chromium nickel alloy, etc.) and for cutting, drilling and grooving of all composite materials and industrial ceramics. In nuclear power station, water jet machining can be used in cutting nuclear fuel pipe and deuterium oxide pipelines, fuel waste reprocessing and removal of concrete structures from atomic reactors. In addition, in the field of medical treatment and civil construction, there are also many applications of water jet machining.

其优越性。

5）当使用细小喷嘴时，加工件上承受的作用力极小，能顺利地切割质地柔软的材料（如橡胶、纸、木头、石棉、复合材料等）和容易变形的构件（如飞机蒙皮的蜂窝夹层、海绵构造等）。

3. 水射流加工的应用及发展趋势

近年来，由于水射流加工的独特优点越来越明显，加工超高压系统的技术日趋完善，水射流加工技术已迈入高科技行列，世界上工业先进的国家都在竞相开发。

水射流加工的应用十分广泛，几乎适用于加工所有材料，除钢铁、铝、铜等金属材料外，还可以加工特别硬脆、很薄、很软或切屑飞扬的非金属材料，如塑料、皮革、纸张、布匹、木材、石棉、大理石、玻璃、陶瓷、橡胶和复合材料等。

在汽车工业中，瑞典沃尔沃汽车公司是第一个运用水射流加工的。在20世纪90年代初，水射流切割已在美国的汽车工业中广泛采用，用来切割石棉制动片、橡胶基地毯、复合材料板、玻璃纤维增强塑料等材料，加工的零部件有底盘、车厢、仪表盘、罩盖、车门等。在航空工业中，水射流加工已广泛用于纤维增强复合材料和钛合金的切割以及用来去除发动机涡轮盘、机匣、缸体、火焰筒中孔缘、沟槽、螺纹、交叉孔和盲孔上的毛刺，而不会引起其表层组织的变化。在航天工业中，用以切割高级复合材料、蜂窝状夹层板、钛合金元件和印刷电路板等，可提高疲劳寿命。在机械加工业中，可用于各种金属（如钛、铝、钼、铜、铁、铬镍合金等）和各种复合材料、工业陶瓷等材料的切割、打孔和切槽等加工。在核电站中，水射流加工可用于核燃料管的切割，重水管线的切割，燃料废物的再处理以及原子反应堆屏蔽混凝土建筑的切除等。另外，在医疗领域、土木建筑业中也有水射流加工的应用。

水射流加工技术的发展有以下几个趋势：

1）提高水射流加工机的可靠性和寿命，尤其是其中关键零部件高压泵、高压软管、接头和喷嘴的寿命，故要采取一系列防磨损的措施，能使高压泵和阀体的寿命达到6000h，喷嘴的寿命达到2000h。

2）优化工艺参数，进一步提高效率，减少磨料消耗和降低能耗，以使成本更有竞争力。

3）发展智能化控制，使工艺参数能在加工过程中自适应调整，以提高加工精度，用于制作有一定精度要求的零件，达到其技术经济效果可与等离子体和激光加工相媲美的程度。

4）不断扩大水射流加工的应用范围，由二维的切割和去毛刺加工发展到孔加工和三维型面的加工。同时还能根据加工件的材料及板材厚度，自动提供最佳的切割工艺参数，有利于该项工艺的推广。

3.5.4　超声波加工

1. 超声波加工基本原理

人耳能感受的声波频率为16~16 000 Hz，频率超过16 000 Hz的声波称为超声波。

超声波加工（Ultrasonic Machining，UM）是利用工具端面做超声频振动，通过磨料悬浮液加工脆性材料的一种成形加工方法。超声波加工原理如图3-20所示。加工时在工具1与工件2之间加入液体（工作液）与磨料混合的悬浮液3，并使工具以很小的力 F 轻轻压在工件上。高频电源7作用于磁致伸缩换能器6产生16 000 Hz以上的超声频纵向振动，并借助

There are several trends in the development of water jet machining technology:

1) To improve the reliability and life of water jet machine, especially the service life of key components like high pressure pump, high pressure hose, joint and nozzle the, so a series of measures should be employed to prevent wear, making the service life of high pressure pump and valve up to 6, 000 hours, the nozzle up to 2, 000 hours.

2) Optimizing the process parameters, further improving the efficiency, reducing abrasive consumption and energy consumption, aims to make the cost more competitive.

3) Developing intelligent control makes the process parameters be adjusted adaptively in the machining, in order to improve the machining accuracy and make use of processing a certain precision parts. Thus, its technical and economic effects can achieve a comparable level of comparing with the plasma machining and laser machining.

4) Expanding continually the application scope of the water jet machining is to make it develop from the two dimensional cutting and deburring to the machining of the hole and the three-dimensional surface. At the same time, according to the materials of machining parts and the thicknesses of the plate, the best cutting parameters are automatically provided, which is beneficial to the popularization of the process.

3.5.4　Ultrasonic Machining (USM)

1. Basic Principles of Ultrasonic Machining

The frequency of sound wave that people can feel is in the range of 16 Hz to 16, 000 Hz. The sound wave whose frequency is beyond 16, 000Hz is called ultrasound.

Ultrasonic machining is a kind of forming process method which is to use ultrasonic vibration of the tool end and machine brittle materials through suspension containing abrasive. Principle of ultrasonic machining is shown in Fig. 3-20. While processing, a suspension 3 mixed liquid (working fluid) and abrasive is added between the tool 1 and the workpiece 2, and the tool with a small force F is gently pressed on the workpiece. High frequency power supply 7 acts on magnetostrictive transducer 6 to produce ultrasonic longitudinal vibration more than 1, 600 Hz. Then the ultrasonic longitudinal vibration is amplified to 0. 05-0. 1 mm with the help of horn 4, 5 amplitude, driving the tool end face for ultrasonic vibration to drive abrasives in the suspension greatly impact and constantly polish the machined surface in large velocity and acceleration. In such a way, the material machined surface is crushed into fine particles and fell out of the workpiece. Although the material each falling is few, there is still a certain machining speed owing to the number of hits per second up to 16, 000 times or more.

At the same time, the working fluid generates the high frequency, alternating positive and negative hydraulic shock wave and cavitation effect caused by the ultrasonic vibration of the tool end, making the working fluid into the micro cracks of the processed material to intensify the mechanical damage action. It is so-called cavitation that when the tool which ends with a large acceleration leaves the workpiece surface, the negative pressure and local vacuum are formed in the machining gap, resulting in a lot of micro cavities formed in the working fluid. When the tool end is close to the workpiece surface with a large acceleration, the cavity is closed, causing the largely strong hydraulic shock wave, which can strengthen the processing process. In addition, the positive and negative hydraulic impact also forces the suspension circulation in the processing gap, so that the abrasive parti-

于变幅杆 4、5 把振幅放大到 0.05~0.1 mm，驱动工具端面做超声振动，促使悬浮液中的磨料以很大的速度和加速度不断地撞击、抛磨被加工表面，把被加工表面的材料粉碎成很细的微粒，从工件上被打落下来。虽然每次打击下来的材料很少，但由于每秒钟打击次数多达16 000 次以上，所以仍有一定的加工速度。

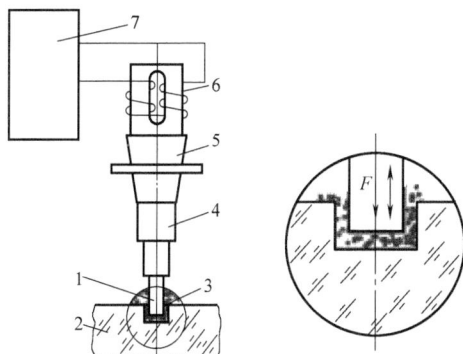

Fig. 3-20 Principle of ultrasonic machining 超声波加工原理

1—Tool 工具　2—Workpiece 工件　3—Abrasive suspension 磨料悬浮液
4、5—Horn 变幅杆　6—Transducer 换能器　7—High frequency power 高频电源

与此同时，工作液受工具端面超声振动作用而产生的高频、交变的液压正负冲击波和"空化"作用，促使工作液钻入被加工材料的微裂缝处，加剧了机械破坏作用。所谓空化作用，是指当工具端面以很大的加速度离开工件表面时，加工间隙内形成负压和局部真空，在工作液体内形成很多微空腔。当工具端面以很大的加速度接近工件表面时，空腔闭合，引起极强的液压冲击波，可以强化加工过程。此外，正负交变的液压冲击也使悬浮液在加工间隙中强迫循环，使变钝了的磨粒及时得到更新。工具逐渐伸入被加工材料中，工具形状便复现在工件上，直至达到所要求的尺寸。

由此可见，超声波加工是磨粒在超声振动作用下的机械撞击和抛磨作用以及超声空化作用的综合结果，其中，磨粒的机械撞击起主要作用。越是脆性材料，受冲击作用遭受的破坏越大，越易于超声加工。超声波适合于加工各种硬脆材料，特别是不导电的非金属材料。

2. 超声波加工装置组成

超声波加工装置一般包括高频电源、超声振动系统、机床和磨料工作液循环系统等几个部分。

（1）高频电源　高频电源也称超声波发生器，其作用是将工频交流电转变为有一定功率输出的超声频电振荡，以提供工具端面往复振动和去除被加工材料的能量。

（2）超声振动系统　该系统由磁致伸缩换能器、变幅杆及工具组成。换能器是将高频电振荡转换成机械振动。由于磁致伸缩的变形量很小，其振幅不超过 0.005~0.01 mm，不足以直接用来加工，因此必须通过一个上粗下细的振幅扩大棒（变幅杆）将振幅扩大至0.01~0.15 mm。超声波的机械振动经变幅杆放大后即传给工具，使悬浮液以一定的能量冲击工件。

（3）机床　超声波加工机床的结构比较简单，包括机架和移动工作台。机架支撑振动系统等部件，移动工作台维持加工过程的进行。

cles to be replaced in time. Gradually the tool enters into the material being processed, tool shape is reproduced the workpiece, until it reaches the required size.

Consequently the ultrasonic machining is the comprehensive result of the mechanical impact, polishing of abrasive particles and ultrasonic cavitation under ultrasonic vibration, mainly depending on the impact effect of the abrasive. Especially brittle material, the damage suffered by the impact is greater, so the brittle material is easier by the ultrasonic processing. Ultrasonic is suitable for processing all kinds of hard and brittle materials, particular non-conductive nonmetallic materials.

2. Components of Ultrasonic Machining Device

Ultrasonic machining device generally includes high frequency power supply, ultrasonic vibration system, machine tool body, circulation system of abrasive working fluid and other parts.

(1) High frequency power   High frequency power also called ultrasonic generator, whose function is to transform alternating current into ultrasonic frequency oscillation with a certain power output, in order to provide energy for reciprocating vibration of tool end face and removing of processed materials.

(2) Ultrasonic vibration system   Ultrasonic vibration system is composed of magnetostrictive transducer, horn and tool. Transducer is to convert high frequency electrical oscillation to mechanical vibration. The deformation of the magnetostriction is very small, resulted in the amplitude less than 0.005-0.01 mm, which is not directly used for processing. Therefore, the amplitude must be amplified to 0.01-0.15 mm by an amplitude amplifier (horn) with larger of the upper part than its down part. The mechanical vibration of the ultrasonic wave is amplified to the tool through the horn, making the suspension impact the workpiece with certain energy.

(3) Machine tool   Structure of the ultrasonic machine tool is relatively simple, including the machine frame and the movable worktable. The frame supports the vibration system and other components, and the moving worktable maintains the process.

(4) Abrasive working fluid and its circulation system   The common working fluid has water, kerosene, machine oil, etc. The abrasives including boron carbide, silicon carbide, alumina, etc., are mixed by centrifugal pump and are poured into the work area, which ensures the abrasive suspension and renewal.

3. Characteristics and Applications of Ultrasonic Machining

Ultrasonic machining has the following characteristics:

1) It is suitable for processing all kinds of hard and brittle materials, especially non-conductive nonmetal materials, such as glass, ceramics (alumina, silicon oxide, etc.), quartz, germanium, silicon, agate, jewelry, diamond, etc. For conductive hard metal materials such as hardened steel, cemented carbide, etc., can also be processed, but the low productivity is rather low.

2) Because the tool can be made more complex shape using softer material, there is no need to make the tool and the workpiece complicated motion. The structure of ultrasonic machine tool is relatively simple, only need one direction of feed with light press, so the operation and maintenance of ultrasonic machine tool are convenient.

3) Because the removal of the processing material relies on the instantaneous local impact by minimal abrasive, the macroscopic cutting force of the workpiece surface is very small. Owing to the cutting stress and the cutting heat is small, the workpiece cannot cause deformation and burn, the surface roughness Ra records good, up to 0.1-1 μm. The ultrasonic machining can produce thin-

（4）磨料工作液及其循环系统　工作液常用的有水、煤油、机油等，将碳化硼、碳化硅、氧化铝等磨料通过离心泵搅拌悬浮后注入工作区，以保证磨料的悬浮和更新。

3. 超声波加工的特点及应用

超声波加工具有以下特点：

1）适合于加工各种硬脆材料，特别是不导电的非金属材料，如玻璃、陶瓷（氧化铝、氮化硅等）、石英、锗、硅、玛瑙、宝石、金刚石等。对于导电的硬质金属材料如淬火钢、硬质合金等，也能进行加工，但加工生产率较低。

2）由于工具可用较软的材料做成较复杂的形状，故不需要使工具和工件做比较复杂的相对运动，因此超声加工机床的结构比较简单，只需一个方向轻压进给，操作、维修方便。

3）由于去除加工材料是靠极小磨料瞬时局部的撞击作用，故工件表面的宏观切削力很小，切削应力、切削热很小，不会引起变形及烧伤，表面粗糙度值 Ra 也较低，可达 0.1～1 μm，而且可以加工薄壁、窄缝、低刚度零件。

超声波加工在工业生产中的应用越来越广泛，常用于型孔与型腔的超声加工、一些淬火钢、硬质合金冲模，拉丝模，塑料模具型腔的最终抛磨光整加工、超声清洗、超声切割，以及超声波复合加工，如超声电火花加工、超声电解、超声振动切削等。

3.6 快速原型制造技术

3.6.1 概述

1. 快速原型制造技术产生背景

进入 20 世纪 80 年代，市场需求已由卖方市场转化为买方市场并日趋全球化。空前激烈的市场竞争迫使制造企业必须以更快的速度设计、制造出性能价格比高并满足人们需求的产品。制造领域正在面临两大挑战：①大幅减少产品开发时间；② 提高小批量多品种产品制造的柔性。因此，产品开发的速度和制造技术的柔性就成为赢得竞争的关键问题。CAD 和 CAM 已经大大改善了传统的产品设计和制造，但在 CAD 与 CAM 真正集成快速开发新产品方面仍存在许多障碍。CAD 与 CAM 在以下两个方面仍存在鸿沟：一是快速生成三维模型和样机，二是经济有效生产具有复杂几何形状的模型和模具。从技术发展角度看，计算机技术、CAD、材料科学、数控技术、激光技术等的发展与普及为新的制造技术的产生奠定了基础。为了能够大大地减少样件、模型和样机的开发时间，一些制造企业开始应用快速原型方法制造复杂模型和零部件样机。快速原型制造（Rapid Prototyping Manufacturing，RPM）技术就是在这种社会背景下于 20 世纪 80 年代后期在美国问世的。应用快速原型制造技术可以快速生成三维模型和样机，经济有效地生产具有复杂形状的模型和样机。

RPM 技术与虚拟制造技术一起，被列为未来制造业的两大支柱技术。有人称 RPM 技术是继 NC 技术之后的又一次技术革命。它借助于计算机技术、激光技术、数控技术、精密传动技术等现代手段将 CAD 与 CAM 集成于一体，根据在计算机上构造的产品三维模型，能在很短的时间内直接制造出产品的样品，无须使用传统制造中的刀具、夹具和模具，从而缩短了产品开发周期，加快了产品更新换代的速度，降低了企业投资新产品的风险。

walled, narrow and low stiffness part.

The ultrasonic machining has been applied more and more extensively in industrial production. It is commonly used in the ultrasonic machining of profiled hole and cavity, some hardened steel, stamping die and wire-drawing die of cemented carbide, and used in the final polishing and grinding, ultrasonic cleaning and ultrasonic cutting of plastic mold cavity, as well as ultrasonic composite machining, such as ultrasonic EDM, ultrasonic electrolysis, ultrasonic vibration cutting, etc.

3.6 Rapid Prototyping and Manufacturing Technology

3.6.1 Introduction

1. Background of Rapid Prototyping Manufacturing (RPM) Technology

In 1980s, market demand has been transformed from a seller's market to a buyer's market and is increasingly globalized. Unprecedented intensive market competition compelled the manufacturing enterprises to have to produce the product that has a high performance price ratio and meets the people' needs at a faster rate. Manufacturing field is facing to two important challenging tasks: ①Substantial reduction of product development time; ②improvement on flexibility for manufacturing low volume products and variety of types of products. Therefore, the product development speed and the flexibility of manufacturing technology has become the key to win the competition. CAD and CAM have significantly improved the traditional product design and manufacturing. However, there are a number of obstacles in the real integration of CAD with CAM for the rapid development of new products. The gap between CAD and CAM remains unfilled in the following aspects: One is the rapid creation of 3D models and prototypes; the other is the cost-effective production of patterns and moulds with complex configuration.

The RPM technology and the virtual manufacturing technology are named as two major pillars of the future manufacturing industry. Some people thought the RPM technology as another technological revolution after NC technology. It integrates CAD with CAM with the help of computer technology, laser technology, NC technology, precise transmission technology and other modern means. Based on the 3D product model constructed in computer, it is able to manufacture directly the sample of product in very a short time without use the tools, fixtures and moulds in traditional manufacturing, thus shortening the product development cycle, speeding up the upgrading of products, and reducing the risk of new product investment.

2. Basic Principle of RPM

As the RPM technology breaks through traditional machining modes such as the "forced forming" and the "removal forming", applies the additive way little by little to fabricate parts. It is named as "additive manufacturing" or "growth manufacturing", and also named as "3D printing" technology.

RPM technology is a comprehensive technology which integrates CAD, NC with material science, mechanical engineering, electronic technology and laser technology. It adopts the manufacturing principle of "software dispersing-material accumulating" to complete the forming process of the workpiece. Its working procedure is shown in Fig. 3-21. Its working principle is as follows.

2. RPM 技术的基本原理

快速原型制造突破了传统的"受迫成形"和"去除成形"的加工模式，采用一点一点添加的方式制造零件，有学者称这种新技术为"增材制造"或"生长型制造"，还有学者统称其为"三维打印"技术。

RPM 技术是集 CAD、数控技术、材料科学、机械工程、电子技术和激光技术于一体的综合技术。它采用"软件离散-材料堆积"的制造原理完成零件的成形过程。RPM 工作过程如图 3-21 所示。其工作原理如下：

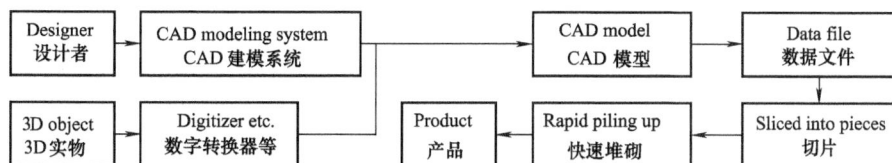

Fig. 3-21 RPM working procedure RPM 工作过程

样件是通过添加材料而不是去除材料制作的，即零件首先运用几何建模技术建立零件的实体模型。然后从数学上把三维模型分层切片。对每一片都要生成固化或粘接轨迹。这些固化或粘接的轨迹被直接用于指使机器通过固化或粘接一行行材料来生产零件。当一层做好后，再用同样的方式在原先的一层上再造出另一层。这样从下至上逐层堆积便可以制造出所设计的模型。总之，快速成型环节由两部分组成，即数据准备和模型生产。

RPM 的工作流程大致分为三个阶段：

1）前处理。首先应用三维造型软件（如 Pro/E、UG、CATIA 等）或逆向工程技术构造产品的三维模型，然后再对三维模型进行近似处理。

2）快速原型。首先用配置在快速原型（Rapid Prototyping，RP）设备上的切片软件沿成形制件的高度方向每隔一定距离（多为 0.1 mm）从 CAD 模型上依次截取平面轮廓信息；然后 RP 设备上的激光头或喷射头在数控装置的控制下按截面轮廓信息相对于 X-Y 平面工作台运动，进行选择性激光扫描（实现固化、切割或烧结）或者进行选择性喷射（喷射热熔材料或黏结剂）。材料用物理或化学方法逐层成形并相互黏结。每成形一层，工作台便下移一个切片厚度。这样一层层堆积便构成三维实体制件。

3）后处理。为改善制件的性能，往往需要进行后处理，如去除支撑，修磨至产品要求，在纸质制件的表面涂覆一层金属、陶瓷或高分子材料，以提高制件表面的机械强度、耐磨性和防潮性等。

3. RPM 的特点

RPM 最主要的特征就是由 CAD 模型直接驱动快速制造任意复杂形状三维实体零件。与传统的加工方法相比，具有以下共同特点：

1）制造过程柔性化。成形过程无须专用工具和模具，使得产品的制造过程几乎与零件的复杂程度无关。

2）产品开发快速化。从 CAD 设计到产品的成形完成只需几小时或几十小时，即便是大型的较复杂的零件只需要上百小时即可完成。RP 技术与其他制造技术集成后，新产品开发的时间和费用将节约 10%~50%。产品的单价几乎与批量无关，特别适合于新产品的开发和单件小批量生产。

The prototype part is produced by adding materials rather than removing materials, that is, a part is first modeled by a geometric modeler such as a solid modeler. And then the 3D model is mathematically sectioned (sliced) into a series of parallel cross-selection pieces. For each piece, the curing or binding paths are generated. These curing or binding paths are directly used to instruct the machine for producing the part by solidifying or binding a line of material. After a layer is built, a new layer is built on the previous one in the same way. Thus, the model is built layer by layer from the bottom to top. In summary, the rapid prototyping activities consist of two parts: data preparation and model production.

The workflow of RPM is divided into three stages on the whole:

1) Pre-treatment. 3D model of the product is firstly constructed by means of such 3D modeling softwares as Pro/E, UG, CATIA or reverse engineering technology, and then the approximate processing of 3D model is performed.

2) Rapid prototyping (RP). The face profile information is collected successively every certain distance (0.1 mm in the most cases) from the CAD model along the height of the forming part by means of the slicing software installed in the RP equipment. Then, the laser head or jetting nozzle of the RP equipment is moved relative to X-Y worktable based on the section profile information under the control of the NC device to perform selective laser scanning (solidifying, cutting or sintering) or selective jetting (to jet hot-melt material or binder). The materials are formed layer by layer and bond each other by using physical or chemical methods. When a layer is formed, the worktable will be moved down a slice thickness. The 3D solid part can be formed by piling up layer by layer.

3) Post-treatment. In order to improve the performance of the part, it is often necessary to carry out the post-treatment on the product, such as removing the support, polishing the product to the requirement, coating a layer of metal, ceramic or polymer materials on the surface of paper product so as to improve the mechanical strength, abrasive resistance and moisture resistance of the product.

3. Characteristics of RPM

The most important feature of RPM is to make the 3D solid part with arbitrary complex shape under the direct drive of CAD model. Comparing with traditional machining methods, RPM methods have the common characteristics as follows:

1) Flexibility of manufacturing process. Without need for special tooling and mold in the forming process, the manufacturing process of product has almost nothing to do with the complexity of the parts.

2) Rapid development of product. It takes only a few hours or a few hours from the product CAD to the forming finish. The forming of large-sized complex part can be completed only a hundred hours. Integrated the RP technology with other manufacturing technologies, the new product development time and cost will save 10%-50%. The unit price of the product has nothing to do with the batch. RPM is suitable for the development of new products and small batch production especially.

3) Using "dispersing-accumulating" principle instead of "removing" principle in product manufacturing. RPM can be used for manufacturing the 3D geometric solids with arbitrary complexity.

4) Complete digitization of manufacturing processes. Based on the computer software and NC technology, high integration of CAD and CAM and real machining without drawing are realized. The forming process does not need or need less manual intervention.

5) Widespread materials. As different RPM technologies have different forming ways, they use different materials. Such materials as metal, paper, plastic, resin, paraffin wax, ceramics and so on

3）产品制造采用"离散-堆积"原理而不是"去除"原理。RPM 可以用于制造任意复杂的三维几何实体。

4）制造过程的完全数字化。以计算机软件和数控技术为基础，实现了 CAD、CAM 的高度集成和真正的无图样加工。成形过程中无须或少需人工干预。

5）材料来源广泛。由于各种 RPM 工艺的成形方式不同，因而使用的材料也不相同，如金属、纸、塑料、树脂、石蜡、陶瓷等都得到了很好的应用。

6）发展的可持续性。RPM 中剩余的材料可继续使用，大大提高了材料的利用率。

3.6.2　典型的 RPM 工艺

1. 立体光刻

立体光刻机（SLA）是由美国 3D Systems 公司的查尔斯·赫尔发明的。1988 年，3D Systems 公司推出世界上第一台商品化样机 SLA—250。它被认为是用途最广的成型机械，所用材料是液态光固化树脂。在光子的照射下，小分子被聚合成大分子。零件就是根据这种原理在如图 3-22 所示的装有液态树脂的大桶内制造出来的。

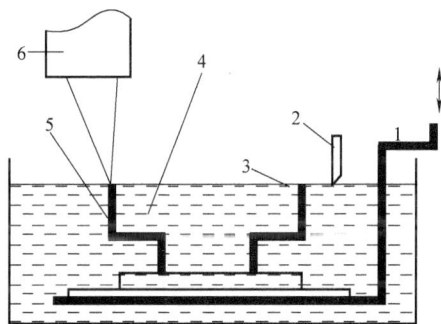

图 3-22　Stereo lithography principle 立体光刻的工艺原理图
1—Elevator 升降台　2—Flatting device 刮平器　3—Liquid surface 液面　4—Liquid photo-curable
resin 光敏树脂　5—Formed parts 成型零件　6—Laser 激光器

立体光刻机利用激光束在液态光敏树脂池的表面上扫描分层截面。激光束不像数控机床的刀具那样沿轮廓或 z 字形移动，而是沿平行线扫描。在树脂桶内的升降台刚好位于液面之下，其深度在光吸收限内。利用偏转镜使激光束沿 X 轴和 Y 轴水平偏转，这样，激光束横扫 树脂表面从而形成一薄层的固态实体。当一层固化后，升降台下降一个用户规定的距离，并在刚固化好的树脂表面涂上一层新的液态树脂。刮平器把铺在先前凝固层上的黏稠聚合树脂摊开。重复此过程，模型就这样自下而上一层一层地被制造出来了。在各层凝固完成时，制件的固化率大约是 95%。为使制件完全固化，仍然需要后期固化处理。

SLA 法成形精度较高，制件结构清晰且表面光滑，适合制作结构复杂和精细的制件。但制件韧性较差，设备投资较大，需要支撑以防止变形，液态树脂有一定的毒性。

2. 层合实体制造

层合实体制造（Laminated Object Manufacturing，LOM）又称分层实体造型（Slicing Solid Manufacturing，SSM），是美国 Helisys 公司的 Michael Feygin 于 1986 年研制成功的，1987 年获得美国专利。1990 年 Helisys 公司开发了世界上第一台商业机型 LOM—1015。目前基于

have found good applications.

6) Sustainability of development. Remained material in RPM can be used again, increasing material utilization greatly.

3.6.2　Typical RPM Technologies

1. Stereo Lithography

Stereo lithography apparatus (SLA) was invented in 1984 by Charles Hull of 3D systems Inc. in the United States. In 1988, 3D systems Inc. made the first commercial prototype named as SLA-250 in the world. It is considered as the most widely used prototyping machine. The material used is liquid photo-curable resin. Under the initiation of photons, small molecules are polymerized into large molecules. Based on this principle, the part is built in a vat of liquid resin as shown in Fig. 3-22.

The SLA machine creates the prototype by tracing layer cross-sections on the surface of the liquid photopolymer pool with a laser beam. Unlike the contouring or zigzag cutter movement used in CNC machining, the beam traces in parallel lines. An elevator table in the resin vat rests just below the liquid surface whose depth is the light absorption limit. The laser beam is deflected horizontally in X and Y axes by galvanometer-driven mirrors so that it moves across the surface of the resin to produce a thin solid pattern. After a layer is built, the elevator drops a user-specified distance and a new coating of liquid resin covers the solidified layer. A wiper helps spread the viscous polymer over the previous one. In this way, the model is build layer by layer from bottom to top. When all layers are completed, the prototype is about 95% cured. Post-curing is needed to completely solidify the prototype.

SLA method has high forming accuracy, clear part structure and smooth surface. It suitable for making the fine parts with complex structure. But SLA method requires larger equipment investment; the toughness of the part made by this method is poor; it needs the braces to support the part to prevent deformation; the liquid resin has certain toxicity.

2. Laminated object manufacturing (LOM)

Laminated object manufacturing (LOM) is also named as slicing solid manufacturing (SSM), which was developed successfully in 1987 by Michael Feygin of Helisys company. It was patented in 1988. In 1990, Helisys Corp. developed the first commercial machine in the world, whose model is LOM-1015. At present, there are more than 30 kinds manufacturing processes based on LOM.

The operation principle of LOM process is shown in Fig. 3-23. It produces the parts from bonded paper plastic, metal or composite sheet stock coated with thermosol on the back in advance. The laser beam cuts the sheet material on the worktable based on the layering profile information collected by the software, and the heated pressing roller presses the sheet material so as to bond the layer of sheet material to a stack of previously formed laminations. Then a laser beam follows the contour of the part cross-section generated by CAD to cut it to the required shape. The laser beam cuts contour of the part cross-section and its frame on the fresh layer bonded just now, and cuts the area in between the contour of cross-section and the outside frame into net shape for easy to remove the excess material of every sheet in post processing. After a layer of sheet material is cut by laser beam, the worktable together with the formed parts on it drops a distance equal to the thickness of the sheet material, thus making the formed part separate from the banding sheet material. The feeding device turns the material supplying shaft and the material collector and move the material band, sending the

LOM 的制造工艺已达 30 多种。

LOM 工艺原理如图 3-23 所示。它以单面事先涂有热溶胶的纸、金属箔、塑料膜、陶瓷膜等片材为原料制作零件。激光束按切片软件截取的分层轮廓信息切割工作台上的片材，热压辊热压片材，使之与下面已成形的工件粘接。然后，激光束按照由 CAD 生成的零件截面轮廓把它切割成所需的形状。激光在刚粘接的新层上切割出零件截面轮廓和外框，并在截面轮廓与外框之间多余的区域内切割出网格以便在后处理时容易去除这部分多余的片材。

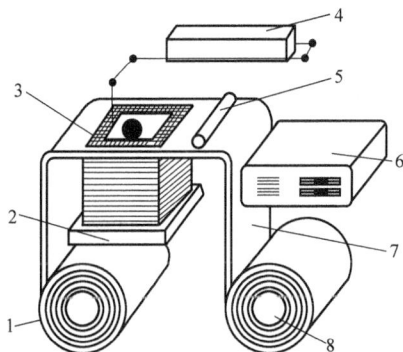

Fig. 3-23 Operation principle of LOM LOM 工艺原理

1—Material collector 收料器 2—Elevator 升降台 3—Machining surface 加工平面
4—CO₂ Laser CO$_2$ 激光器 5—Heated pressing roller 热压辊 6—Computer 计算机
7—Material band 料带 8—Material supplying shaft 供料轴

激光切割完成一层的截面后，工作台与其上面的已成形工件一起下降一个片材厚度，与带状片材分离。送料机构转动收料辊和送料辊使料带移动，把新一层片材送到加工区域。热压辊热压片材，工件高度增加一个料厚，再在新层上进行激光切割。如此反复直至零件的所有截面粘接、切割完毕，最终得到分层制造的实体零件。

LOM 工艺的优点是：①材料适应性强；②只需切割零件轮廓线，成形厚壁零件的速度较快，易于制造大型零件；③不需要支撑；④成形过程中不存在材料相变，成形后的制件无残余应力，因此制件不易产生翘曲变形。缺点是层间结合紧密性差。

3. 选择性激光烧结

选择性激光烧结（SLS）方法是由美国得克萨斯大学奥斯汀分校机械工程系的 C. R. Dechard 于 1989 年首先研制出来的，同年获美国专利。DTM 公司 1992 年首先推出了 SLS 商品化产品 "烧结站 2000 系统"，这是一种可以替代液相固化系统的选择性激光烧结系统。SLS 的原理与 SLA 十分相像，主要区别在于所使用的材料及其状态。SLS 使用粉末状的材料，这是该项技术的主要优点之一。因为理论上任何可熔的粉末都可以用来制造模型，这样的模型可以做实用的原型元件。

SLS 工艺原理如图 3-24 所示。它采用 CO$_2$ 激光束去烧结一层一层的粉末状材料而不是液体材料。在 SLS 过程中，逆时针旋转的滚筒机构把一薄层粉末铺撒在工作位置，先将粉末状材料预热到略低于其熔点的温度。然后用激光束扫描粉末表面的截面，把粉末加热到烧结温度，激光束扫描到的粉末被烧结。未被激光束扫描到的区域仍原样保留不动，可用作下一层粉末的支撑，这有助于减少制件的扭曲变形。当一层被扫描烧结完毕后，工作台下移一个

new sheet material into the machining area. The heated pressing roller presses the sheet material, the thickness of the sheet is added to the height of the formed parts. The laser beam cuts the new layer of sheet material again. The manufacturing process repeated until all cross-sections of the workpiece are bonded and cut. Finally, the solid workpiece made by means of LOM method can be obtained.

3. Selective Laser Sintering (SLS)

Selective laser sintering method was developed firstly in 1989 by C. R. Dechard at the Mechanical Engineering Department of Texas University at Austin and was patented in the same year. In 1992, DTM Corp. offers a commercial SLS product— "sintering station 2000 system". This is an alternative to liquid-curing systems. The principle of SLS is similar very much to that of SLA. The chief difference between them lies in the material and its status they use. SLS uses the powder materials, which is one of the chief advantages of this technology. Theoretically, all fusible powders can be used for making pattern, and this pattern can be used as the practical prototyping element.

Fig. 3-24 shows the operation principle of SLS. SLS uses a carbon dioxide laser to sinter successive layers of powder instead of liquid. In SLS processes, a thin layer of powder is applied by a counter-rotating roller mechanism onto the work place. The powder material is preheated to a temperature slightly below its melting point. The laser beam traces the cross-section on the powder surface to heat up the powder to the sintering temperature so that the powder scanned by the laser is bonded. The powder that is not scanned by the laser will remain in place to serve as the support to the next layer of powder, which aids in reducing the distortion of part. When a layer of the cross-section is completed, the worktable drops a distance equal to the thickness of the sheet material, and the piston for supplying powder material moves up correspondingly. The leveling drum levels another layer of powder over the sintered one again for the next pass. In this way, the 3D solid part can be piled up layer by layer. After all sintering work is finished, the excess powder should be removed from the sintered part. The required part can be obtained through smoothing and drying.

The advantages of LOM process are as follows: ①Many materials like metal, ceramics, plastics and composite material can be used, and the utilization of material is high; ②the brace is not required, the parts with complex structure can be made.

4. Fused Deposition Modeling (FDM)

Fused deposition modeling was developed successfully in 1988 by Dr. Scott Crump of America, and in 1991, Stratasys Inc. in the United States offered a commercial equipment FDM-1000. Later, Stratasys Inc. offered FDM-1650, FDM-2000, FDM-3000, FDM-8000 and FDM Quantum etc. one after another. In recent years, 3D Systems Inc. in the United States has developed multi-jet manufacture (MJM) technology on the basis of FDM technology, which can use multiple jets for modeling simultaneously so as to improve the modeling speed greatly.

FDM system is composed of jetting nozzle, filament feeding mechanism, motion mechanism, heated forming chamber and worktable, in which jetting nozzle is the most complex part in structure. In an FDM process as shown in Fig. 3-25, a spool of thermoplastic filament is fed by filament feeding mechanism into a heated FDM extrusion head. The movement of the FDM head is controlled by computers. Inside the flying extrusion head, the filament is melted into liquid (1℃ above the melting temperature) by a resistant heater. The head traces an exact outline of each cross-section layer of the parts. As the head moves horizontally in the X and Y axes, the thermoplastic material is extruded out of a nozzle by a precise pump. The material solidifies in 1/10 second as it is directed on

片层厚度，而供粉活塞则相应上移。铺粉滚筒再次将铺在已烧结平面上的粉末铺平供下一次激光烧结。如此反复便一层一层堆积出三维实体制件。全部烧结后去掉多余的粉末，再进行修光、烘干后便可获得所要求的制件。

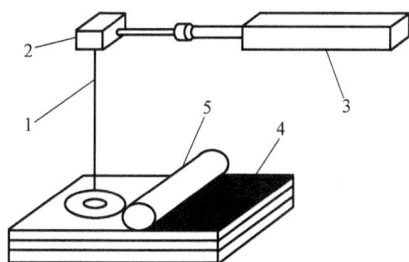

Fig. 3-24　Operation principle of SLS　SLS 工艺原理

1—CO_2 laser beam CO_2 激光束　2—Scanning mirror 扫描镜　3—CO_2 laser CO_2 激光器

4—Powder 粉末　5—Leveling drum 平整滚筒

SLS 工艺的优点是：①可以采用金属、陶瓷、塑料、复合材料等多种材料，且材料利用率高；②不需要支撑，故可制作形状复杂的零件。其缺点是成形速度较慢，成形精度和表面质量较差。

4. 熔融沉积成形

熔融沉积成形（FDM）工艺由美国的 Scott Crump 博士于 1988 年研制成功，并于 1991 年由美国的 Stratasys 公司率先推出商品化设备 FDM—1000。之后又相继推出了 FDM—1650、FDM—2000、FDM—3000、FDM—8000 和 FDM Quantum 等机型。近年来，美国 3D Systems 公司在 FDM 技术的基础上开发了多喷头（Multi-Jet Manufacture，MJM）技术，可使用多个喷头同时造型，大大提高了造型速度。

FDM 系统主要由喷头、供丝机构、运动机构、加热成形室和工作台等 5 个部分组成，而喷头是结构最复杂的部分。在如图 3-25 所示的熔融沉积成形过程中，一卷热熔性丝材由供丝机构送入加热的 FDM 挤压头。FDM 挤压头的运动由计算机控制。在飞速移动的挤压头内，热熔性丝材被电阻加热器融化成液态（高于熔点 1℃）。喷头沿着零件每一个截面层的准确轮廓移动。当喷头沿 X 轴、Y 轴水平移动时，熔融的材料在一个精密泵的作用下从喷嘴挤出。当熔融的材料对准工作位置时，会在 0.1s 内固化。当一层材料固化完成后，挤压头沿着 Z 轴方向向上移动（或工作台下降）一个程序设定的距离再沉积固化出另一新的薄层。每一层都通过加热粘接到先前固化了的一层上。如此一层层沉积固化便可堆积出三维实体制件。

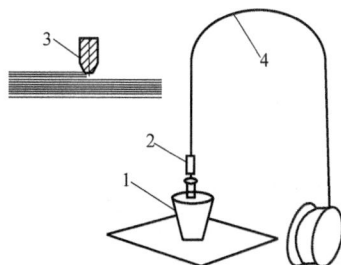

Fig. 3-25　FDM process principle　FDM 的工作原理

1—Formed part 成形制件　2、3—Jetting nozzle 喷头　4—Filament 料丝

to the workplace. After one layer is completed, the extrusion head moves up (or the worktable moves down) a programmed distance in Z direction for deposition solidifying the next new layer. Each layer is bonded to the previous layer through thermal heating. In this way, 3D solid parts can be piled up through the deposition solidifying layer by layer.

The advantages of FDM process are as follows: ①wide material resources, such as ABS engineering plastics, wax, polyethylene, ceramics, nylon, etc; ②as the process does not use laser device, it is simple in use and maintenance and has low cost; ③ rapid forming rate; ④if the water-solubility support material is used, the brace is removed easily and rapidly; ⑤the temperature is relative low (60-300℃), and there is not dust in the whole forming process, and there is not toxic chemical gas, leak of laser and liquid polymer which would arise from previous several methods.

5. Three-Dimensional Printing (3DP)

Three-dimensional printing method was developed by Emanual Sachs of Massachusetts Institute of Technology in 1989, and was patented in the name of three-dimensional printing. Later, it is commercialized by Soligen Inc. in the United States in the name of DSPC (direct shell production casting), which is used for manufacturing the ceramic shell and core in casting.

(1) Operation principle and characteristics of 3DP   Operation principle of 3DP is just like that of a 2D printer on the desk in the past. Its process is similar to the selective laser sintering, which also uses such powder materials as ceramic powder and metallic powder. The difference is that it uses the liquid bonding material injected by an ink-jet printing head to "print" the section of part on the powder, instead of sintering the powder by laser. The specific technological process of 3DP is as follows. After the sintering of the first layer is finished, the forming cylinder drops a distance (equal to the thickness of a layer: 0.013-0.1 mm). The powder-supply cylinder moves up a height and pushes a lot of powder out. The powder is sent to the forming cylinder, paved and compacted. Under the control of the computer, the nozzle jets the bonder selectively to construct the sectional layer in light of the forming data of the next structural section. The excess powder would be collected by powder collecting device in powder paving process. In this way, sending powder, paving powder and jetting bonder go round and round, a 3D powder object formed by bonding layer by layer can be obtained finally.

3DP has many advantages. It has rapid forming speed and wide material resource. If the color materials are added to the bonder, the color parts can be printed. This is one of the most competitive characteristics of the method. The places without sprayed binder are dry powder, which play the role of support and is removed more easily after the forming process is finished. 3DP is especially suitable for making the prototypes with complex shape in inner cavity. The shortcomings of 3DP are that the parts formed by bonding has low strength, and can only be used as conceptual model and is difficult to carry out functional test.

(2) 3DP of organ   Now, many patients who need an organ transplant have to wait very a long time to get the appropriate biological organ. 3DP technology is developing to the direction of the "manufacturing" biological organs. In North Carolina's Wake Forest University of the United States, some scientists want to change the biological cells into "ink", and are seeking the possibility to print the human bones and organs. They believe that the human organ will be "printed" from nothing some day.

In fact, there are many scientists who have already been seeking the possibility of this technology all over the world. This technology even has new names, such as biological printing, organ printing,

FDM 工艺的优点是：①加工材料范围广，如 ABS 工程塑料、蜡、聚乙烯、陶瓷和尼龙等；②因不用激光器件，故使用、维护简单，成本较低；③成形速度快；④当采用水溶性支撑材料时，支撑去除方便快捷；⑤整个成形过程温度为 60～300℃，并且不会产生粉尘，也不存在前几种方法出现的有毒化学气体、激光和液态聚合物的泄漏。

5. 三维打印

三维打印方法是美国麻省理工学院的 Emanual Sachs 等人 1989 年研制的，并申请了 3DP（Three-Dimensional Printing）专利。后被美国的 Soligen 公司以 DSPC（Direct Shell Production Casting）名义商品化，用以制造铸造用的陶瓷壳体和芯子。

（1）三维打印工作原理及其特点　三维打印的工作原理就像一台过去的桌面二维打印机。其工艺过程与选择性激光烧结（SLS）工艺类似，也是采用粉末材料成形，如陶瓷粉末、金属粉末。所不同的是材料粉末不是通过激光烧结连接起来的，而是使用一个喷墨打印头喷射液体黏结剂，将零件的截面"印刷"在材料粉末上面。具体工艺过程如下：当一层黏结完毕后，成型缸下降一个距离（等于层厚为 0.013～0.1 mm），供粉缸上升一高度，推出若干粉末，并被铺粉辊推到成型缸，铺平并被压实。喷头在计算机控制下，按下一个构造截面的成形数据有选择地喷射黏结剂构造层面。铺粉时，多余的粉末将被集粉装置收集。如此周而复始地送粉、铺粉和喷射黏结剂，最终完成一个三维粉体的黏结。

三维打印的优点是：成型速度快，材料来源较广；在黏结剂中添加颜料，可以打印彩色零件，这是该方法最具竞争力的特点之一；未被喷射黏结剂的地方为干粉，在成形过程中起支撑作用，成形结束后比较容易去除。三维打印特别适合制作内腔形状复杂的原型。其缺点是：用黏结剂粘接的零件强度较低，只能作为概念模型，难以进行功能性试验。

（2）器官的三维打印　现在，很多需要器官移植的病人要等待很长时间才能等来合适的生物器官。三维打印技术也正在向"制造"生物器官的方向发展。在美国北卡罗来纳州的维克森林大学，一些科学家欲让生物细胞变成"墨水"，正在探寻打印人体骨骼和器官的可能性。他们相信某一天，人体器官也可以从无到有被"打印"出来。

事实上，全球各处有很多科学家已经在探寻这种技术的可能性了。这种技术甚至有了新的名字：生物印刷、器官印刷、计算机辅助组织工程学或生物制造。美国克莱姆森大学的一位研究人员这样描述生物打印技术。生物打印机有点像传统的喷墨打印机，只不过原先的喷墨现在变成了细胞组织和一种特殊的化学成分，而原先打印出来的纸，现在变成了"培养皿"。这种化学成分可以让培养皿中的液体变成果冻般的胶体，然后将细胞"打印"在胶体上面。机器可以反复添加液体、化学成分和细胞，一层层地把组织层制造出来，最终创造三维的生物机体。

目前，已有科学家能制造出大约 2 in（1 in = 0.0254 m）厚度的组织。但是如果没有足够的营养，细胞会死亡。2012 年 3 月，美国一家公司的三维生物印刷机，已经可以生产出血管。预计未来 10 年内，科学家就可以生产出更复杂的器官，如心脏、牙齿、骨骼等。

6. 直接激光制造

直接激光制造技术，也称选择性激光熔接（Selective Laser Melting, SLM），它是 20 世纪 90 年代在快速成型技术的基础上，结合激光熔覆技术发展起来的一种无模快速制造技术。直接激光制造与选择性激光烧结工艺不同。选择性激光烧结工艺采用的是间接法，即用激光熔化低熔点高分子聚合物或金属粉末以粘接高熔点的金属粉末。这样制造的元件是多孔组

computer aided tissue engineering or bio-manufacturing and so on. A researcher of Clemson University described biological printing technology in this way. The biological printer was something like the traditional ink jet printer, only that the original ink jet became a cell tissue and a kind of special chemical composition now; and originally printed paper became the culture dish now. This kind of chemical composition can make the liquid in the culture dish change into the jelly like gel, and then the cells are printed on the jelly. The machine can add liquid, chemical composition and cell again and again, the tissue layers are made layer by layer. Finally, 3D living organism can be created.

At present, the scientists have been able to make the tissue about 2 inches thick. But if there is not enough nutrition, the organs would die. In 2012, the 3D biological printer of a company in America had produced the blood vessel. It is expected that the scientists will be able to manufacture more sophisticated organs (such as heart, tooth, skeleton and so on) in the next ten years.

6. Direct Laser Fabrication (DLF)

Direct laser fabrication is also named as selective laser melting (SLM). It is a rapid prototyping manufacturing technology without mould developed in 1990s by combining laser cladding technology on the basis of RP technology. Direct laser fabrication is different from the selective laser sintering (SLS). The SLS uses the indirect method, that is, the high molecular polymer or metalic powder with low melting point is melted by laser to splice the metalic powder with high melting point. The components made in this way present in porous tissue with lower density. In post processing, such treatments as thermally degradable polymer, sintering for second time and metallic cementation are required. The components have not only limited accuracy and strength, but longer post processing time. As the metalic powder of SLM is in the completely melted state, the components made by it have higher accuracy, good density and mechanical property. They can satisfy the engineering requirements after a certain precision machining. Another advantage of SLM is that it has obvious superiority in the feasibility to handle pure nonferrous metals.

The operation principle of DLF technology is shown in Fig. 3-26. Firstly, the 3D model of component is cut into slice in computer, thus the data file of each slice can be obtained. The worktable system with 3D motion is mainly composed of a precise X-Y worktable, a precise Z worktable and stepping motors of themselves. In forming operation, the powder materials are sent at a certain speed by means of conveying appliance to the position of laser focus and are melted. The plane scanning motion of NC X-Y worktable and the vertical motion of powder conveying appliance are controled by the slicing CAD software. A cladding section can be obtained by laser cladding point by point and line by line.

After cladding a layer, the laser head goes up (or the table goes down) along the Z axis to a certain height (equal to the layering thickness), the second layer is cladded again by laser and bonded metallurgically to the first layer. In this way, the required 3D component can be obtained by continuous stacking. The method does not need any bonding agent and can make the metal parts with 100% density microstructure. As the method can be used to rapidly manufacture the metal parts with compact near-net shape with the help of laser cladding method, it is also called laser engineering net shaping (LENS). This unparalleled advantage makes direct laser fabrication have huge application value in such key industrial fields as avigation, aerospace, shipbuilding, moulds which involve in the national competitiveness.

织，致密度较低。在后处理中，还需要热降解聚合物、二次烧结及渗金属处理等，不仅制件的精度、强度有限，且后处理时间较长。而 SLM 的金属粉末为完全熔化状态，所以它制成的零件精度高、致密性好、力学性能良好，只需要一定的精加工即可满足使用要求。SLM 的另一个优点是它在处理有色纯金属的可行性上具有很明显的优势。

直接激光制造技术的工作原理如图 3-26 所示。首先在计算机上对零件的三维 CAD 模型进行切片分层，得到每一层切片的数据文件。三维运动工作台系统主要由精密 X-Y 轴工作台和精密 Z 轴工作台及其驱动步进电机组成。成型时，将粉状材料以一定的控制速度由输送装置送到激光焦点所在的位置融化。通过分层 CAD 文件控制 X-Y 轴二维数控工作台做平面扫描运动和垂直方向的送粉机构，实现逐点逐线激光熔覆，获得一个熔覆截面。

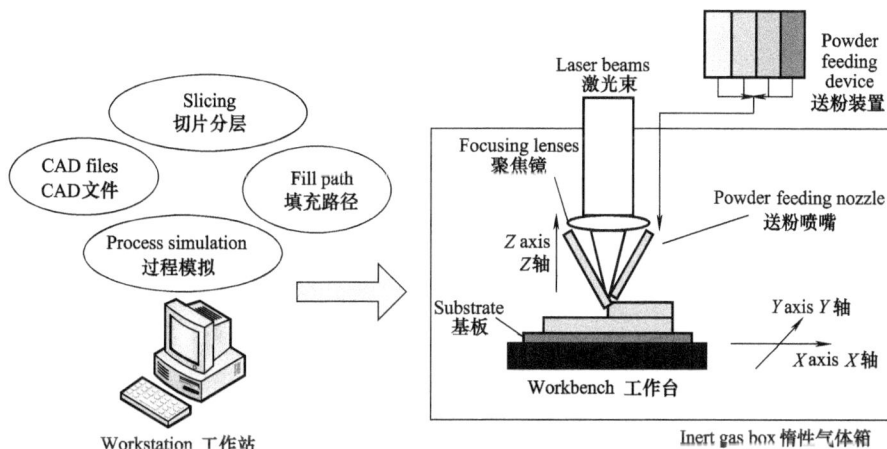

Fig. 3-26　Operation principle of direct laser fabrication 直接激光制造技术的工作原理

一层熔覆过后，激光头沿 Z 轴上升（或工作台沿 Z 轴下降）一定高度（等于一个分层厚度），再激光熔覆第 2 层，使第 2 层与第 1 层冶金结合在一起。如此下去不断地层叠，即可获得所需的三维零件。这种方法不需要任何黏结剂，可以制造 100% 密度显微结构的金属零件。由于这种方法能够借助激光熔覆方法快速制造出致密的近净形金属零件，因此也被称为激光工程净形制造（Laser Engineering Net Shaping，LENS）。正是这种无可比拟的优势，使得激光直接制造技术在航空、航天、造船、模具等关乎国家竞争力的重要工业领域内具有极大的应用价值。

3.6.3　RPM 技术的应用

1. 快速产品开发

RPM 在快速产品开发（Rapid product development，RPD）方面的应用如图 3-27 所示。RPM 在产品开发中不受复杂形状的限制，可以迅速地将显示于计算机屏幕上的设计变为可进一步评估的实物。根据原型，可对设计的正确性、造型的合理性、可装配性和干涉等进行具体的检验。对形状较复杂而贵重的零件（如模具），如果仅依据 CAD 模型不经原型阶段就进行加工制造，这种简化的做法风险极大，往往需要多次试制才能成功。它不仅延误开发的进度，而且往往需要花费更多的资金。一般来说，产品开发采用 RPM 技术，可减少产品开发成本的 30% ~ 70%，减少开发时间的 50%。

3. 6. 3 Application of RPM Technology

1. Rapid Product Development (RPD)

The applications of RPM in RPD are shown in Fig. 3-27. In the product development, RPM are not restricted by complex shape and can change rapidly the design on the screen of the computer into solid object for further evaluation. Based on the prototype, the correctness of design, model rationality, assemblability, interference and so on can be inspected specifically. For the valuable parts with complex shape (such as the mould), if the manufacture is carried out based only on the CAD model without rapid prototyping, this simplified approach has great risk, often it needs trial-manufacturing many times to get success. It not only delays the progress of development, but often needs to spend much money. Generally speaking, the application in product development can reduce the product development cost by 30%-70% and the product development time by 50%.

2. Applications of RP Technology in the Medical Domain

The applications of RP technology are just unfolding not only in manufacturing industry, but are full of life in the biomedical domain. The skeletons and internal organs of the human body have the extreme complex structure. RP is almost the only way to reproduce really the internal organs of the human body and reflect the characteristics of the lesion. Based on the data acquired from CT scan or MRI magnetic resonance, solid model of human organ made by means of RP method can assist the doctor to make diagnosis and determine the treatment plan; human artificial limb made by means of RP technology can achieve maximum degree of anastomosis with the binding site, thus shortening the operation time and reducing the postoperative complications. In recent years, the domestic and foreign scholars are more interested in studying how to make the biomaterials into human organs through RP technology, where the study of artificial bone has made gratifying achievements. For instance, using the injection method, Tsinghua University has made the biomaterials into the cell carrier scaffold of large artificial bone through material accumulating and shaping under the low temperature environment. Animal experiments prove that the scaffold can be effectively degraded. Some experts even said that although RP technology appeared firstly in the manufacturing industry, its most exciting application will be in the biomedical field.

3. 6. 4 Development Trend of RPM Technology

1. Development of Rapid System

With the acceleration of CNC programming and tool path generation, RPM has lost part of the speed advantage. Therefore, RPM developers strive to improve the speed of RPM machine under the premise of not sacrificing accuracy. For example, the high-speed deflection system and plane displacement system with high precision are studied so as to improve the laser scanning speed; dual laser is used in SLS apparatus; variable laser spot size is used in the SLA 7000 machine, i. e. using the laser spot with small diameter to generate the boundary with high accuracy, and using the laser spot with large diameter to fast fill inside. Other approaches to increase shaping speed are the application of variable thickness slice, rapid post processing, improving the speed of data file processing and so on.

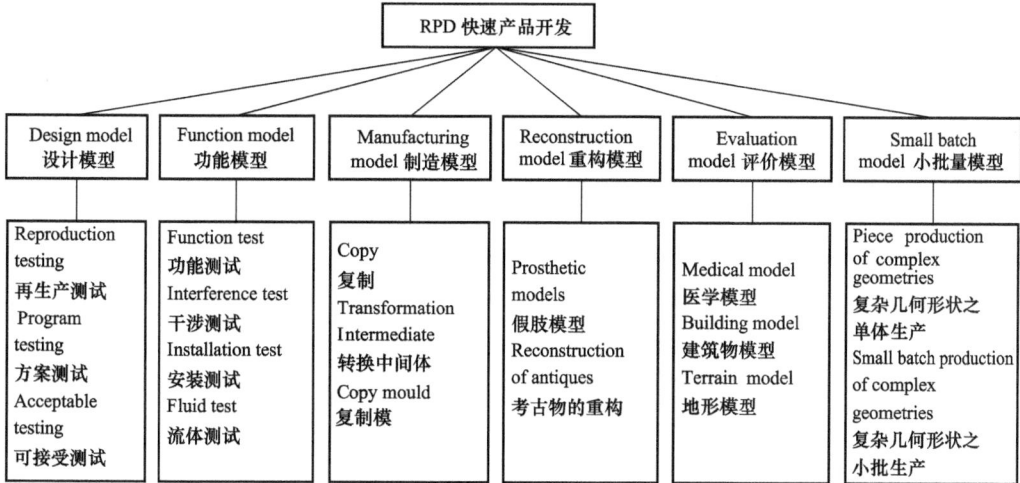

Fig. 3-27 Applications of RPM in RPD RPM 在 RPD 方面的应用

2. RP 技术在医学领域中的应用

RP 技术不仅在制造业领域的应用方兴未艾，在生物医学领域的应用也充满生机。人体的骨骼和内部器官具有极其复杂的结构。要真实地复制人体内部的器官构造，反映病变特征，快速成型几乎是唯一的方法。以 CT 扫描或 MRI 磁共振数据为基础，利用 RP 方法快速制作的人体器官实体模型可以帮助医生进行诊断和确定治疗方案；借助 RP 技术制作的人体假肢还能与结合部位实现最大程度的吻合，从而缩短手术时间，减少术后并发症。近几年国内外学者更加热衷于研究将生物材料快速成型为人工器官的课题，其中人工骨的研究已取得可喜的成果。例如，清华大学采用喷射方法，将生物材料在低温环境下堆积成形，制成多孔大段人工骨的细胞载体支架。经过动物实验证明该支架能有效降解。有专家甚至说，虽然 RP 技术最初出现在制造行业，但它最激动人心的应用将是在生物医学领域。

3.6.4 RPM 技术的发展趋势

1. 快速系统的研制

随着 CNC 编程和刀具路径生成的加快，RPM 失去了一部分速度优势。为此 RPM 开发商在不牺牲精度的前提下，竭力提高 RPM 成形机的速度。如研究快速高精度偏转系统及平面位移系统，提高了激光扫描速度；在选择性激光烧结设备中采用了双重激光；在 SLA 7000 设备中采用变化的激光光斑尺寸，用小直径的光斑生成高精度的边界，而用大直径的光斑在内部快速填充。其他提高成形精度的方法还有：变厚度切片的应用；快速后处理工艺；提高数据文件处理的速度等。

2. 新材料的研究

成形材料是 RPM 技术发展的关键环节。金属、陶瓷和复合材料的应用代表了新材料在 RPM 领域的研究进展，因为这些材料更适合于制造各种功能的零件，更符合工程的实际需要。在 LOM 工艺中出现了玻璃纤维和碳纤维加强薄膜。在 SLS 技术中开发了可直接烧结的材料，如铜合金、钢、硬质合金（WC/Co）和陶瓷（SiC，Al_2O_3）。德国 EOS 公司和日本

2. Research on New Materials

Forming material is the key to the development of RPM technology. The applications of metals, ceramics and composites represents the research progress of new materials in the field of RPM, because these materials are more suitable for the manufacture of various functional parts and more conformable to the actual needs of the engineering project. Glass fiber and carbon fiber reinforced film were found in the LOM process. Materials that can be sintered directly were developed in the SLS technology, such as copper alloy, steel, cemented carbide (WC/Co) and ceramics (SiC, Al_2O_3). German EOS company and Japan CMIT company added the ceramic powder to the epoxy photopolymer to make the material which can be used for direct patternmaking rapidly. Michigan University in the United States carried out the metal welding by means of high power laser to shape directly the steel mould. Combining the layer-by-layer accumulation with the 5-axis NC machining, Stanford University uses the laser to complete the direct metal sintering and acquires the accuracy similar to NC machining.

3. Desktop and Networking of RPM System

With the continuous development of computer technology, information technology, multimedia technology and mechatronic technology, desktop manufacturing system (DMS) based on RPM technology will appear. Like the printer and drawing machine, it will be used as the peripheral equipment of computer and will really become 3D printer or 3D fax machine. RPM equipment will be made gradually into economical, popular, easy-to-use, green, universal peripheral equipment of the computer.

With the development and popularization of information superhighway, full sharing of resources and equipment can be realized. The on-line personalized design platform provided by some enterprises through network can make those enterprises that do not have the ability to develop product or rapid prototyping equipment use the resources and devices in network. The enterprises with the ability of RPM carry out the product development and production so as to realize remote manufacturing.

4. Rapid Prototyping and Micro/Nano Fabrication

At present, it is difficult for the micro machining technology to machine 3D irregular microstructures, and further increase in the ratio of depth to width is limited. But the RP technology based on the dispersion-accumulation manufacturing principle can manufacture the components with any complex shape. There were the reports on the fabrication of micro structures by RPM process. For instance, the University of Southern California in the United States manufactured a 290 μm wide metal chain according to LOM principle, and the shape of each layer was achieved by means of electrochemical method; using laser technology, Japanese scientist made a cattle with 10 μm long and 7 μm high from synthetic resin.

5. Biofabrication and Growth Forming

Biofabrication can be defined as the production of complex living and non-living biological products from raw materials such as living cells, molecules, extracellular matrices, and biomaterials. Biofabrication technology combines the biotechnology, biomedical science and manufacturing science to solve the human health problems. How to manufacture the "biological parts" which are able to change or reproduce the living body or its part of the function is the issue studied by biofabrication. Take the artificial bone as an example.

CMIT 公司在环氧光敏树脂中添加陶瓷粉，可快速直接制模。美国密歇根大学采用大功率激光器进行金属熔焊使钢模具直接成型；斯坦福大学用逐层累加与五坐标数控加工结合方法，用激光将金属直接烧结成型，可获得与数控加工相近的精度。

3. RPM 系统的桌面化和网络化

随着计算机技术、信息技术、多媒体技术、机电一体化技术的不断发展，将会出现基于 RPM 技术的桌面制造系统（Desktop Manufacturing System，DMS）。其将与打印机、绘图机一样作为计算机的外围设备来使用，真正成为三维立体打印机或三维传真机，逐步使 RPM 设备变成经济型、大众化、易使用、绿色环保、通用化的计算机外围设备。

随着信息高速公路的发展和普及，可实现资源和设备的充分共享。通过网络，企业提供在线个性化设计的平台，可使不具备产品开发能力或快速成型设备的企业充分使用网络上的共享资源和设备。具备快速成型制造能力的公司进行产品开发和成形制作，从而实现远程制造（Remote Manufacturing）。

4. 快速原型与微纳米制造

目前，常用的微加工技术难于加工三维异形微结构，深宽比的进一步增加受到了限制。而快速原型根据离散/堆积的制造原理，能够制造任意复杂形状的构件。已有采用 RPM 工艺制备微细结构的报道，如美国南加州大学采用 LOM 原理加工出一条 290 μm 宽的金属链，其每一层的形状通过电化学方法获得；日本科学家采用激光技术，用合成树脂制成了长 10 μm、高 7 μm 的牛。

5. 生物制造和生长成形

生物制造可以定义为利用活细胞、分子、细胞外基质和生物材料等原材料从事活性与非活性生物产品的生产过程。生物制造技术将生物技术和生物医学与制造科学相结合，解决了人类的健康保健问题。如何制造能够改变或者复现生命体或其一部分功能的"生物零件"，正是生物制造要研究的问题。

以人工骨替代骨骼为例，首先对人体的骨骼进行 CT 扫描，然后进行骨骼内部结构的仿生 CAD 建模和骨骼外腔的三维造型。利用常温固化的羟基磷化石等生物相容性和生物可降解性较好的材料，在 RP 设备上制出具有生物活性的人工替代骨，在成形过程中置入骨生长因子。这种方法有望解决目前人工替代骨加工周期长、生物相容性和生物可降解性不好、内部微孔结构不可控的问题。

此外，当处理的材料，如细胞、基因片段等具有活性并携带着一定的生长信息时，必须对生长成形进行研究。生长成形要解决的问题是如何在快速原型的信息处理中考虑每个材料单元所发生的生长情况，解决生长机理并充分利用和模拟生长现象，提炼出最少和最关键的材料单元的控制量来制造出最终的产品。

3.7　绿色制造技术

3.7.1　绿色制造概述

1. 绿色制造技术的产生背景

尽管各类制造业对经济繁荣做出了重大贡献，但它们同时每年也产生了大约 55 亿 t 无

Firstly, CT machine is used to scan the human bones. And then, bionic CAD modeling of the internal structure of skeleton and 3D modeling of the external cavity of the skeleton are carried out. By the use of good biocompatible and biodegradable materials solidified at normal temperature, such as hydroxyl phosphorus fossil, the artificial bones with living are fabricated by means of the RP machine. The bone growth factor is put in the forming process. This method is expected to solve the problems, such as long processing cycle of artificial bone, poor biocompatibility and biodegradability, uncontrollable internal microporous structure and so on.

Besides, when such materials as cells, gene segments etc. with the biological activity and some growth information are processed, the growth forming have to be studied. The problems to be solved in growth forming are to consider the growth situation generated in each material unit in the information processing of RP, to solve the growth mechanism and fully use and simulate growth phenomenon, and to refine the least and the most critical controlled quantity of material unit to manufacture the final product.

3.7 Green Manufacturing Technology

3.7.1 Introduction to Green Manufacturing

1. Background Green Manufacturing Technology

Although manufacturing industries make a significant contribution to economic prosperity, they also generate approximately 5.5 billion tons of nonhazardous waste and 0.7 billion tons of hazardous waste each year. Industrial countries are beginning to face one of the consequences of the rapid development. Wide diffusion of consumer goods and shortening of product lifecycles have caused an increasing quantity of used products being discarded. In Europe, 800, 000 tons of old television sets, computer equipment, radios, and measuring devices, and 3 million tons of automobile equipment are thrown into the national garbage center each year. In the United States, the municipal solid waste (MSW) generated by households and industrial establishments is about 4 pounds per person each day. Historically, much effort focused on the proper treatment and disposal of toxic and hazardous waste from industries. Unfortunately, this reactive environmental protection approach cannot completely solve the problems of potential toxic or hazardous material releasing from products or the waste stream into the environment. Facing this environmental problem, both the governments and industrial companies are making more strict regulations to promote environmental friendly products and technology. One example is that the governments of Germany and the U. S. require that manufacturers take responsibility for the disposal of their products. The other example is the establishment of ISO14000 series of Environmental Management Systems (EMS) standards. All of these regulations intend to minimize the environmental impact of products. Since 1990s, under the impetus of green tide and sustainable development, green manufacturing technology has developed rapidly and been widely used in developed countries.

2. Connotation of Green Manufacturing Technology

Green manufacturing (GM) is also named as environmentally conscious design and manufacturing

害废物和 7 亿 t 有害废物。工业国家正面临快速发展的后果。消费品广泛扩散和产品生命周期的缩短造成越来越多的废旧产品被丢弃。在欧洲，每年有 800 000 t 旧电视机、计算机设备、收音机、测量设备，以及 300 万 t 汽车设备被扔进国家的垃圾中心。在美国，家庭和工业场所产生的城市生活垃圾平均每人每天大约为 4 lb（1 lb ≈ 0.454 kg）。从历史上看，大量的工作都集中在如何对各行各业产生的有毒有害废物进行适当处理和清理上。遗憾的是这种应付式的环保方法不能完全解决从产品或废水释放的潜在有毒或有害物质进入环境的问题。面对这一环境问题，各国政府都在制定更严格的法规来促进环保产品和技术的发展。一个例子是德国和美国政府要求制造商对其产品的处置负责。另一个例子就是确立了 ISO14000 系列环境管理体系（EMS）标准。所有这些法规旨在尽量减少产品对环境的影响。自 20 世纪 90 年代以来，绿色制造技术在绿色浪潮和可持续发展思想的推动下迅速发展，并在发达国家得到了广泛的应用。

2. 绿色制造技术的内涵

绿色制造又称环境意识设计与制造或面向环境的制造。绿色制造是对设计与制造的一种思考方式，重点研究如何最富有成效地利用原材料和自然资源，如何把对工人和自然环境的不利影响降到最低程度。

在污染控制方面，与传统的"末端治理"相反，绿色制造采用积极主动的方法把产品在设计和制造期间对环境的影响降到最低程度，从而提高产品在环境意识市场的竞争力。一般有两种绿色制造方法。在第一种方法中（零废品生命周期），认为产品在全生命周期内对环境的影响可以减少到零。循环绝对是可持续的，产品的设计、制造、使用及报废处理均对环境无影响。这种方法的重点是创建一个尽可能可持续的产品周期。可持续生产意味着在产品设计、生产、销售、使用和报废处理过程中对环境和职业健康造成的危害极小（或没有），而且消耗的资源（材料和能源）也极少。第二种方法（增量式废品生命周期）是基于这样一个前提，即当前的加工周期存在一定的负面影响。这种影响可以通过在称为增量式废品生命周期控制技术方面的某些改进进行减少或清除。这种方法旨在通过清洁技术减少有毒材料的负面影响。"清洁技术"是一种用于消除或显著减少有害废物产生源头的方法。

图 3-28 所示为一种绿色制造的体系结构。绿色制造的体系结构中包括的具体内容有：

1）两个过程。一是指具体的制造过程即物料转化过程。这是一个充分利用资源，减少环境污染，实现具体绿色制造的过程；二是指在构思、设计、制造、装配、包装、运输、销售、售后服务以及回收整个产品生命周期中每个环节均充分考虑资源和环境问题，以实现最大限度地优化利用资源和减少环境污染的广义绿色制造过程。

2）三项内容。三项内容包括绿色资源、绿色生产过程和绿色产品。

3）三条途径。一是改变观念，树立良好的环境保护意识，加强立法和宣传教育；二是针对具体产品的环境问题，采用绿色设计和绿色制造工艺解决所出现的问题；三是加强管理，利用市场机制和法律手段，促进绿色技术、绿色产品的发展。

4）两个目标。即资源优化和环境保护。通过资源综合利用、短缺资源的代用、可再生资源的利用、二次能源的利用及节能降耗措施延缓资源能源的枯竭，实现持续利用；减少废料和污染物的生成和排放，降低整个生产活动给人类带来的风险，最终实现经济效益和环境效益的最优化。

（ECD&M）or manufacturing for environment （MFE）. Green manufacturing is a way of thinking about design and manufacturing, which focuses on the most efficient and productive use of raw materials and natural resources, and minimizes the adverse impacts on workers and the natural environment.

As opposed to the traditional "end-of-pipe" treatment for pollution control, green manufacturing is proactive approach to minimize the product's environmental impact during its design and manufacturing, and thus to increase the product's competitiveness in the environmentally conscious market place. There are two general approaches to green manufacturing. **In the first approach** （zero-wasted lifecycle）, it is assumed that the environmental impact of a product during its lifecycle can be reduced to zero. The cycle can be absolutely sustainable, and the product may be designed, manufactured, used, and disposed without affecting the environment. The emphasis in this approach is to create a product cycle that is as sustainable as possible. **Sustainable production** means that products are designed, produced, distributed, used and disposed with minimal （or none）environmental and occupational health damages, and with minimal use of resources （material and energy）. **The second approach** （incremental waste control lifecycle）is based on the premise that there is a certain amount of negative impact from the current process cycle. This impact can be reduced or cleaned based on some improvement in technology that is named as incremental waste lifecycle control. This approach is to reduce the negative impact of hazardous materials through clean technology. A "cleaner technology" is a source reduction or recycling method applied to eliminate or significantly reduce hazardous waste generation.

A green manufacturing architecture is shown in Fig. 3-28. The specific contents included in the green manufacturing architecture are as follows:

1）Two processes. One is the specific manufacturing process, that is, material conversion process. This is a process to implement green manufacturing by making full use of resources and reducing environment pollution. The other means the generalized green manufacturing process in which the resource and environment problems of every link in whole product life cycle including conception, design, manufacturing, assembly, packaging, transportation, sales, after-sale service and recycling should be fully considered so as to maximize the use of resources and reduce environment pollution.

2）Three items. Three items include green resources, green production and green products.

3）Three ways. One way is to change concepts, set up a good environment protection consciousness? and reflect in concrete action, these can be achieved by strengthening legislation, publicity and education. Second, aim at environment issue of specific product, adopt green design and green manufacturing process, establish green degree evaluation system and solve problems. Third, enhance management, take advantage of market mechanism and legal mean to promote green technology development and green product.

4）Two objects. Two objects include resource comprehensive utilization and environment protection. Achieve sustainable utilization through resource comprehensive utilization, shortage resource alternative, renewable resource utilization, secondary energy utilization and energy saving measures to slow the depletion of energy resource. Reduce generation and emissions of waste and pollutant, improve environment compatibility of industrial product in production process, reduce risk of entire

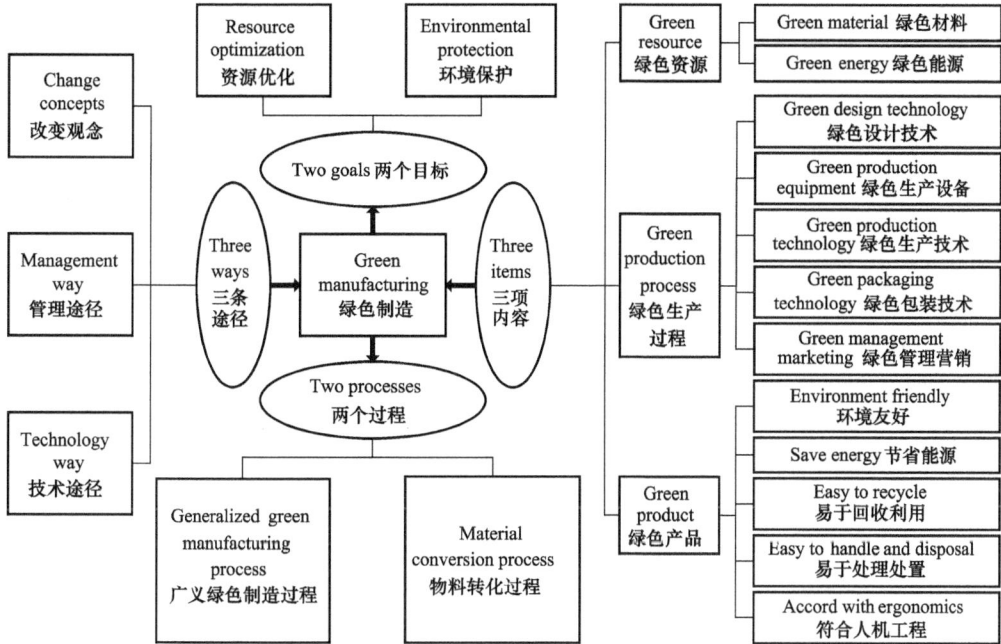

Fig. 3-28　Green manufacturing architecture 绿色制造的体系结构

3.7.2　绿色产品设计

1. 绿色产品设计的概念和特点

绿色产品设计是污染防治的必然产物。绿色产品是指在产品使用和处置过程中能够减轻环境负担的产品，这种产品更具有市场吸引力。绿色产品设计是指绿色工程设计。绿色工程设计的目标是要人们了解设计决策如何影响产品的环境兼容性。

绿色产品设计是绿色制造的关键，因为产品在设计阶段就基本确定了采用何种材料、资源以及何种加工方式，同时也确定了产品在整个生命周期中的环境属性。正如 Boothroyd 在福特汽车公司做的报告中指出的那样，尽管设计费用仅占产品全部成本的 5 %左右，但却决定了 80%~90%的产品生命周期的全部消耗。

传统设计方法是以企业的发展战略和获取企业自身最大经济利益为出发点，主要考虑产品的功能、质量和成本等基本属性。因而，传统的产品设计仅涉及产品生命周期中的市场分析、产品设计、工艺设计、制造和销售以及售后服务等几个环节，如图 3-29 所示。

绿色设计也称面向环境的设计（Design for Environment，DFE），是将产品全生命周期中的设计、制造、使用、回收及再生等各个环节作为有机整体。绿色设计的基本流程及设计目标如图 3-30 所示。

2. 绿色设计备受关注的 4 个目标

1）面向污染防治的设计。它是指制造商和消费者一开始就采取的行动，以防止废物的产生（即使用较少的材料去实现相同的功能，或设计耐用的产品延长产品的使用寿命）。

2）面向更好的材料管理设计。它是指允许产品零部件或材料在其最高增值应用中被回收和再利用而采取的行动（即设计易于拆解成组成材料的产品，或采用无须分离而被整体

production activity, realize optimization of economic benefit and environment benefit.

3.7.2 Green Product Design

1. Concept and Characteristics of Green Product Design

Green product design is expanded from pollution prevention. Green products—products that can reduce the burden on the environment during use and disposal have additional marketing appeal. Green product design refers to green engineering design. The aim of green engineering design is to develop an understanding of how design decisions affect a product's environmental compatibility.

Green product design is the key to the green manufacturing, because it is basically determined to adopt what materials, resources and methods of processing in the design phase. And the environmental attributes of the product is also determined in the life cycle. As Boothroyd pointed out in the report of Ford Motor Company, although the design cost only accounts for about 5% of the total product cost, it determines the overall consumption in the 80% to 90% of product life cycle.

Based on the development strategy and the maximum economic benefit of the enterprise, traditional design method considers primarily the function, quality, cost and other basic properties of the product. Therefore, the traditional product design is only involved in the market analysis, product design, process design, manufacturing, sale and after-sale service etc. in the product life cycle, as shown in Fig. 3-29.

Green design, also called design for environment (DFE), takes such links in the product life cycle as product design, manufacture, use, recycling and regeneration and so on as a whole system. The objectives and basic process of green design are shown in Fig. 3-30.

2. Four Focused Objectives of Green Design

1) Design for pollution prevention. It refers to activities by manufacturers and consumers that prevent the generation of waste in the first place (that is, using less material to perform the same function, or designing durable products to extend the product service life).

2) Design for better materials management. It refers to activities that allow product components or materials to be recovered and reused in their highest value-added application (that is, designing products that can be readily disassembled into constituent materials, or using materials that can be recycled together without the need for separation).

3) Design for re-manufacturing and recycling. It refers to reducing virgin material extraction rates, waste generated from raw material separation and processing, and energy uses associated with manufacturing. It can also divert residual material from municipal waste, relieving pressure on over-burdened landfills.

4) Design for composting and incineration. It refers to making products entirely out of biodegradable materials. For example, starch-based polymers (which are inherently biodegradable) and film can substitute for plastic in a variety of applications.

3. Principles of Green Design

To achieve the green design should follow three principles at least. Fig. 3-31 shows the three essential factors of green products.

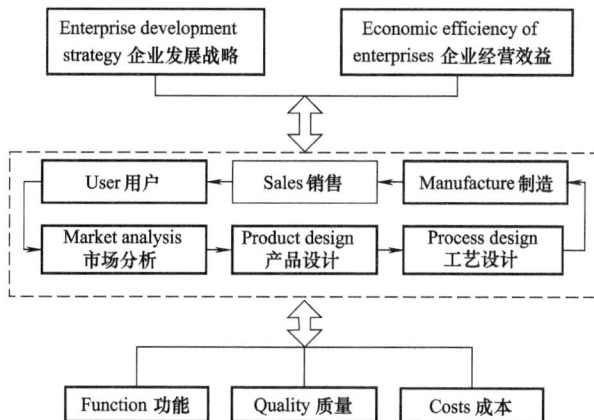

Fig. 3-29 Goals and process of traditional design method 传统设计方法的目标及流程

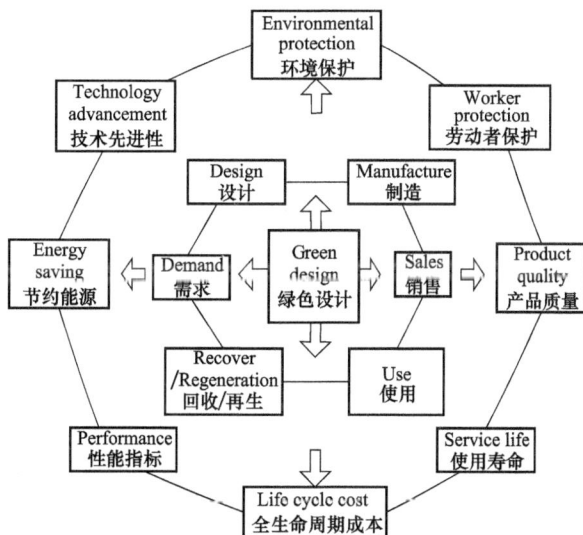

Fig. 3-30 Objectives and basic process of green design 绿色设计的基本流程及设计目标

再利用的材料）。

3）面向再制造和回收利用的设计。它是指减少原材料的提取率，减少在原料分离和处理过程中产生的废物，以及减少与制造有关的能源消耗。例如，也可以从市政垃圾中分离出残留的材料以缓解不堪重负的垃圾填埋场的压力。

4）面向堆制肥料和焚化的设计。它是指完全由生物可降解材料制造产品。例如，淀粉基聚合物（本质上就是可降解的）和胶片可以在各种应用场合代替塑料。

3. 绿色设计原则

实现绿色设计至少要遵循 3 个原则，即通常说的 "3R"（Reuse "再利用"、Recycle "回收和循环"、Reduce "减少"）。图 3-31 所示为绿色产品应具备的三要素。

1）再利用。要求产品及其零部件和附件外包装能够被反复使用。这就要求设计师在对产品进行设计时，尽可能使零部件结构简单化和标准化。这样不但能节约资源，而且由于是

1) Reuse. It is required that the product and its components, and the external packing of accessory can be used repeatedly. This asks for the designers to simplify and standardize the structure of components as far as possible in the product design. This will not only save resources, but the standard components can also be recycled. Manufacturers should try to extend the use of the product rather than to replace it very quickly.

2) Recycle. It is required that the manufactured product can be turned into available resources again after completing its functions, rather than the unrecoverable garbage. There are two cases of recycling. One is the primary recycling, that is, the recycled waste is used to produce the same type of new products; the other is the secondary recycling, that is, the waste resources are changed into the raw materials for other products. Thus, the primary recycling can better save natural resources.

3) Reduce resource consumption. It is required that less raw materials and energy are used to achieve an established production or consumption so as to save resources and reduce pollution from the source. We should choose such clean, renewable primary energy as solar energy, wind energy as far as possible, rather than gasoline and other non-renewable secondary energy.

The three principles mentioned above are focused on the impact of the product on the environment. How to implement green design in product design There are many kinds of methods in which the important methods are green material design, green product structure design, green energy consumption design, green packaging design, and green manufacturing process design and so on.

4. Key Methods to Implement Green Design

(1) Green material design   Green material, also called eco-material, refers to the material and products that have the good performance, less consumption on the resources and energy, small pollution on the ecological and environmental, low energy consumption, low noise, non-toxic and harmless to the environment in the process of preparation and production. It also includes the materials and manufactured goods what is harmful to human and environment, but can take appropriate measures to reduce or eliminate. Green material design follows the following 4 principles:

1) Try to use environmentally-friendly materials and to avoid using toxic, hazardous and radiation properties of materials. Green material design highlights the selection of environmentally-friendly materials. The bamboo notebook made in ASUS Corporation is just a typical example, as shown in Fig. 3-32. The shell of the notebook is made of bamboo, instead of the ABS plastics which is not conducive to recycling and would causes serious environmental pollution. IMPRINT chair made in Danmark, as shown in Fig. 3-33, is another example, which uses a kind of tree fiber material to replace plastics so as to relieve the burden on the natural resource and environment.

2) To simplify the surface processing of the product. Often plating, painting and other processing are carried out on the surface of some products to pursue surface color and metal texture. These surface treatments would produce sewage and toxic gases and are not conducive to recycling, thus having a great impact on the environment. Currently, many factories which attach importance to environmental protection have adopted other technological means. For example, the common used plastic housings in electronic products after they are formed in the injection mold usually have the smooth and bright appearance, which is achieved through using mirror surface treatment to the mold cavity.

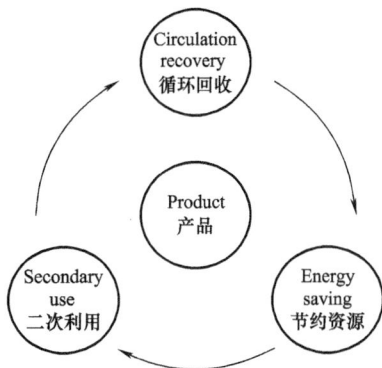

Fig. 3-31　Three essential factors of green product 绿色产品应具备的三要素

标准件，还可以对其进行回收再利用。制造商应该尽量延长产品的使用期，而不是非常快地更新换代。

2）回收和循环使用。要求生产出来的物品在完成其功能后能重新变成可以利用的资源，而不是不可恢复的垃圾。再循环有两种情况：一是原级再循环，即废品被循环用来生产同种类型的新产品；另一种是次级再循环，即将废物资源转化为其他产品的原料。由此可见，原级再循环能够更好地节省自然资源。

3）减少资源消耗。要求用较少的原料和能源投入来达到既定的生产或消费目的，进而从源头就注意节约资源和减少污染。尽可能选用太阳能、风能等清洁、可再生的一次能源，而不是汽油等不可再生的二次能源，以有效地缓解能源危机。

以上 3 种原则都着重注意了产品对环境的影响。该如何在产品设计中实施绿色设计呢？方法多种多样，其中重要的有绿色材料设计、产品绿色结构设计、绿色能耗设计、绿色包装设计、绿色制造过程设计等。

4. 实施绿色设计的关键方法

（1）绿色材料设计　绿色材料（Green Material）又被称为生态材料（Eco-material），是指具有良好的使用性能，并对资源和能源消耗少，对生态和环境污染小，在制备、生产过程中能耗低、噪声小、无毒性并对环境无害的材料和材料制品，也包括那些对人类、环境有危害，但采取适当的措施后就可以减少或消除的材料及制成品。绿色材料设计遵循以下 4 个原则：

1）尽量使用环保性材料，避免选用有毒、有害和有辐射特性的材料。绿色材料设计主要凸显在环保材料的选材方面。华硕公司推出的竹韵笔记本式计算机就是一个典型例子，如图 3-32 所示。其外壳采用了环保竹材，取代了对环境污染严重、不利于回收的 ABS 塑料。还有丹麦的 IMPRINT 椅子，如图 3-33 所示，用一种树木纤维材料代替塑料，减轻对自然资源和环境的负担。

2）简化产品的表面工艺。有些产品为了追求表面色彩和金属质感，往往会对表面采用电镀、喷漆等处理。这些表面处理会产生污水和有毒气体，而且不利于回收，从而对环境造成了极大影响。目前很多重视环保的工厂都采用了其他工艺手段。例如，电子产品中常见的塑胶外壳，可以通过对注塑模具型腔采用镜面处理来实现塑胶品成型后的光滑明亮的外观。

3）尽量使用单一的材料类型。使用单一的材料可以极大地便利产品的回收。

3) Try to use a single material. The use of a single material can greatly facilitate the recovery of products.

4) Materials used by green products should be easy to recycle, reuse, re-manufacture or degrade.

(2) Green structure design　In addition to meet the basic requirements of ordinary products, the disassembly and recyclability are mainly considered in green structure design. In the ECD&M literature, many researchers emphasize the importance of recycling end-of-life (EOL) products and the role of product disassembly for effective recycling.

1) Design for recovering and recycling (DFR). Recycling is defined as recovering materials or components of a used product to make them available for new products. Another definition is the use of product design to facilitate the recovery and reuse of materials in the product. These definitions infer "closing the loop" of materials and components after usage by reusing them for raw materials or secondary materials at different stages of the product's lifecycle. One of the purposes of disassembly is recycling, while the purpose of DFR includes mainly resource recovery and reuse.

The contents of DFR are as follows: ① Recycling performance analysis of the components. Whether the materials of the components in the EOL products can be recycled depends upon its performance degradation; ②identification and symbol of recyclable materials; ③recycling method; ④analysis of structure processability of the recycling parts; ⑤economic analysis and evaluation of recyclability.

Two engineering problems associated with design for recyclability are dismantling techniques and recycling costs. Dismantling required the knowledge of the destination or recycling possibility of the component parts disassembled. It is not possible or economical to recycle a product completely, therefore, the aim of recycling is to maximize the recycle resources and to minimize the mass and pollution potential of the remaining products.

2) Design for disassembly (DFD). The main contents of DFD include design for dismountable product and design for dismountable process. The purpose of the design for dismountable product is to ensure that the product has a good dismountable property by considering the difficulty of disassembly in the product concept forming stages. Specifically, it involves three contents: ①to reuse the product components directly or indirectly; ②to recycle the components, especially the electronic components; ③to recycle the materials. The purpose of the design for dismountable process is to determine the optimum disassembly path and to choose the best disassembly tool for manual and automatic disassembly. Manual disassembly or automatic disassembly has two situations: one is to disassemble the product into single component; the other is the selective disassembly based on the final state of the product, selective demolition.

(3) Design for green energy consumption　Design for green energy consumption means to design rational product structure, function, manufacturing process or to use new technology, new theory in order to minimize the energy consumption and energy loss of the product in use. The reduction of energy consumption means the reduction of environment pollution. Green energy consumption design is set about mainly from two aspects: ①to use green renewable energy, such as solar energy, wind energy etc. as far as possible; ② to save energy and reduce emission.

4）绿色产品所用材料要具有可回收性，易于再利用、回收、再制造或易于降解。

Fig. 3-32　Bamboo notebook 竹韵笔记本式计算机

Fig. 3-33　IMPRINT chair IMPRINT 椅子

（2）绿色结构设计　绿色结构设计除满足普通产品的基本要求外，主要考虑的是结构的易拆卸性与可回收性。在有关绿色制造的文献中，许多研究人员都强调废弃产品回收再利用的重要性以及产品可拆卸性对高效回收的作用。

1）面向回收的设计（Design for Recovering and Recycling，DFR）。回收再利用被定义为回收用过产品的材料或零部件以使它们可用于新产品。它的另一个定义是运用产品设计以促使产品中材料的回收和再利用。这两个定义意喻着通过把用过的材料和零部件再用作产品生命周期不同阶段的原级或次级材料而形成的闭环。产品拆卸的目的之一是回收，而 DFR 的目的主要包括资源回收和再用。

DFR 的内容包括：①零部件的回收性能分析。产品报废后，其零部件材料能否回收取决于其性能的退化情况。②可回收材料的识别及标志。③回收处理工艺方法。④回收零部件的结构工艺性分析。⑤可回收性的经济分析与评价。

与面向回收设计相关的工程问题是拆解技术和回收成本。拆卸需要了解拆卸零部件的目的或回收的可能性。全部回收某种产品是不可能的或不经济的，因此，回收的目的是使资源回收达到最大化和使残留物品的质量和污染潜力最小化。

2）面向拆卸性设计（Design for Disassembly，DFD）。DFD 的主要内容包括可拆卸产品设计和可拆卸工艺设计。可拆卸产品设计的目的是通过在产品概念形成阶段充分考虑可拆卸的难易程度，以此来保证产品具有良好的可拆卸性能。具体涉及 3 个方面的内容：①产品零部件的直接或者间接再用；②元器件的回收，尤其是电子元器件的回收；③材料的回收。可拆卸工艺设计的目的是为手工拆卸、自动拆卸确定最佳拆卸路径和选择最佳拆卸工具。手工拆卸或自动拆卸有两种情况：①将产品拆成单个的零部件；②依据产品的最终状态，有选择地拆卸。

（3）绿色能耗设计　绿色能耗设计就是设计合理的产品结构、功能、制造工艺或利用新技术、新理论使产品在使用中消耗能量最少、能量损失最少。减少能源消耗即意味着减少对环境的污染。绿色能耗设计主要从两个方面着手：①尽可能使用绿色可再生能源，如太阳能、风能等；②要节能减排。

（4）绿色包装设计　绿色包装是指符合环保要求的包装。绿色包装最重要的是要使用可分解、降解的包装材料。另外，要尽量简化包装。简化包装并不意味着随意地去减少材料，还需确保产品的防护性和运输性。

（5）绿色制造过程设计　绿色制造是一个综合考虑环境影响和资源效益的现代化制造模式。其目标是使产品从设计、制造、包装、运输、使用到报废处理的整个产品生命周期

（4）Design for green packaging　Green packaging means the packaging which meets the environmental protection requirements. The most important thing for green packaging is to use the decomposable, degradable packaging materials. Besides, the packaging should be simplified as far as possible. Simplifying packaging does not necessarily mean reducing the material at will. It also should ensure the product protection and transportation.

（5）Design for green manufacturing process　Green manufacturing is a modern manufacturing mode considering synthetically environmental impact and resource efficiency. The goal is to make the product in the product life cycle from the design, manufacturing, packaging, transport, use to disposal have the minimal impact on the environment, the maximum resource utilization, and make the economic benefit and social benefit of enterprise coordination and optimization.

3.7.3　Green Machining

1. Introduction to Green Machining

Under the precondition of ensuring the quality, cost and reliability, function and energy utilization of the products, green machining refers to the machining process which can make full use of resources and relive the harmful effects on the environment. Its connotation is to realize the high quality, low consumption, high efficiency and clean production in the machining process. Green machining can be divided into two types of machining technologies: one is the resource-saving type, the other is the environment-protecting type.

Considering from saving resources, green machining technology is mainly used in three aspects, i. e. the less chip or chipless machining, dry machining and new unconventional machining. In machining, green machining generally refers to dry cutting and dry grinding. Cutting fluids have to be used in conventional machining such as turning, milling, grinding and so on. Dry machining can obtain the clean and pollution-free chip, thus saving large amount of money paid on the cutting fluid and its disposal.

2. Key Technologies of Green Machining

Whether the green machining can be achieved mainly depends on three factors, that is, material selection, machining methods and machining equipment. Here the technical issues of machining equipment are discussed.

（1）Cutting tool technology　Dry machining requires that the cutting tool's material should have good red hardness, good wear resistance, heat shock resistance and anti-adhesion. The geometrical parameters and structure design of the dry cutting tool should meet the requirements for chip breaking and chip removal. Especially, the chip breaking should be solved when machining ductile material. At present, the 3D design technology for the chip breaker of lathe cutter has been relatively mature, the chip breaker can be rapidly designed for different workpiece materials and cutting variables, and the chip breaking ability and the ability to control chip flow direction can be improved greatly. The development of tool materials makes the blade bear higher temperature, thus reducing the requirement for lubrication. Vacuum or jet system can improve the chip removal condition. The manufacturing of complex tool can solve the problem on chip removal in the enclosure space.

中，对环境的影响最小，资源利用率最高，并使企业经济效益和社会效益协调优化。

3.7.3 绿色加工

1. 绿色加工概述

绿色加工（Green Machining，GM）是指在保证产品质量、成本、可靠性、功能和能量利用率的前提下，充分利用资源，尽量减轻加工过程对环境产生有害影响。其内涵是指在加工过程中实现优质、低耗、高效及清洁化。绿色加工可分为两种加工技术：一种是资源节约型的，另一种是环保型的。

从节约资源方面考虑，绿色加工技术主要应用在少无切屑加工、干式加工、新型特种加工3个方面。在机械加工中，绿色加工主要是指干切削和干磨削。常规的加工方法，如车削、铣削、磨削等，都采用切削液。干式加工可获得洁净无污染的切屑，从而节省切削液及其处理的大量费用。

2. 绿色加工的关键技术

能否实现绿色加工主要取决于选材、加工方法、加工设备等3个方面的因素。这里着种讨论加工设备方面的技术问题。

（1）刀具技术　干式加工要求刀具材料具有很高的红硬性、良好的耐磨性、耐热冲击和抗黏结性。刀具的几何参数和结构设计要满足干切削对断屑和排屑的要求。加工韧性材料时尤其要解决好断屑问题。目前车刀三维曲面断屑槽的设计技术已经比较成熟，可针对不同的工件材料和切削用量很快设计出相应的断屑槽，并能大大提高切屑折断能力和对切屑流动方向的控制能力。刀具材料的发展使刀片可承受更高的温度，从而降低了对润滑的要求。真空或喷气系统可以改善排屑条件。复杂刀具的制造可解决封闭空间的排屑问题。

（2）机床技术　干加工对机床提出了更高的要求。干式加工在切削区域会产生大量的切削热。如果不及时散热，会使机床受热不均而产生热变形，从而影响工件的加工精度。因此机床应配置循环冷却系统，并在机床结构上采取良好的隔热措施。干切削时产生的切屑是干燥的，因此应尽可能将干切削机床设计成立式或倾斜式床身，以利于切屑的快速排出。

（3）辅助设备技术　辅助设备主要包括夹具、量具等。辅助设备的绿色技术主要是指在选用辅助设备时，尽量满足低成本、低能耗、少污染、可回收的原则。

3. 干切削技术

在切削（含磨削）过程中，不使用冷却润滑液或使用极少量的冷却润滑 MQL（小于 50 ml/加工小时），且加工质量和加工时间与湿式切削相当或更好的切削技术称为干式切削。只有当所有工序都实现干切削后，才称为实用化的干式切削。然而，在干切削条件下，切削液在加工中的冷却、润滑、冲洗、防锈等作用将不复存在。如何在没有切削液的条件下创造与湿切削相同或近似的切削条件，这就要求人们去研究干式切削机理，从刀具技术、机床结构、工件材料和工艺过程等方面采取一系列的措施。

（1）干切削刀具　如前所述，干切削刀具是实施干切削的重要条件之一。分别从刀具材料、刀具涂层和刀具几何形状3个方面进行分析。

1）采用新型的刀具材料。目前用于干式切削的刀具材料主要有超细硬质合金、陶瓷、立方氮化硼（CBN）和聚晶金刚石（PCD）等超硬度材料。超细硬质合金比普通硬质合金具有更好的韧性、耐磨性和红硬性，可制作大前角的深孔钻头和刀片，用于干式铣削和干式

（2）Machine tool technology   The dry machining puts forward higher requirements for the machine tool. Dry machining would produce a large amount of cutting heat in the cutting area. If the cutting heat is not dissipated in time, it would cause machine tool to arise thermal deformation because of uneven heating, thus influencing on the machining accuracy of the workpiece. Therefore, the machine tool should be equipped with circulating cooling system, and good heat insulating measures are taken on the structure of machine tools. As the chips produced in dry cutting is dry, the machine tool with vertical or inclined bed should be used as far as possible in order to facilitate the rapid discharge of chips.

（3）Auxiliary equipment technology   Auxiliary equipment mainly includes fixtures, measuring tools and so on. Green technology of auxiliary equipment means mainly that the selection of the auxiliary equipment should follow such principles as the low cost, low energy consumption, less pollution, recyclable as far as possible.

3. Dry Cutting Technology

The cutting/grinding technology which doesn't use or uses minimum quantity liquid (MQL, less than 50 ml/h) and whose machining quality and machining time are equivalent to or even better than the wet cutting is called dry cutting. Only when the dry cutting is performed in all working operations can the cutting be named as practical dry cutting.

However, under the dry cutting conditions, the cooling, lubrication, washing and anti-rust and other functions of cutting fluid will no longer exist in the machining process. How to create cutting conditions similar to wet cutting under the conditions of not use cutting fluid requires people to research the dry cutting mechanism and to take a series of measures from cutting tool technology, machine tool structure, workpiece material, and technological process and other aspects.

（1）Dry cutting tool   As mentioned above, dry cutting tool is one of the important conditions to carry out dry cutting. The following three aspects—cutting tool material, cutting tool coating and cutting tool geometry are analyzed respectively.

1）To use new tool materials. Currently, the cutting tool materials used for dry cutting are ultrafine cemented carbide, ceramics, cubic boron nitride (CBN) and polycrystalline diamond (PCD) and other super-hard materials. The ultrafine cemented carbides have better toughness, wear resistance and red hardness than ordinary cemented carbides, which can be used for making the deep hole drills and blades with large rake angle and for dry milling and dry drilling operations. Ceramic tool has good red hardness and very suits for general dry cutting. However, due to its brittle properties, it is not suitable for discontinuous cutting like dry milling. The hardness of CBN is very high, next to diamond, and has good chemical stability at a high temperature. Using CBN tool for cutting cast iron can greatly improve the cutting speed; for cutting the hardened steel in turning can obtain the machining quality similar to grinding. PCD tool has very high hardness and good thermal conductivity and is suitable for dry machining copper, aluminum and its alloy. But PCD tool is easy to produce carbonization when the cutting temperature is higher than 700℃.

2）Tool coating technology. The coating treatment of cutting tools is an important way to improve the performance of the cutting tools. The coating of cutting tools is divided into two catego-

钻削。陶瓷刀具材料具有很好的红硬性，很适合于一般的干切削。但由于其性能较脆，不适合于干铣削等断续切削。立方氮化硼的硬度很高，仅次于金刚石，且具有良好的高温化学稳定性。采用 CBN 刀具加工铸铁，可大大提高切削速度；用于加工淬火钢，可以"以车代磨"。PCD 刀具具有非常高的硬度和热导率，比较适用于干式加工铜、铝及其合金工件。但在切削温度高于 700℃ 时易出现炭化现象。

2）采用刀具涂层技术。对刀具进行涂层处理，是提高刀具性能的重要途径。涂层刀具分两大类：一类是"硬"涂层刀具，如 TiN、TiC 和 Al_2O_3 等涂层刀具。这类刀具表面硬度高，耐磨性好。其中 TiC 涂层刀具抗后刀面磨损的能力特别强，而 TiN 涂层刀具则有较高的抗"月牙洼"磨损能力。另一类是"软"涂层刀具，如 MoS_2、WS 等涂层刀具。这类涂层刀具也称为"自润滑刀具"，因为它与工件材料的摩擦系数很低，从而能有效减少切削力和降低切削温度。例如，瑞士开发的"MOVIC"涂层丝锥，刀具表面涂覆一层 MoS_2。切削实验表明未涂层丝锥只能加工 20 个螺孔，用 TiAlN 涂层丝锥时可加工 1000 个螺孔，而用 MoS_2 涂层的丝锥可加工 4000 个螺孔。高速钢和硬质合金刀具经过物理气相沉积（Physical Vapor Deposition，PVD）涂层处理后，可以用于干切削。原来只适用于进行铸铁干切削的 CBN 刀具，在经过涂层处理后也可用来加工钢、铝合金和其他超硬合金。

从机理上讲，涂层的功能类似于切削液的功能。它产生一层保护层，把刀具与切削热隔离开来，从而能在较长的时间内保持刀刃的坚硬和锋利。表面光滑的涂层还有助于减小摩擦，从而减少切削热，保护刀具材料不受化学反应的作用。TiAlN 涂层和 MoS_2 软涂层还可交替涂覆，形成一种硬度高、耐磨性好和摩擦系数小得多的涂层刀具。

目前已开发出的纳米涂层（Nanocoating）刀具。纳米涂层采用多种材料的不同组合（如金属/金属组合、金属/陶瓷组合、陶瓷/陶瓷组合、固体润滑剂/金属组合等），以满足不同的要求。纳米涂层可使刀具的硬度和韧性显著增加。纳米涂层刀具具有优异的抗摩擦、磨损及自润滑性能。

3）刀具几何形状设计。干切削刀具常以"月牙洼"磨损为主要失效原因，月牙洼是由于加工中没有切削液，刀具和切屑接触区域的温度升高所致。因此，通常应使刀具有较大的前角和刃倾角。但前角增大会影响刀刃强度，此时应配以适宜的负倒棱。这种刀具结构可使刀尖和刃口具有足够的材料以承受切削力，减轻冲击和月牙洼扩展对刀具的不利影响。

（2）干切削加工工艺 在高速干切削方面，美国 Makino 公司提出"红月牙"（Red Crescent）干切工艺。其切削机理是高的切削速度使切削区附近工件材料达到红热状态，导致工件材料的屈服强度明显下降，从而提高材料去除率。实现"红月牙"干切工艺的关键在于刀具。目前主要采用 PCBN 和陶瓷等刀具来实现干切工艺。如用 PCBN 刀具干车削铸铁盘时切削速度可达 1000 m/min。当然，选用什么刀具材料还要视工件材料而定。虽然上述 PCBN 很适合进行高速干切削，但主要是对高硬度黑色金属和表面热喷涂的硬质材料进行干切削。金刚石刀具与铁元素有很强的化学亲和力，故不能用来加工黑色金属。

干切削通常是在大气氛围中进行，但在特殊气体氛围中（如氮气、冷风或干式静电冷却）而不使用切削液进行的切削也取得了良好的效果。

1）在氮气氛围中进行干切削——吹氮加工。吹氮加工使用的氮气可借助氮气生成装置除去空气中的氧、水分和 CO_2 而获得，然后经由喷嘴吹向切削区。由于氮气是不可燃气体，吹氮加工很适合加工易燃的镁合金。更重要的是氮气氛围还可抑制刀具的氧化磨损，保护刀

ries. One is "hard" coated cutting tools, such as the cutting tools coated with TiN, TiC and Al_2O_3. This kind of cutting tools has high surface hardness and good wear resistance. The cutting tools coated with TiC have particularly strong ability to resist the flank wear, while the cutting tools coated with TiN have higher ability to resist the "crater" wear. The other is "soft" coated cutting tools, such as the cutting tools coated with MoS_2, WS. This type of coated cutting tools is also known as "self lubricating cutter", since it has very low friction coefficient to the workpiece material. Therefore, it can effectively reduce the cutting force and reduce the cutting temperature. Take the "MOVIC" coating tap developed by Swiss for example, the surface of the tap coated with a layer of MoS_2. Cutting experiments show that the tap without coating can be used to machine 20 screw hole only; the tap coated with TiAlN is able to machine 1, 000 screw holes; while the tap coated with MoS_2 is able to machine 4, 000 screw holes. Through physical vapor deposition (PVD) coating, the cutting tools made of HSS and carbide can be used in dry cutting. Originally, the CBN cutting tool is suited only for dry cutting cast iron, after coating treatment, it can also be used for cutting steels, aluminum alloys and other super-hard alloys.

From the mechanism, the function of coating is similar to that of cutting fluid. The coating produces a protective layer which isolates the cutting tool from the cutting heat, thus keeping the cutting edge hard and sharp for a longer period of time. Smooth coating surface also helps to reduce friction and cutting heat so as to protect the tool material from chemical reaction. TiAlN and MoS_2 can also be coated alternately, thus forming a multi-layer cutting tool with high hardness, good wear resistance and small friction coefficient.

At present, the nano-coating cutting tools have been developed. Nano-coating adopts different combinations with variety of materials (such as metal/ceramics combination, ceramic/metal combination, metal/ceramic combination, solid lubricant/metal combination, etc.) in order to meet the different requirements. Nano-coating can significantly increase the hardness and toughness of the cutting tools. Nano-coating cutting tools have the excellent wear resistance and self-lubricating performance.

3) Geometrical design of cutting tool. Usually, the "crater" abrasion is the main reason for the failure of dry cutting tools. The crater is caused by the high temperature in the contact area between the tool and chip because of no cutting fluid in machining. Therefore, the cutting tool should has a larger rake angle and cutting edge inclination angle. But the increase of the rake angle will affect the strength of cutting edge. Therefore, the cutting tools with large rake angle usually have the first face land with appropriate negative rake angle. This tool structure makes the tool nose and cutting edge have enough volume of material to bear the cutting force and to reduce the adverse effects of impact and crater extension on the cutting tools.

(2) Dry cutting technology In high speed dry cutting, Makino company in the United States puts forward to a dry cutting technology called "Red Crescent". Its cutting mechanism is that high cutting speed makes the workpiece material near the cutting zone get to hot and red state, which causes the yield strength of the workpiece to decrease obviously, thus improving the material removal rate. The key to achieve "Red Crescent" lies in the cutting tool. At present, the cutting tools made of PCBN or ceramics are mainly used for dry cutting. For example, when the PCBN tool is used for dry

具涂层和防止切屑粘连到刀具,从而提高刀具的寿命。

2)干式静电冷却技术。该技术的基本原理是通过电离器将压缩空气离子化和臭氧化,然后经由喷嘴送至切削区,在切削点周围形成特殊气体氛围。这种冷却技术不仅能有效降低切削区的温度,更重要的是能在刀具与切屑、刀具与工件接触面上形成具有润滑作用的氧化薄膜。大量切削试验表明,在多数情况下,采用干式静电冷却技术的刀具寿命与湿切削时相当或超过;在少数情况下,也能达到湿切削时刀具寿命的80%~90%。

3)冷风干切削。冷风干切削的工作原理是:让低温冷风射流机生成的干燥低温冷风(-30~$-50℃$,有时也混入极微量的植物油)喷射到切削点,对刀具的前、后刀面实施冷却,同时引发被加工材料的低温脆性,使切削过程较为容易,并相应改善刀具磨损状况。冷风切削系统主要由空气压缩机、低温冷风射流机、微量油雾化器、喷射器、刀具等组成。

4)高速干切削。高速切削具有切削效率高、切削力小、加工精度高、加工过程稳定以及可以加工各种难加工材料等特点。随着高速机床技术的不断发展,切削速度和切削功率急剧提高,使得金属切除率大大增加。同时,切削液的用量也越来越大。但高速切削时切削液实际上很难到达切削区。也就是说,大量的切削液根本起不到实际的冷却作用。这不仅增加了制造成本,还加重了切削液对资源、环境等方面的负面影响。

高速干切削技术是将高速切削技术与干切削技术有机地融合而形成的一项新兴先进制造技术。切削技术、刀具材料和刀具设计技术的发展,使高速干切削的实施成为可能。采用高速干切削技术可以获得高效率、高精度、高柔性,同时又限制使用切削液,消除了切削液带来的负面影响。因此高速干切削技术是符合可持续发展要求的绿色制造技术。

复习题与习题

3-1 什么是先进制造工艺技术?其主要内容是什么?

3-2 何谓超精密加工技术?根据加工特点,它可分为哪几类?

3-3 超精密加工的关键技术有哪些?

3-4 超精密加工对机床设备和环境有何要求?

3-5 超精密切削刀具应具备哪些主要条件?

3-6 超精密磨削一般采用什么类型的砂轮?这些砂轮又如何修整?

3-7 与常规机电系统相比,微机电系统(MEMS)具有哪些特点?

3-8 何谓微细加工?目前有哪些微细加工方法?

3-9 微细加工与一般尺度的加工主要有哪些不同?

3-10 为何要实施高速切削加工?

3-11 实施高速切削加工的关键技术有哪些?

3-12 为什么要采用特种加工?特种加工有何特点?

3-13 现代特种加工技术主要有哪些?各适应什么场合?

3-14 试述快速原型制造技术的基本原理。其有何工艺特点?

3-15 快速原型制造技术有何工艺特点?

3-16 试述层合实体制造(LOM)的工艺原理。它有何特点?

3-17 试述选择性激光烧结(SLS)的工艺原理。它有何特点?

3-18 选择性激光熔接(SLM)与选择性激光烧结相比有何特点?

3-19 绿色制造的定义与内涵是什么?简述绿色制造的特点。

turning the cast iron plate, the cutting speed can reach to 1,000 m/min. Of course, what kind of tool material is used depends on the material to be cut. Although the PCBN is very suitable for high speed dry cutting, it is mainly used for cutting the ferrous metals with high hardness and the hard materials with thermal spraying. The diamond tools have a strong chemical affinity, therefore, it can't be used for cutting ferrous metals.

Dry cutting is usually carried out in the air atmosphere. But the cutting operations performed in the special gas atmosphere (such as nitrogen, cooling air or dry electrostatic cooling) without cutting fluids have also achieved good effect.

1) Dry cutting in nitrogen atmosphere—blowing nitrogen machining. In blowing nitrogen machining, the nitrogen is generated by means of nitrogen generating device by which the oxygen, water and CO_2 in the air are removed from the air, and then is blown through the nozzle to the cutting zone. As nitrogen is not flammable, blowing nitrogen machining is very suitable for machining the flammable magnesium alloy. It is more important that the nitrogen atmosphere can also inhibit the oxidation wear of cutting tool, protect coating and prevent the chip from adhesion to the cutting tool, thus improving the tool life.

2) Dry electrostatic cooling technology. The basic principle of this technology is that the compressed air is ionized and ozonized by the ionizer. Then the air is sent through the nozzle to the cutting zone and forms a special gas atmosphere around the cutting point. This cooling technology cannot only reduce the temperature of cutting area, but more important is that it can form the oxide film which plays the role of lubrication on the contact surfaces between the tool and chip, the tool and workpiece. A large number of cutting experiments show that in most cases, the tool life under dry electrostatic cooling condition equals to or exceeds that in wet cutting; in a small number of cases, it can also reach 80%-90% of the tool life in wet cutting.

3) Cold air dry cutting. The working principle of cold air cutting is that the dry cold air (−30 ℃ to −50 ℃, sometimes mixed with minimum quantity of vegetable oil) generated by the cryogenic cold air jet machine is sprayed into the cutting point to cool the rake face and the flank of cutting tool. Simultaneously, the cold air would cause the low temperature brittleness of the material to be cut, make the cutting easy, and correspondingly improve the tool wear condition. The cold air cutting system is mainly composed of air compressor, low temperature cold air jet machine, minimum quantity oil atomizer, injector, cutting tool and so on.

4) High-speed dry cutting. High-speed cutting has many characteristics, such as high cutting efficiency, small cutting force, high machining accuracy, cutting heat concentration, stable cutting process, and the ability to cut various difficult-to-cut materials, etc. With the continuous development of high-speed machine tool technology, the rapid increase of cutting speed and cutting power makes the metal removal rate increase greatly. At the same time, the amount of cutting fluid is also increasing. In fact, the cutting fluid is very difficult to enter the cutting zone in high-speed cutting. That is to say, a large amount of cutting fluid does not exert the actual cooling effect. This increases not only the manufacturing cost, but also the negative impact of cutting fluid on resources, environment and other aspects.

3-20 试从污染控制方面简述两种绿色制造方法。

3-21 绿色制造涉及的三个问题、三项内容各是什么？

3-22 绿色制造的目标是什么？

3-23 绿色设计包含哪些主要内容？绿色设计的基本内涵是什么？

3-24 何谓绿色材料？在绿色设计中选择材料的原则是什么？

3-25 绿色加工的关键技术有哪些？

3-26 干式切削的关键技术有哪些？

High-speed dry cutting technology is a new advanced manufacturing technology formed by the organic integration of high-speed cutting technology with dry cutting technology. The development of cutting technology, tool material and the design technology of cutting tool made it possible to implement high-speed dry cutting. The use of high-speed dry cutting technology can obtain high efficiency, high precision, and high flexibility. At the same time, it can restrict the use of cutting fluid, thus eliminating the negative impact of the cutting fluid. Therefore, high-speed dry cutting technology is a green manufacturing technology met the requirements of the sustainable development.

Review Questions and Problems

3-1　What is advanced manufacturing technology? What are its main contents?

3-2　What is ultra-precision machining technology? According to the machining characteristics, which categories can it be divided into?

3-3　What are the key technologies of ultra-precision machining?

3-4　What are the requirements of ultra-precision machining for machine tools, equipment and environment?

3-5　What main conditions should the ultra-precision cutting tools have?

3-6　What type of grinding wheel is used in ultra-precision grinding? How are these wheels dressed?

3-7　Compared with conventional electro-mechanical systems, what are the characteristics of micro electro-mechanical systems (MEMS)?

3-8　What is micro machining? What micro machining methods are there at present?

3-9　What are the main differences between micro machining and general machining?

3-10　Why do we implement high-speed machining?

3-11　What are the key technologies in the implementation of high-speed machining?

3-12　Why do we need unconventional machining? What characteristics does the unconventional machining have?

3-13　What are the major modern unconventional machining technologies? What situations do they adapt?

3-14　Try to state the basic principle of rapid prototyping technology. What technological characteristics does it have?

3-15　What technological characteristics does the rapid prototyping & manufacturing technology have?

3-16　Try to state the technological principle of laminated solid manufacturing (LOM). What characteristics does it have?

3-17　Try to state the technological principle of selective laser sintering (SLS). What characteristics does it have?

3-18　Compared with selective laser sintering, what characteristics does the selective laser melting (SLM) have?

3-19　What is the definition and connotation of green manufacturing? Briefly describe the features of green manufacturing.

3-20　Briefly describe two green manufacturing methods from pollution control.

3-21　What are the three issues and the three contents involved in green manufacturing?

3-22　What is the goal of green manufacturing?

3-23　What main elements does the green design include? What is the basic connotation of green design?

3-24　What is green material? What are the principles of choosing materials in green design?

3-25　What are the key technologies of green machining?

3-26　What are the key technologies of dry cutting?

Chapter 4   Manufacturing Automation Technology

第4章　制造自动化技术

4.1 Introduction to Manufacturing Automation

Manufacturing automation is the goal that people pursue in the long-term production activities. The manufacturing automation technology not only can significantly improve productivity and product quality, reduce manufacturing cost, improve economic efficiency, but also can effectively improve working conditions, greatly enhance the enterprise competitive power in the international market. Manufacturing automation technology is an important symbol of the development of manufacturing industry, it represents the level of advanced manufacturing technology and reflects the level of national science and technology.

4.1.1 Connotation of Manufacturing Automation

"Automation" was proposed by Mr. D. S. Harder of GM in 1936. Its core means "automatic completion of a specific job". At that time, the specific job that Mr. Harder said refers to the automatic handling between the machines. The functional objective of automation is to replace the physical labor.

In narrow sense, manufacturing automation means the automation of product machining and assembly inspection process in production workshop, including automatic machining, automatic loading/unloading of parts, automatic part handling, automatic cleaning and inspection of parts and products, automatic chip breaking and removal, assembly automation, automatic machine fault diagnosis and so on. In broad sense, manufacturing automation means the automation of whole product manufacturing process as well as the integrated automation of each link, such as product design automation, enterprise management automation, machining process automation and quality control automation and so on.

The goal of manufacturing automation (TQCSE) is to improve the response speed of manufacturing enterprises to the rapidly changing market and the competitiveness of manufacturing industry. the development of manufacturing automation will be flexible, integrated, agile, intelligent, and globalization to meet the requirements of rapid market changes. Based on the national conditions, the development of manufacturing automation in China uses the appropriate automation technology with human-computer combination, the higher degree of automation equipment (such as CNC machine tools, industrial robots etc.) is integrated effectively with the low level of automation equipment. On this basis, taking the human as the center and the computer as the important tool, the automated manufacturing system with the characteristics of flexibility, intelligence, integration, quick response and fast reorganization can be achieved.

4.1.2 Development of Manufacturing Automation Technology

1. Development of Manufacturing Automation Technology

The development of manufacturing automation has experienced a long process (see Tab. 4-1). Reviewing the history, the development process of manufacturing automation can be divided into three stages: rigid automation, flexible automation and integrated automation.

4.1 制造自动化概述

制造自动化是人类在长期的生产活动中不断追求的目标。采用制造自动化技术不仅可以显著提高劳动生产率、大幅度提高产品质量、降低制造成本、提高经济效益，还可有效地改善劳动条件，大大提高企业在国际市场的竞争力。制造自动化技术是制造业发展的重要标志，它代表着先进制造技术的水平，也体现了一个国家科技水平的高低。

4.1.1 制造自动化的内涵

"Automation"一词是美国通用汽车公司 D. S. 哈德先生于 1936 年提出来的，其核心含义是"自动地去完成特定的作业"。当时哈德先生所说的特定作业，是指零件在机器之间的自动搬运。自动化功能的目标是代替人的体力劳动。

在"狭义制造"概念下，制造自动化的含义是生产车间内产品的机械加工和装配检验过程的自动化，包括切削加工自动化、工件装卸自动化、工件运储自动化、零件与产品清洁及检验自动化、断屑与排屑自动化、装配自动化、机器故障诊断自动化等。而在"广义制造"概念下，制造自动化则包含了产品设计自动化、企业管理自动化、加工过程自动化和质量控制自动化等产品制造全过程以及各个环节综合集成自动化。

制造自动化的目标（TQCSE）更主要是提高制造企业对瞬息万变的市场的响应速度，提高制造业的竞争能力。制造自动化的发展将以其柔性化、集成化、敏捷化、智能化、全球化的特征来满足市场快速变化的要求。立足国情，我国制造自动化的发展采用人机结合的适度自动化技术，将自动化程度较高的设备（如数控机床、工业机器人）和自动化程度较低的设备有效地集成起来。在此基础上，可以实现以人为中心，以计算机为重要工具，具有柔性化、智能化、集成化、快速响应和快速重组的制造自动化系统。

4.1.2 制造自动化技术的发展

1. 制造自动化技术的发展历程

制造自动化的发展经历了一个漫长的发展过程（见表4-1）。回顾历史，可将制造自动化的发展历程分为刚性自动化、柔性自动化和综合自动化 3 个发展阶段。

（1）刚性自动化 半自动和自动机床、组合机床、组合机床自动线的问世，解决了单一品种大批量生产自动化问题。其主要特点是生产效率高、加工品种单一。这个阶段于 20世纪 50 年代达到了顶峰。

（2）柔性自动化 为满足多品种、小批量甚至单件生产自动化的需要，出现了一系列柔性制造自动化技术，如数控技术（NC）、计算机数控（CNC）、柔性制造单元（FMC）、柔性制造系统（FMS）等。

（3）综合自动化 随着计算机及其应用技术的迅速发展，各项单元自动化技术逐渐成熟。为充分利用资源，发挥综合效益，自 20 世纪 80 年代以来以计算机为中心的综合自动化得到了发展，如计算机集成制造系统（CIMS）、并行工程（CE）、精益生产（LP）、敏捷制造（AM）等先进制造模式得到了发展和应用。

（1）Rigid automation   The emergence of the semi-automatic and automatic machine tool, modular machine tool and transfer line composed of modular machine tool solves the automation problem in large volume production of single variety. This stage had reached its peak in 1950s.

（2）Flexible automation   In order to meet the needs of the automation of many varieties, small batch and even job production, a series of flexible manufacturing automation technologies, such as NC technology, computer numerical control（CNC）, flexible manufacturing cell（FMC）, flexible manufacturing system（FMS）and so on have emerged.

（3）Integrated automation   With the rapid development of computer application technology, the unit automation technologies become gradually perfect. In order to make full use of resources and produce integrated benefits, the comprehensive automation based on computers has been developed since 1980s, such as computer integrated manufacturing system（CIMS）, concurrent engineering（CE）, lean production（LP）, agile manufacturing（AM）and other advanced manufacturing modes have been developed and applied.

2. Development Trend of Manufacturing Automation Technology

The development trend of manufacturing automation technology is manufacturing agility, manufacturing network, manufacturing virtualization, manufacturing intelligence, manufacturing globalization and green manufacturing.

（1）Manufacturing agility   Agility is the inevitable trend of manufacturing environment and manufacturing process for manufacturing activities in 21th century. Agility of manufacturing environment and manufacturing process includes three aspects：①flexibility, such as flexible mechanical equipment, flexible process, and operation system flexibility；②reconstruction ability, such as to realize the rapid reorganization of the system, to constitute dynamic alliance；③rapid integrated manufacturing technology, such as rapid prototyping manufacturing（RPM）technology.

（2）Manufacturing network   Manufacturing network, especially Internet/Intranet based manufacturing has become an important development trend. The network inside manufacturing environment can realize the integration of the manufacturing process. The network of manufacturing environment and the whole manufacturing enterprises can realize the integration of the manufacturing environment and engineering design, management information system and other subsystems in the manufacturing environment and the manufacturing enterprise. The network of enterprise to enterprise can realize the sharing, combination and optimal utilization of enterprise resources. Remote manufacturing can be implemented through network.

（3）Manufacturing virtualization   It mainly means the virtual manufacturing. Integrating modern manufacturing technology, computer graphics, parallel engineering, artificial intelligence, virtual reality technology, multimedia technology and many other high technologies as a whole. Virtual manufacturing is an integrated system technology formed by multidisciplinary knowledge based on the manufacturing technology and system modeling and simulation technology supported by and computer technology. By establishing the system model, the real manufacturing environment and its manufacturing process are presented in the virtual environment that supported by the computer and related technology. The real manufacturing environment and all activities of its manufacturing process and

Tab. 4-1 Development of manufacturing automation 制造自动化发展历程

Time 时间	Some important events in the development of manufacturing automation 制造自动化发展中的一些重大事件
1900	Electro-hydraulic profiling machine（Italy）电液仿形机床（意大利）
1913	Progressive assembly line（America Ford）流水装配线（美国福特）
1920	Robot（Czechoslovakia）机器人（捷克斯洛伐克）
1924	Automatic production line（England）自动生产线（英国）
1936	Automation（America）自动化（美国）
1947	Remote machine hand（America）遥控机械手（美国）
1952	3-axis NC vertical milling machine（America MIT）三轴数控立式铣床（美国 MIT）
1958	Automatic programming tool（America）自动编程系统（美国）
1958	Machine center（America）加工中心（美国）
1959	Industrial robot（Polar coordinates）（America）工业机器人（极坐标型）（美国）
1960	Adaptive control milling machine（America）自适应控制铣床（美国）
1962	Industrial robot（Cylindrical coordinates）（America）工业机器人（圆柱坐标型）（美国）
1965	Direct numerical control（DNC）（America）计算机直接数字控制（美国）
1967	CAD/CAM software（America）CAD/CAM 软件（美国）
1970	FMS patent（England）FMS 专利（英国）
1973	Computer integrated manufacturing（America Harrington）计算机集成制造（美国哈林顿）
1980	Manufacturing automation protocol（America）制造自动化协议（MAP）（美国）
1989	Lean production（Japan）精益生产（日本）
1991	Research on intelligent manufacturing system（Japan，America，European Community）智能制造系统 IMS 研究（日本、美国、欧共体）
1991	Agile manufacturing，virtual manufacturing（America）敏捷制造，虚拟制造（美国）
1994	Advanced manufacturing technology program（America）先进制造技术计划（美国）
2012	Industrial Internet（America）工业互联网（美国）
2013	Industrial 4.0（Germany）工业 4.0（德国）
2015	Made in China 2025 中国制造 2025

2. 制造自动化技术的发展趋势

制造自动化技术的发展趋势主要是制造敏捷化、制造网络化、制造虚拟化、制造智能化、制造全球化和制造绿色化。

（1）制造敏捷化 敏捷化是制造环境和制造过程面向 21 世纪制造活动的必然趋势。制造环境和制造过程的敏捷化包括 3 个方面的内容：①柔性，如柔性机械装备、柔性工艺过程、系统运行柔性等；②重构能力，如能实现系统的快速重组，组成动态联盟；③快速化的集成制造工艺，如快速原型制造 RPM 技术。

（2）制造网络化 制造的网络化，特别是基于互联网或内联网的制造已成为重要的发展趋势。制造环境内部的网络化可以实现制造过程的集成；制造环境与整个制造企业的网络化可以实现制造环境与企业中工程设计、管理信息系统等各子系统的集成；企业与企业间的网络化可以实现企业间的资源共享、组合与优化利用。通过网络，实现异地制造。

（3）制造虚拟化 制造虚拟化主要是指虚拟制造。它是以制造技术和计算机技术支持的系统建模技术和仿真技术为基础，集现代制造工艺、计算机图形学、并行工程、人工智能、虚拟现实技术和多媒体技术等多种高新技术为一体，由多学科知识形成的一种综合系统技术。它将现实制造环境及其制造过程通过建立系统模型呈现到计算机及其相关技术所支撑的虚拟环境中，在虚拟环境下模拟现实制造环境及其制造过程的一切活动和产品制造全过

product manufacturing process are simulated in the virtual environment. The product manufacturing and behavior of manufacturing system are predicted and evaluated. Virtual manufacturing is a key technology to realize agile manufacturing.

(4) Manufacturing intelligence   Intelligence is the further development and extension of the manufacturing system on the basis of flexibility and integration. GE of America put forward the industrial Internet in 2012 and payed attention to the influence of software, network, big data etc. on industrial automation. Germany in 2013 proposed "Industry 4. 0" with the purpose to realize "intelligent factory" based on the cyber physical system (CPS). In 2015, China proposed "Made in China 2025", emphasizing the integration of industrialization and information. What they have in common is that the real world is closely connected through the Internet. The advanced computing power in net space is applied effectively to the real world, thus, all data in the manufacturing process relevant to design, development and production will be collected by the sensor and analyzed, forming the intelligent production system with self disciplined operation. At present, the intelligent complete manufacturing system, also called holonic manufacturing system (HMS) is put forward in the intelligent manufacturing system (IMS). HMS is made up of intelligent complete units, and its bottom equipment has the characteristics of openness, self-discipline, cooperation, flexibility, knowable, easy to integration and good robustness.

(5) Manufacturing globalization   The development and implementation of intelligent manufacturing system planning and agile manufacturing strategy promote the globalization of manufacturing industry. With the appearance of "global network" and "global market", "global competition" and "global management", the research and application of globalization manufacturing are developed rapidly, which includes mainly the following contents: The global network for product sale is being formed; the international cooperation in the product design and development and transnational product manufacturing; the reorganization and integration of manufacturing enterprises in the world, such as the dynamic alliance; trans-region and trans-nation coordination, sharing and optimized utilization of manufacturing resources and so on.

4. 2   Modern CNC Machining Technology

Numerical control technology means a kind of automation technology used to control the equipment operation and machining process by means of digital signal. The numerical control technology is the foundation of manufacturing process automation, and the core of the automatic flexible system, and the important part of the modern integrated manufacturing system. NC machining means the part machining process performed automatically on NC machine tool in light of the prepared program under the control of CNC system. NC technology has improved the function, efficiency, reliability of the mechanical equipment and product quality to a new level, thus making the traditional manufacturing industry undergo profound changes.

4. 2. 1   Introduction to CNC Machine Tool

1. Definition of CNC Machine Tool

Numerical control machine tools is a kind of machine tool using computer to control automatical-

程，并对产品制造及制造系统的行为进行预测和评价。虚拟制造是实现敏捷制造的关键技术。

（4）制造智能化　智能化是制造系统在柔性化和集成化基础上进一步的发展和延伸。当前，智能制造已成为各制造大国的研究热点。美国 GE 公司于 2012 年提出了工业互联网，注重软件、网络、大数据等对于工业自动化的影响。德国 2013 年提出工业 4.0，基于"信息物理融合系统（Cyber Physical System，CPS）"实现"智能工厂"。我国于 2015 年提出"中国制造 2025"，强调工业化和信息化的融合。它们的共同之处在于通过互联网紧密连接现实世界，将网络空间的高级计算能力有效运用于现实世界中，从而在生产制造过程中，与设计、开发、生产有关的所有数据将通过传感器采集并进行分析，形成可自律操作的智能生产系统。目前，在智能制造系统（Intelligent Manufacturing System，IMS）计划中提出了智能完备制造系统，也称全能制造系统（Holonic Manufacturing System，HMS）。HMS 是由智能完备单元复合而成，其底层设备具有开放、自律、合作、适应柔性、可知、易集成和鲁棒性好等特性。

（5）制造全球化　智能制造系统计划和敏捷制造战略的发展和实施，促进制造业的全球化。随着"网络全球化""市场全球化""竞争全球化""经营全球化"的出现，全球化制造的研究和应用发展迅速，包括以下主要内容：产品销售的全球网络正在形成；产品设计和开发的国际合作及产品制造的跨国化；制造企业在世界范围内的重组与集成，如动态联盟；制造资源的跨地区、跨国家的协调、共享和优化利用等。

4.2　现代数控加工技术

数控技术是指用数字化信号对设备运行及其加工过程进行控制的一种自动化技术。数控技术是实现制造过程自动化的基础，是自动化柔性系统的核心，也是现代集成制造系统的重要组成部分。数控加工是指数控机床在数控系统的控制下，自动地按预先编制的程序对机械零件进行加工的过程。数控技术把机械装备的功能、效率、可靠性和产品质量提高到一个新水平，使传统的制造业发生了深刻的变化。

4.2.1　数控机床概述

1. 数控机床的定义

数控机床是用计算机通过数字信息来自动控制机械加工的机床。数控机床可以通过编制程序，即数字（代码）指令来自动完成机床各个坐标的协调运动，正确地控制机床运动部件的位移量，并且按加工的动作顺序要求自动控制机床各个部件的动作（如主轴转速、进给速度、换刀、工件夹紧放松、工件交换、切削液开关等）。它是集计算机应用技术、自动控制、精密测量、微电子技术、机械加工技术于一体的一种具有高效率、高精度、高柔性和高自动化的数控设备。

2. 数控机床的组成

数控机床的基本组成包括加工程序、输入/输出装置、数控装置、伺服系统、辅助控制装置、反馈系统及机床本体。图 4-1 所示为一般数控机床的组成。

（1）程序载体　程序载体目前是指软盘、磁盘和 U 盘；

ly machining by means of the digital information. By means of part programming, i. e. digital (code) instructions, the CNC machine tool can complete automatically the coordinating motions of various coordinates, correctly control the displacement of the machine moving parts, and control automatically the actions of each machine component (such as RPM of the spindle, feed rate, tool changing, clamping and loosening of the workpiece, part exchange, switching on/off coolant, etc.) in light of the requirements of the machining sequence. Numerical control machine tools is a kind of numerical control equipment with high efficiency, high precision, high flexibility and high automation, which integrates the computer application technology, automatic control, precision measurement, microelectronics technology with machining technology.

2. Elements of CNC Machine Tool

The basic elements of CNC machine tools include machining programs, I/O devices, numerical control devices, servo systems, auxiliary control devices, feedback systems and machine tools. The elements of a general CNC machine tool is shown in Fig. 4-1.

(1) Program carrier   Currently, it refers to the floppy disk, disk and U disk.

(2) Input device   For the modern CNC machine tools, the part program is input by manual mode, DNC network communication and RS232 serial communication.

(3) CNC system   CNC system is the core of machine tool to carry out automatic machining and mainly composed of the operating system, the main control system, programmable controller, various types of I/O interface and other components. The main functions of CNC system are as follows: multi-coordinate control and interpolation of many functions; to input, edit and modify a variety of programs; information conversion; compensation; selection of machining methods; display; self-diagnose; communication and networking.

(4) Servo system   Servo system is the implementation part of CNC system, and is mainly composed of the servo motor, drive control system and position detection feedback device and so on. Servo system, the executive components and mechanical transmission components of the machine tool constitute the feed system of CNC machine tool. Servo system is used to control the feed rate, direction and displacement of the executive components based on the speed and displacement commands sent by CNC device.

(5) Main drive system   It is one of the main components used for transmitting during cutting operation. It is mainly composed of spindle drive control system, spindle motor and mechanical transmission mechanisms.

(6) Strong electricity control device   The strong electricity control device is the control system between the NC device and the mechanical and hydraulic components of the machine tool. It is mainly composed of various intermediate relays, contactors, transformers, power switches, terminals and various types of electrical protection components. Its main roles are to receive such commands as changing the speed of main motion, tool selection and exchange, auxiliary device action and other signals sent by NC device, and through necessary compilation, logic judgment and the power amplifying, to drive directly the corresponding electrical, hydraulic, pneumatic and mechanical components so as to complete the actions specified by the commands.

（2）输入装置　输入装置主要用于输入数控加工程序。对于现代数控机床，数控加工程序可以通过 MDI 方式、DNC 网络通信、RS232 串口通信输入。

（3）数控系统　数控系统是机床实现自动加工的核心，主要由操作系统、主控制系统、可编程控制器、各类 I/O 接口等组成。其主要功能有：多坐标控制和多种函数的插补，多种程序的输入、编辑和修改，信息转换，补偿，加工方法的选择，显示，自诊断，通信和联网。

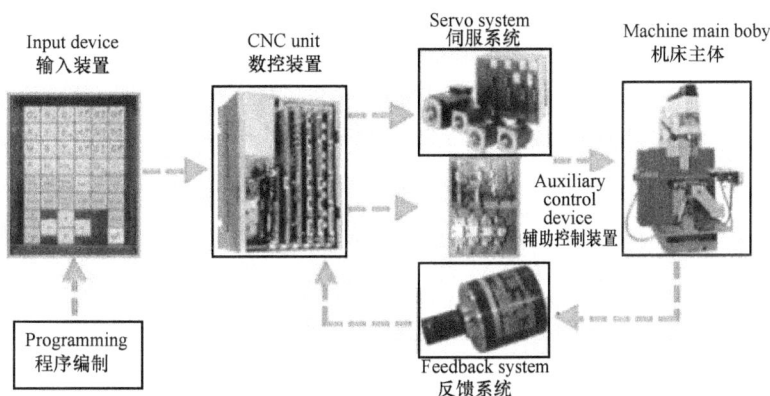

Fig. 4-1　Elements of CNC machine tool 数控机床的组成

（4）伺服系统　伺服系统是数控系统的执行部分，主要由伺服电动机、驱动控制系统及位置检测反馈装置等组成。它与机床上的执行部件和机械传动部件组成数控机床的进给系统。它根据数控装置发来的速度和位移指令控制执行部件的进给速度、方向和位移。

（5）主传动系统　主传动系统是机床切削加工时传递扭矩的主要部件之一。它主要由主轴驱动控制系统、主轴电动机及机械传动机构等组成。

（6）强电控制装置　强电控制装置是介于数控装置和机床的机械、液压部件之间的控制系统，主要由各种中间继电器、接触器、变压器、电源开关、接线端子和各类电气保护元器件等构成。其主要作用是接收数控装置发出的主运动变速、刀具选择交换、辅助装置动作等指令信号，经必要的编译、逻辑判断、功率放大后直接驱动相应的电器、液压、气动和机械部件，以完成指令所规定的动作。

（7）辅助装置　辅助装置主要包括刀具自动交换装置（Automatic Tool Changer，ATC）、工件自动交换装置（Automatic Pallet Changer，APC）、工件夹紧机构、回转工作台、液压控制系统、冷却润滑装置、排屑装置、过载与限位保护装置等。

（8）机床本体　机床本体是指数控机床机械结构实体。与普通机床相比具有如下特征：①采用高性能主传动及主轴部件；②进给传动采用高效传动件。一般采用滚珠丝杠副、直线滚动导轨副等；③具有较完善的刀具自动交换和管理系统；④具有工件自动交换和工件夹紧机构；⑤床身机架具有很高的动、静刚度。

4.2.2　五轴联动数控机床

长期以来，大型精密的高档五轴联动数控机床一直被一些工业发达国家视为重要的战略

（7）Auxiliary devices　Auxiliary devices include mainly automatic tool changer（ATC）, automatic pallet changer（APC）, workpiece clamping mechanism, rotary table, hydraulic control system, coolant and lubrication devices, chip removal device, overload and limiting position protection devices and so on.

（8）Machine body　It refers to the mechanical structure entity of CNC machine tool. Compared with the ordinary machine tools, CNC machine tool has the following characteristics：①Main drive and spindle components with high performance are used；②high efficiency transmission parts are used in feed drive system. Generally, the ball screw pair and linear rolling guide way are used；③there is a more perfect automatic tool exchange and management system；④there are automatic workpiece exchange mechanism and clamping mechanism；⑤bed frame has a high dynamic and static rigidity.

4.2.2　5-axis Linkage CNC Machine Tool

For a long time, large, precise and high-level 5-axis linkage CNC machine tools have been regarded by some industrialized countries as the important strategic materials. "Toshiba event" happened in the late 1980s when Toshiba company of Japan exported illegally the large 5-axis CNC milling machine to the Soviet Union shows that the 5-axis CNC machine tool has a significant influence on a country' s aerospace, military, scientific research, precision instruments and other industries, and on the rapid development of the national economy.

The movement of the numerical control axes which can simultaneously participate in the interpolation is called linkage. The number of linkage axes is usually used to measure the ability of a CNC machine tool to machine surfaces. 5-axis linkage refers to the control of any two rotary axes of A, B, C coordinate axis in addition to controlling the three linear axes linkage. That is, the movements of 5 axes can be controlled at the same time. 5-axis linkage CNC machine tool is a high-tech, high precision machine tool used specially for machining complex surfaces.

1. Structural Types of 5-axis Linkage CNC Machine Tools

5-axis CNC machine tools have three linear motion axes（i. e. X axis, Y axis and Z axis）, in addition, there are also at least two rotary axes to achieve rotary feed motions. In order to machine curve surfaces, the spindle head and worktable can be multi-axis linkage to achieve continuous rotary feed motion. 5-axis linkage CNC machine tools have various structural layout. From the view of kinematic design, suppose the transmission chain starts from the workpiece to the cutting tool, the linear motion is represented by L, and the rotary motion is represented by R. Then, 5-axis machining centers with three linear axes and two rotary axes have seven kinds of motion combinations：RRLLL, LRRLL, LLRRL, LLLRR, RLRLL, RLLRL and RLLLR. The most common motion combinations are LLLRR, RRLLL and RLLLR. The three motion combinations and their typical configurations are shown in Fig. 4-2.

Fig. 4-2a shows a gantry machining center with moving girder. Its motion combination is LLLRR. The workpiece is mounted on a stationary worktable. The cross girder moves along the guideways（Y axis）on the top of the columns located on the left and right sides of the machine tool. The

物资。20世纪80年代末日本东芝公司向苏联出口大型五轴联动数控机床的"东芝事件"表明，五轴联动数控机床对一个国家的航空航天、军事、科研、精密器械等行业，对国民经济的迅速发展都有着举足轻重的影响力。

把可以同时参与插补的数控轴移动称为联动。联动轴数的多少通常用来衡量数控机床对曲面加工的能力。五轴联动是指除了控制 X、Y、Z 3个直线坐标轴联动外，还同时控制围绕这3个直线坐标轴旋转的 A、B、C 坐标轴中的两个坐标轴，即形成同时控制5个轴联动。五轴联动数控机床是一种科技含量高、精密度高，专门用于加工复杂曲面的机床。

1. 五轴联动数控机床的结构类型

五轴联动数控机床除了具有 X、Y 和 Z 这3个直线运动坐标轴外，还有至少两个旋转坐标轴实现旋转进给运动。为了进行曲面加工，主轴头和工作台可以多轴联动实现连续回转进给运动。五轴联动数控机床的结构配置多种多样。从运动设计的角度，假定传动链从工件开始到刀具，直线运动以 L 表示，回转运动以 R 表示，具有3个移动轴和2个回转轴的五轴加工中心的运动组合共有7种：即 RRLLL、LRRLL、LLRRL、LLLRR、RLRLL、RLLRL 和 RLLLR。其中最常见的运动组合有 LLLRR、RRLLL 和 RLLLR。这3种运动组合和配置形式如图4-2所示。

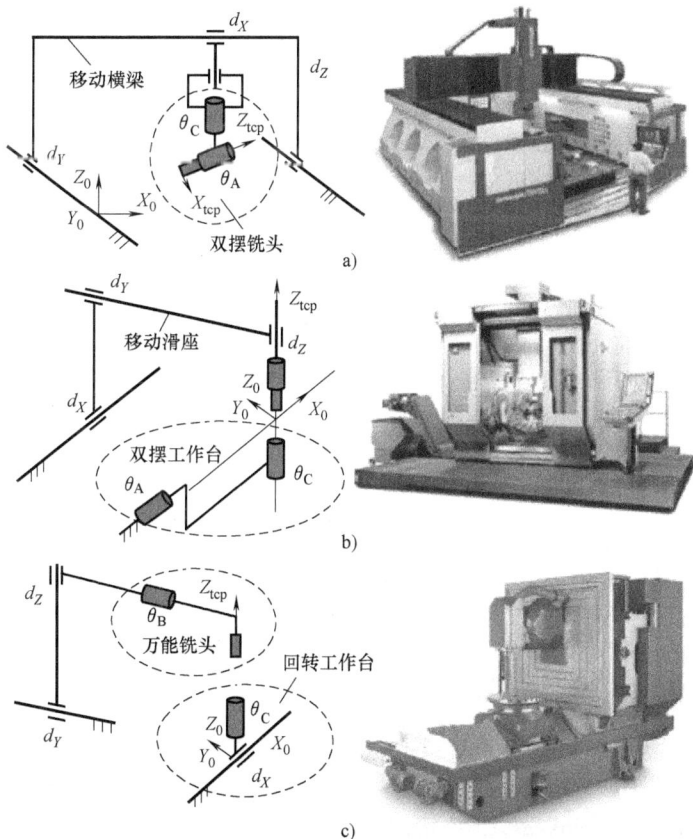

Fig. 4-2　Motion combination and layout of 5-axis linkage CNC machine tool

五轴联动数控机床的运动组合和配置形式

a) LLLRR　b) RRLLL　c) RLLLR

spindle slide moves along cross girder (*X* axis), the spindle ram moves up and down (*Z* axis). The milling head can wiggle in two directions to realize the rotations of *A* axis and *C* axis.

Fig. 4-2b shows a vertical machining center. Its motion combination is RRLLL. The workpiece is fixed on the worktable wiggled around *A* axis and *C* axis. The cross girder moves along the columns (*X* axis) on both left and right sides, the spindle slide moves along the *Y* axis, the spindle ram moves up and down along the *Z* axis.

Fig. 4-2c is also a vertical machining center, but its motion combination is RLLLR. The workpiece is fixed on the *C*-axis rotary table, and the table moves along the *X* axis. The spindle slide moves along the *Y* and *Z* axes, and the universal milling head can be rotated in the *B* axis.

2. Advantages of 5-axis CNC Machine Tools

Compared with other machine tools, 5-axis CNC machine tools have many advantages in machining. For example, using 5-axis linkage to machine parts can reduce the number of the fixture. In addition, the use of 5-axis linkage CNC machine tool can save many special cutting tools in the machining process, thus reducing the cost of cutting tool. The 5-axis CNC machine tool can increase the length of the effective cutting edge, which is beneficial for reducing the cutting force, increasing the tool life and reducing the tool cost. The specific advantages are listed in detail in Tab. 4-2.

Tab. 4-2   Advantages of 5-axis linkage CNC machine tool 五轴联动数控机床的优势

Wide applications 更广泛的应用范围	As the position (angle) of the cutting tool relative to the workpiece can be adjusted at any time in the machining of 5-axis CNC machine tool, the machining interference of tool can be avoided. Therefore, the 5-axis linkage CNC machine tool can complete many complex machining tasks which cannot be processed by 3-axis linkage CNC machine tool. 由于五轴联动数控机床在加工过程中刀具相对于工件的位置(角度)可以随时调整,避免了刀具的加工干涉,因此五轴联动数控机床可以完成三轴联动机床不能完成的许多复杂的加工
Better machining quality 更好的加工质量	The 5-axis linkage CNC machine tool can machine the workpiece with complex cavity in one time mounting. And the pose angle of cutting tool of 5-axis CNC machine tool can be adjusted at any time, the machine tool can machine the workpiece at the better angle, which can avoid multiple clamping and greatly improve the machining efficiency and machining quality. 五轴联动数控机床可以在一次安装中完成复杂型腔零件的加工。并且,由于五轴联动数控机床在加工时可以随时调整刀具的位姿角,因此就可以以更好的角度加工工件,这避免了多次装夹,大大提高了加工效率和加工质量
Higher machining efficiency 更高的加工效率	In the traditional machining on 3-axis CNC machine tool, much time is used in the workpiece handling, loading/unloading of the workpiece, setting and so on. The 5-axis linkage CN machine tool can complete the machining tasks which can be completed by several 3-axis CNC machines, thus greatly saving occupied space, material handling time and cost among different machining units, and significantly improving the working efficiency. 在传统三轴数控机床加工过程中,大量的时间被消耗在搬运工件、上下料、安装调整等上面。五轴数控机床可以完成数台三轴数控机床才能完成的加工任务,这大大节省了占地空间和工件在不同加工单元之间运送的时间和花费,显著提升了工作效率

3. Introduction to Several Advanced 5-axis Linkage CNC Machine Tools

(1) Hermle C52 machining center   Fig. 4-3 shows a C52 machining center made in Hermle, Germany. This is a 5-axis linkage CNC machine tool with a cradle type double swing table. The machine tool is suitable for aerospace, mold manufacturing, energy and semiconductor and other industries.

The cradle type double swing table is longitudinal layout, and is driven by servo motor and

图 4-2a 所示为动梁式龙门加工中心，其运动组合采用 LLLRR。工件安装在固定工作台上不动，横梁沿位于左右两侧立柱顶部的导轨移动（Y 轴），主轴滑座沿横梁运动（X 轴），主轴滑枕上下移动（Z 轴）。铣头可在两个方向上摆动以实现 A 轴和 C 轴的转动。

图 4-2b 所示为立式加工中心，其运动组合采用 RRLLL。工件固定在 A 轴和 C 轴双摆工作台上，横梁沿左右两侧立柱移动（X 轴），主轴滑座沿 Y 轴移动，主轴滑枕沿 Z 轴上下移动。

图 4-2c 所示也为立式加工中心，但其运动组合采用 RLLLR。工件固定在 C 轴回转工作台上，工作台沿 X 轴移动。主轴滑座沿 Y 轴和 Z 轴移动，万能铣头可作 B 轴回转。

2. 五轴联动数控机床的优势

与其他机床相比，五轴联动数控机床在零件加工方面具有很多优势。例如，采用五轴联动加工零件可以减少夹具的数量。另外，五轴联动数控机床可在加工中省去许多特殊刀具，从而降低刀具成本。五轴联动数控机床在加工中能增加刀具的有效切削刃长度，减小切削力，提高刀具使用寿命，降低刀具成本。具体优势详见表 4-2。

3. 几种先进的五轴联动数控机床简介

（1）哈默 C52 加工中心　图 4-3 所示为德国哈默（Hermle）公司的 C52 加工中心的总体结构。这是一种摇篮式双摆工作台五轴联动数控加工机床，适用于航空航天、模具制造、能源和半导体等工业。

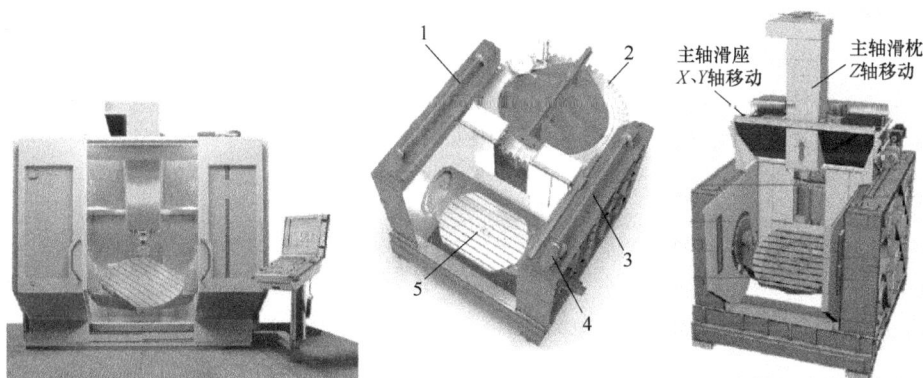

Fig. 4-3　Overall configuration of Hermle C52 machining center 哈默 C52 加工中心的总体结构
1—Linear guide 线性导轨　2—Tool magazine 刀库　3—Ball screw 滚珠丝杠
4—Servo drive 伺服驱动　5—Double swing table 双摆工作台

摇篮式双摆工作台呈纵向布局，采用伺服电动机和无背隙齿轮传动，摆动范围为 +100°/−130°，可以进行五轴联动的立/卧式车削加工或铣削加工。为了提高机床的动态性能，机床移动部件采用轻量化设计。框架式主轴十字滑座在台式床身的顶部，完成 X、Y 方向移动。主轴滑枕做 Z 轴垂直方向移动。该加工中心采用盘式刀库。

由图 4-3 可见，主轴下层的滑鞍由安装在床身两侧壁上的伺服电动机和滚珠丝杠驱动沿三根线性导轨移动，以实现重心驱动，这样可避免移动过程的偏斜，从而提高机床的工作精度。

（2）西田 YMC430-Ⅱ 精密加工中心　日本西田 YMC430-Ⅱ 精密加工中心主要用于加工微小的高精度零件，在结构设计上特别注意提高刚度和减少热变形的影响，其总体结构如

gears without backlash. The swing range of the table is $+100°/-130°$. The vertical/horizontal turning operations or milling operations with 5-axis linkage can be done on the machining center. The moving parts of the machine tool are designed in lightweight in order to improve the dynamics of the machine tool. The cross slide of the frame-type spindle located on the top of bench-type bed can move in the X, Y direction. The spindle ram moves along the Z axis in the vertical direction. The machining center adopts the disc-type tool magazine.

It can be seen from Fig. 4-3 that the slide located in the lower part of the spindle is driven by servo motor and ball screw mounted on the two sides of the bed to move along the three linear guideways to realize the centrobaric drive, thus preventing it from the deflection in the moving process and improving the working accuracy of the machine tool.

(2) YMC430-II precision machining center of Yasda   The YMC430-II precision machining center of Yasda is mainly used for machining the small parts with higher accuracy. The special attention is paid to improve the rigidity and to reduce the influence of thermal deformation in the structural design. The overall layout is shown in Fig. 4-4.

It can be seen from Fig. 4-4 that the integral gantry structure with the H shaped section, symmetrical structure in 4 directions and large section factor can ensure the machine tool structure with high rigidity, high precision and thermal stability. The spindle slide and the spindle unit are arranged in front of H shaped column and equipped with weight balance system in order to ensure the moving accuracy along Z axis. In addition, the spindle unit with symmetrical structure can significantly reduce the offset of tool center point relative the worktable caused by thermal deformation.

The X, Y and Z axes of YMC430-II all are driven by the linear motor, and the linear guideways with high rigidity and high precision are used in order to improve the moving precision and rigidity, simplify mechanical structure, avoid reverse backlash, and ensure the dynamic performance of the machine tool.

4.2.3   Parallel Kinematics Machine Tool

1. Overview of Parallel Kinematic Machine Tool

Parallel kinematic machine tool (PMT), also called virtual axis machine tool, is the product combined the parallel robot technology with the modern CNC machine tool technology. PMT has many characteristics of machine tool and robot. It is the new mechanical and electrical equipment which has not only the agility and flexibility of the robot, but also the rigidity and precision of the machine tool.

In 1994, the Giddings & Lewis of America launched the Variax machining center in International Manufacturing Technology Exhibition (IMTS 94) held in Chicago, as shown in Fig. 4-5.

The supporting parts, such as the bed, guideway, column and cross girder in the traditional machine tool can't be found from the Variax machining center. The chief characteristic of the machining center is that the triangular frames are used to replace the traditional bed and column, and the 6 "legs" which can stretch out and draw back are used to support and connect the upper platform (equipped with the spindles) and the lower platform (equipped with worktable). Each "leg"

图 4-4 所示。

由图 4-4 可见，H 形横截面、左右前后 4 个方向都对称和断面系数大的整体龙门式结构，保证了机床结构的高刚度、高精度和热稳定性。主轴滑座和主轴部件位于 H 形立柱的正前方，具有重量平衡系统，以保证 Z 轴移动的精度。加上结构对称的主轴部件，可显著减少热变形所引起的刀具中心点相对工作台的偏移量。

西田 YMC430-Ⅱ 的 X、Y、Z 轴皆由直线电动机驱动，采用高刚度和高精度的线性导轨，以提高移动精度和刚性，简化机械结构，避免反向间隙，保证机床的动态性能。

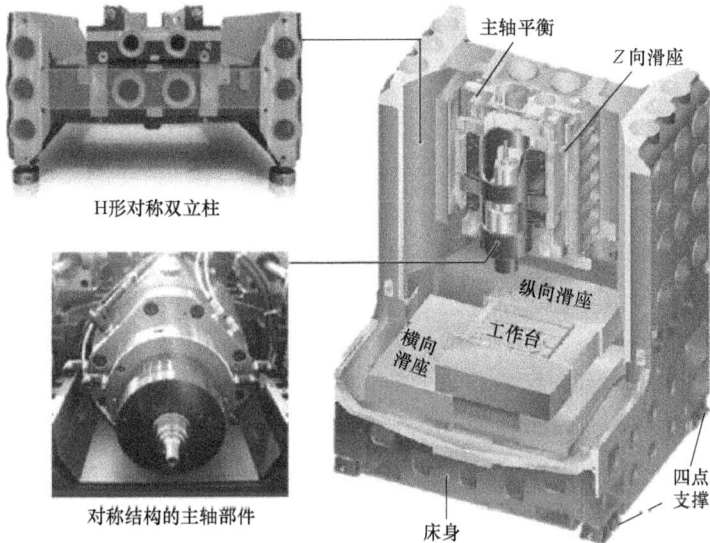

Fig. 4-4　Overall structure of YMC430-Ⅱ precision machining center
西田 YMC430-Ⅱ 精密加工中心的总体结构

4.2.3　并联运动机床

1. 并联运动机床概述

并联机床（Parallel Kinematic Machine tool，PMT）又称虚（拟）轴机床，是并联机器人技术和现代数控机床技术结合的产物。它同时兼顾了机床和机器人的诸多特性，既具有机器人的灵活与柔性，又具有机床的刚度和精度，是集多种功能于一体的新型机电设备。

1994 年，美国 Giddings&Lewis 公司在美国芝加哥国际制造技术博览会上（IMTS 94）推出了 Variax 加工中心，如图 4-5 所示。

在 Variax 加工中心上，根本看不到如传统机床上的床身、导轨、立柱和横梁等支撑部件。其结构的最大特点是采用三角形构架结构取代了传统的床身、立柱等，可以伸缩的 6 条"腿"支撑并连接上平台（装有主轴头）与下平台（装有工作台）。每条"腿"均由各自的伺服电动机与滚珠丝杠驱动。6 条"腿"的伸缩可使装有主轴头的上平台进行 6 坐标轴的运动，从而改变主轴与工件的相对空间位置，满足加工中刀具运动轨迹的要求。

Variax 加工中心的刚度比一般加工中心的刚度高 5 倍。它完全无悬臂结构，各构件只受拉力或压力而无弯曲力矩。因而无须像传统机床那样靠增加质量来提高刚性。6 条"腿"与

is driven respectively by the servomotor and ball screw, and the stretching out and drawing back of the 6 "legs" can move the upper platform along the 6 coordinate axes so as to change the spacial position of the spindle relative to the workpiece and to meet the requirement of the tool path in machining.

The rigidity of the Variax machining center is five times higher than that of the ordinary machining center. It has no cantilever structure, each component bears either tension or pressure only without bending moment. Therefore, it is no need to improve the rigidity by increasing the mass like the traditional machine tools. The 6 "legs" are connected to the upper and lower platforms through the universal joints. Each universal joint can rotates in two coordinate axes so as to avoid the bending moment transmitting to the legs.

2. Comparison of Parallel Kinematics Machine Tool with Traditional Machine Tool

The structure sketch of the parallel kinematics machine tool and the traditional machine tool is shown in Fig. 4-6. The main difference between them are as follows:

The basic characteristics of the traditional machine tool layout is that taking the bed, column, cross girder and so on as the support components, the spindle unit and worktable move along the straight guideways on the support components, and the cutting path of tool bit is formed based on the series kinematics principle with X, Y, Z coordinate motion superposition.

The basic characteristics of the parallel kinematics machine tool layout are that taking the frame of machine tool as a fixed platform, the spacial parallel mechanism is formed by a number of rods. The spindle unit is mounted on the moving platform of the parallel mechanism. By changing the rod length or moving the fulcrum of the rod, the cutting path of the tool bit can be formed based on the parallel kinematics principle.

Compared with traditional machine tool, parallel kinematic machine tool has the following advantages:

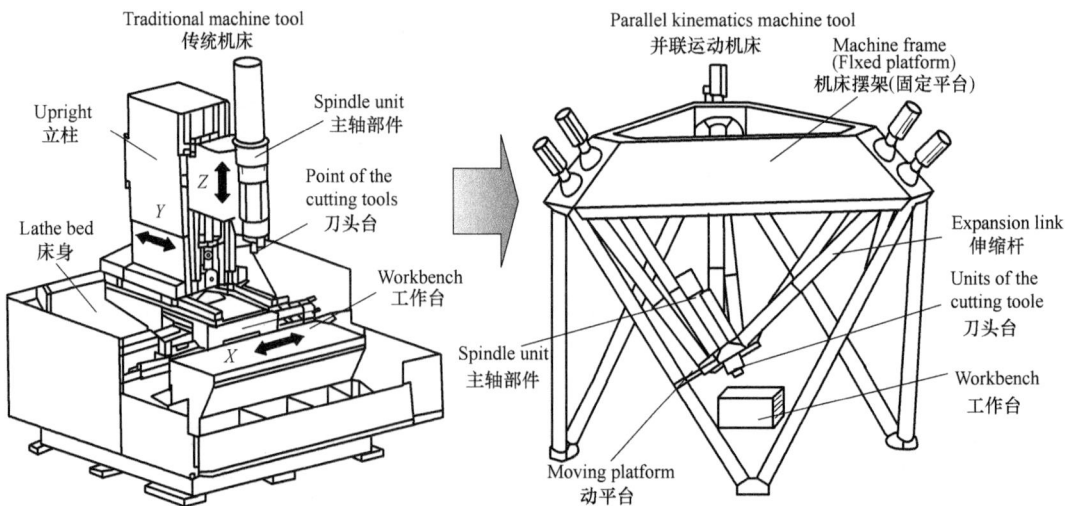

Fig. 4-6 Comparison of the PMT with traditional machine tool
并联运动机床与传统机床的比较

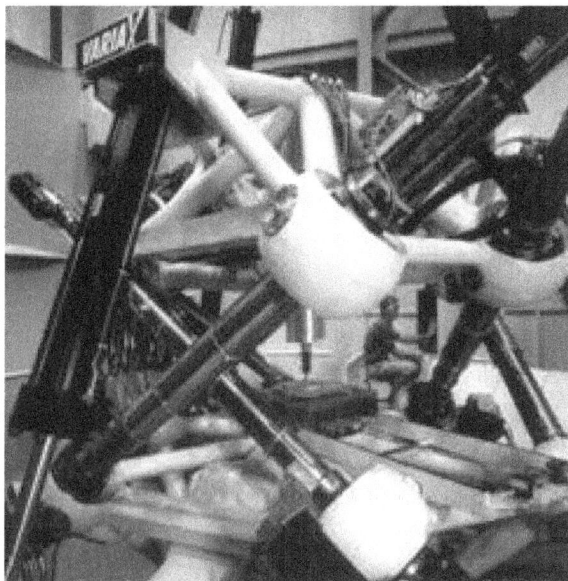

Fig. 4-5   Variax machining center   Variax 加工中心

上、下平台通过万向接头连接，每个万向接头可作两坐标轴的回转运动，以免使弯曲力矩传递到"腿"上。

2. 并联运动机床与传统机床的比较

图 4-6 所示为并联运动机床与传统机床的结构示意图。二者的主要区别表现如下：

传统机床布局的基本特点是以床身、立柱、横梁等作为支承部件，主轴部件和工作台沿支承部件上的直线导轨移动，按照 X、Y、Z 坐标运动叠加的串联运动学原理，形成刀头点的加工轨迹。

并联运动机床布局的基本特点是以机床框架为固定平台的若干杆件组成空间并联机构。主轴部件安装在并联机构的动平台上，改变杆件的长度或移动杆件的支点，按照并联运动学原理形成刀头点的加工轨迹。

与传统机床相比，并联运动机床具有以下优点：

1）刚度重量比大。因采用并联闭环静定或非静定杆系结构，且在准静态情况下，传动构件理论上为仅受拉压载荷的二力杆，故传动机构的单位重量具有很高的承载能力。

2）动态性能好。运动部件惯性的大幅度降低有效地改善了伺服控制器的动态品质，允许动平台获得很高的进给速度和加速度，因而特别适合高速加工。

3）机床结构简单，集成化、模块化程度高。

4）变换坐标系方便。由于没有实体坐标系，机床坐标系与工件坐标系的转换全部靠软件完成，非常方便。

5）使用寿命长。并联运动机床由于没有传统机床导轨，避免了导轨磨损、锈蚀、划伤等现象。

但并联运动机床也存在一些缺点。例如：并联运动机床的驱动杆多，互相牵制，导致工作空间小；每个驱动支路的关节较多，影响了整体刚度；驱动杆的反馈困难，运动精度难以保证。目前并联机床的加工精度还难以和传统高精度机床相媲美。

1) Higher rigidity-weight ratio. As the parallel close-loop structure with statically determinate or statically indeterminate rod systems is used, and under the quasi static condition, the transmission component is theoretically two-force rod subjected to tension and compression load only, the unit weight of the transmission mechanism has a high load capacity.

2) Good dynamic performance. The significant reduction in inertia of the moving parts effectively improves the dynamic quality of the servo controller, and allows the moving platform to achieve high feed rate and acceleration. Therefore, the parallel kinematic machine tools are particularly suitable for high-speed machining.

3) The machine tool has the simple structure, high integration and high modularity.

4) Coordinate transformation is simple. As there is no practical coordinate system, the conversion between the machine coordinate system and the workpiece coordinate system is completed entirely by software very conveniently.

5) Long service life. The parallel kinematic machine tool has no guideways of the traditional machine tool, therefore there is no guideway wear, corrosion, scratch and so on.

However, parallel kinematic machine tool also have some shortcomings. For example, many driving rods in the parallel kinematic machine tool would cause mutual restraint, resulting in a small work space; each driving branch has many joints, which would affect the overall rigidity; it is difficult to guarantee the kinematic accuracy because of the difficulty in the feedback of driving rods. Currently, the parallel kinematic machine tool is difficult to match with the traditional high-precision machine tool in the machining accuracy.

4.2.4　Open Architecture CNC System

1. Requirements of the Development of Modern Digital Equipment for the CNC System

As the traditional CNC system uses a dedicated computer hardware system and a dedicated software system, it is no way to use the latest computer technology in the CNC system, thus, leading to high development cost, poor upgrading capability and extendibility, which seriously restricted the application and development of the CNC technology. High speed, high efficiency, compound, precision, intelligence, environmental protection and so on are the development trend of digital equipment in the world, and network, intelligent, open CNC system are the important assurance to achieve high level of equipment. The core is open, that is, the independence of system modules and operating platform, the inter-operability between the modules of the system and the unity of man-machine interface and communication interface. The open architecture makes the CNC system have a better versatility, flexibility, adaptability and extendibility, and develop in the direction of intelligence and networking.

As mentioned above, the trajectory control of the parallel machine tool is completed by the spatial motion synthesis of several rods. Because of the diversification of its structure and configuration, it is difficult to have a control system meet the requirements of all parallel machine tools. The only way is to use a control platform on which the machine tool developers configure their own hardware and software. Therefore, the control of parallel machine tool requires a completely computer-based,

4.2.4 开放式 CNC 系统

1. 现代数字化装备的发展对数控系统的要求

由于传统的 CNC 系统均采用专用的计算机硬件系统和专用的软件系统，CNC 系统无法使用最新的计算机技术，造成开发费用高，升级能力和可扩展能力都比较差，从而严重制约了数控技术的应用和发展。高速、高效、复合、精密、智能、环保等是世界数字化装备的发展趋势，而网络化、智能化、开放式 CNC 系统是实现高水平装备的保证。其核心是开放式，即系统各模块与运行平台的无关性、系统中各模块之间的互操作性和人机界面及通信接口的统一性。开放式体系结构使数控系统有更好的通用性、柔性、适应性、扩展性，并向智能化、网络化方向发展。

如前所述，并联运动机床的轨迹控制是由若干杆件的空间运动综合完成的。由于其结构和配置形式的多样化，很难有一种控制系统能够适合所有并联运动机床的要求，只能用一种控制平台，由机床开发者自行配置硬件和软件。因此，对于并联运动机床的控制来说，需要一种完全以微机为基础的、和谐的、标准化的软件环境，从而能够根据用户需要实现复杂的控制功能。

2. 开放式 CNC 系统的概念与特征

目前，开放式 CNC 系统还没有一个统一明确的定义，但其含义应包括：符合系统规范的应用程序可运行在多个销售商的不同平台上，可与其他的系统应用程序实现互操作，并且具有一致风格的交互界面。

开放式 CNC 系统的主要特征表现在：①功能模块具有可移植性；②功能相似模块之间可互相替换，并具有可扩展性；③有即插即用功能；④使用标准 I/O 和网络功能，容易实现与其他自动化设备的互操作性。表 4-3 列出了传统专用数控系统与开放式数控系统的比较。

3. 基于 PC 的开放式 CNC 体系结构

（1）PC 嵌入 NC 中　即在传统的非开放式 CNC 上插入一块专门的、开放的个人计算机模板，使传统 CNC 具有 PC 的特性。在这种模式中，CNC 部分与原来的 CNC 一样进行实时控制，PC 承担非实时控制。这种结构改善了 CNC 系统的图形显示、切削仿真、编制和诊断功能，使 CNC 系统具有较好的开放性。典型产品有西门子 840C 数控系统和 FANUC-S16 数控系统等。

（2）NC 嵌入 PC 中　运动控制板或整个 CNC 单元（包括集成的 PLC）插入个人计算机的标准槽中。同样，PC 做非实时处理，实时控制由 CNC 单元或运动控制板承担。利用 PC 强大的 Windows 图形用户界面、多任务处理能力以及良好的软、硬件兼容能力，结合运动控制卡和运动控制软件形成高性能、高灵活性和开放性好的数控系统，从而使用户可以开发自身的应用程序。这种模式正成为以 PC 为基础的 CNC 系统的主流。采用这种结构的典型产品为美国 DeltaTau Data System 公司的 PMAC—NC，其中在 PC 中插入一块 PMAC（Programmable Multi-Axis Controller）可编程多轴运动控制器，PMAC 板执行全部实时任务，包括轮廓加工、插补运算、伺服控制、刀具半径补偿和螺距补偿等。这种模式的 CNC 可实现开放式结构，因而能满足机床制造商和最终用户的各种需求。

harmonious, standardized software environment so as to achieve complex control functions according to the user needs.

2. Concept and Characteristics of Open CNC System

At present, the open CNC system does not have a well-defined definition, but its meaning should include: The application programs conforming to the system specifications can run on the different platforms with multiple vendors, and carry out mutual operation with other system application programs, and have the interactive interface with a consistent style.

The main features of the open CNC system are as follows: ①functional modules have transferability; ②the modules with similar function can be interchangeable and have expansibility; ③with plug and play function; ④by means of standard I/O and network functions, the interoperability with other automation devices is achieved easily. The comparison of the traditional dedicated CNC system and open CNC system are listed in Tab. 4-3.

3. Open CNC Architecture Based on PC

(1) PC is embedded in NC   A special, open personal computer template is inserted onto the traditional non-open CNC so that the traditional CNC has the characteristics of PC. In this mode, the CNC part is the same as the original CNC to carry out real time control, and the PC bears non-real time control. This structure improves the graphical display, cutting simulation, compilation and diagnosis function of the NC system, and makes the CNC system have better openness. Typical products are Siemens 840C CNC system and FANUC-S16 CNC system and so on.

(2) NC is embedded in PC   The motion control board or the entire CNC unit (including the integrated PLC) is inserted into the standard slot of PC. Similarly, PC machine is used for non-real-time processing, real-time control is done by the CNC unit or motion control board. Combined PC's powerful windows graphical user interface, multi-task processing capability and good software and hardware compatibility with the motion control card and motion control software constitutes a CNC system with high performance, high flexibility and good openness. Thus, the users can develop their own application programs in the system. This model is becoming the mainstream of PC-based CNC systems. The typical product with this structure is the PMAC-NC of Delta Tau Data System of America, in which a PMAC (programmable multi-axis controller) is inserted in to PC. PMAC board implements all real-time tasks, including contour processing, interpolation calculation, servo control, tool radius compensation and pitch compensation. This kind of CNC can realize open structure, hence, it can meet the various needs of the machine tool manufacturers and the final users.

(3) Pure PC type (full software type)   This is the latest open CNC architecture. Its CNC software is all installed in the computer, the peripheral connection mainly uses related bus standard of the computer. The users can develop various required functions on the windows platform by means of the open CNC kernel to form a various high-performance CNC systems. Compared with the former two types, the software-based open CNC system has become an important trend in the development of numerical control system through software intelligence replacing complex hardware. Typical products are the Open CNC of MDSI Co. of America, the PA8000 NT of PA Co. of Germany and so on.

Tab. 4-3 Comparison of the traditional CNC system to the open CNC system
传统专用数控系统与开放式数控系统的比较

Comparison item 比较项目	Traditional dedicated CNC system 传统专用数控系统	Open CNC system 开放式数控系统
System structure and flexibility 系统结构及可伸缩性	Dedicated hardware 硬件专用 Dedicated software 软件专用 Poor flexibility 可伸缩性差	Hardware based on PC 基于PC的硬件 Software based on the general operating system 基于通用操作系统的软件 The system can be modified as needed 系统可根据需要进行伸缩
System maintainability 系统可维护性	Need to develop dedicated hardware, difficult to adapt to the increasingly intensive competition requirements 需要开发专用的硬件，难以适应日益激烈的竞争要求	Easy to upgrade to keep up with the development of PC technology 容易升级换代以跟上PC技术发展
Difficulty of software development 软件开发难易性	Development of dedicated software is very difficult 专用软件开发难度大	Written in C language, less development time 用C语言编写，开发时间短
Software transparency 软件的透明性	Software belongs to the CNC manufacturers exclusively. It is difficult for machine tool plants and the users to make secondary development 软件为CNC制造商独占，机床厂、用户厂难以进行二次开发	With open software platform, machine manufacturers and users can develop their own software 具有开放软件平台，机床制造商、用户可开发自己的软件
Development of special dedicated system 特殊专用系统开发	Difficult to develop the special, dedicated system 特殊、专用系统开发困难	With open software platform and C++ and other high-level language, easy to develop 使用开放软件平台和C++等高级语言，容易开发
Networking 联网性	Have to use dedicated hardware and dedicated communication technology (method), high networking costs 须用专用硬件和专用通信技术(方法)，联网成本高	Same as the PC networking technology, low networking cost 与PC联网技术相同，联网成本低
PLC software PLC软件	Have to use the manufacturer's language, difficult to transplant and maintain 须用制造商专用语言，难于移植和维修	With standard PLC, good portability and maintainability 使用符合标准的PLC，可移植性强，可维护性好
Interface 接口	Dedicated interface, only the manufacturer's products can be used 专用接口，只能使用制造商的产品	With the standardized interface, it is easy to connect all kinds of servo, stepper motors and spindle motors 使用标准化接口，容易与各类伺服、步进电机及主轴电机连接
Program capacity 程序容量	Usually, dedicated RAM has only 128KB, high expansion cost; need DNC for the large mold program 专用RAM通常只有128KB，扩容成本高，对大型模具程序需采用DNC	The general RAM with memory 4M or more can be expanded to 64MB and can be equipped with hard drive, which can call in the huge program one time 通用RAM，内存4M以上，可扩至64MB，可配置硬盘，一次性调入巨量程序

4. 3　Flexible Manufacturing Technology

4. 3. 1　Introduction to Flexible Manufacturing

Flexible manufacturing was born in the mid-1960s when the British firm Molins, Ltd. developed its System 24. The System 24 was a real flexible manufac turing system (FMS). However, it was doomed from the outset because automation, integration, and computer control technology had not yet been developed to the point where they could properly support the system. As such, it was eventually discard as unworkable.

Flexible manufacturing remained an academic concept through the remainder of the 1960s and 1970s. However, with the emergence of sophisticated computer control technology in the late 1970s and early 1980s, flexible manufacturing became a viable concept. The first major use of flexible manufacturing in the United States were manufacturers of automobiles, trucks, and tractors.

Flexible manufacturing means using the programmable, multi-function digital control equipment to replace the rigid automation equipment, and using the software which is easy to program, modify, expand, and change to replace the rigid-connection process so as to have the rigid production line realize the flexibility, which can response the market needs rapidly and complete the multi-species, small batch production tasks economically and efficiently. Flexible manufacturing technology is a kind of manufacturing automation technology. It is a kind of advanced, flexible, automated, efficient manufacturing technology by integrating the microelectronic technology, intelligent technology with traditional machining technology.

4. 3. 2　Flexible Manufacturing System

1. Definition and Components of Flexible Manufacturing System

In the modern manufacturing setting, flexibility is an important characteristic. It means that a manufacturing system is versatile and adaptable, while also capable of handling relatively high production run. A flexible manufacturing system is versatile in that it can produce a variety of parts. It is adaptable because it can be quickly modified to produce a completely different line of parts.

An FMS is an individual machine or group of machines served by an automated materials handling system that is computer controlled and has a tool handling capability. Because of its tool handling capability and computer control, such a system can be continually reconfigured to manufacture a wide variety of parts. This is why it is called a flexible manufacturing system.

Fig. 4-7 shows a typical FMS structural block. It consists of nine parts as follows:

1) Central management and control computer. It receives instructions from main computer of the factory and implements scheduling, operation control, material management, system monitoring and network communication for the entire FMS.

2) Logistics control device. Its functions are to implement the centralized management and control for automated warehouses, unmanned conveying vehicles, blanks, semi-finished and finished

（3）纯 PC 型（全软件型） 这是最新的开放式 CNC 体系结构。它的 CNC 软件全部装在计算机中，外围连接主要采用计算机的相关总线标准。用户可在 Windows 平台上，利用开放的 CNC 内核开发所需的各种功能，构成各种类型的高性能的数控系统。与前两种相比，全软件型开放式数控系统通过软件智能替代复杂的硬件，已成为数控系统发展的重要趋势。典型产品有美国 MDSI 公司的 Open CNC 和德国 PA 公司的 PA8000 NT 等。

4.3 柔性制造技术

4.3.1 柔性制造概述

20 世纪 60 年代中期，英国莫林斯公司研制出了自己的系统 24，由此出现了柔性制造（Flexible Manufacturing，FM）的新理念。系统 24 是一个真正的柔性制造系统（Flexible Manufacturing System，FMS）。然而，由于当时的自动化、集成以及计算机等技术尚未发展到能够完全支撑系统运行的程度，系统 24 从一开始就注定是要失败的。最终由于不可行而放弃。

柔性制造在 20 世纪 60 年代末和 70 年代间仍然是一个学术概念。然而，随着先进的计算机控制技术在 20 世纪 70 年代末和 80 年代初的相继问世，柔性制造成为一个可行的概念。柔性制造在美国的首要用途是汽车、货车和拖拉机的制造。

柔性制造是指用可编程、多功能的数字控制设备更换刚性自动化设备，用易编程、易修改、易扩展、易更换的软件控制代替刚性联结的工艺过程，使刚性生产线实现柔性化，能够快速响应市场的需求，经济高效地完成多品种、中小批量的生产任务。柔性制造技术（Flexible Manufacturing Technology，FMT）是一种主要用于多品种小批量或变批量生产的制造自动化技术。它是将微电子技术、智能化技术与传统加工技术融合在一起，具有先进性、柔性化、自动化、效率高的制造技术。

4.3.2 柔性制造系统

1. 柔性制造系统的定义和组成

在现代制造环境中，柔性是一个重要的特征。它意味着一个制造系统是多用途的和适应性强的，同时也具有较高的生产能力。之所以说柔性制造系统是多用途的，就在于它能够生产各种各样的零件。之所以说它的适应性强，就在于它在快速调整后能够生产完全不同的零件。

柔性制造系统是由配备一个计算机控制的自动物料运储系统的单台机床或一组机床构成的，并具有刀具运储能力。正是由于其刀具运储能力和计算机控制，这种系统可以进行不断调整以制造种类繁多的零件。这就是称它为柔性制造系统的缘故。

图 4-7 所示为一种较典型的 FMS 结构框图，它由以下 9 个部分组成：

1）中央管理和控制计算机。它接收来自工厂主计算机的指令，对整个 FMS 实行计划调度、运行控制、物料管理、系统监控和网络通信等。

2）物流控制装置。它对自动化仓库、无人输送台车、毛坯、半成品和成品、夹具、刀具等实行集中管理和控制。

products, fixtures, cutting tools and so on.

3) Automated warehouse. Its functions are to automatically call or store the blanks, semi-finished and finished products and so on.

4) Automated guided vehicle (AGV). It is used to transport the workpieces, cutting tools, fixtures and so on between the machine tools, machine tool and automated warehouse, machine tool and central tool magazine. It can be orbital or trackless.

5) Manufacturing cell. It consists of several different types of CNC machine tools and industrial robots. The CNC machine tool also includes machining center or FMC.

6) Central tool magazine. It is the centralized storage area of the tools.

7) Fixture station. Its function is to implement the setting, maintenance and storage of the fixtures.

8) Information transmission network. It is the communication system of FMS.

9) Pallet. It plays the role of transfer buffering from the AGV to the manufacturing unit.

A typical FMS layout is shown in Fig. 4-8.

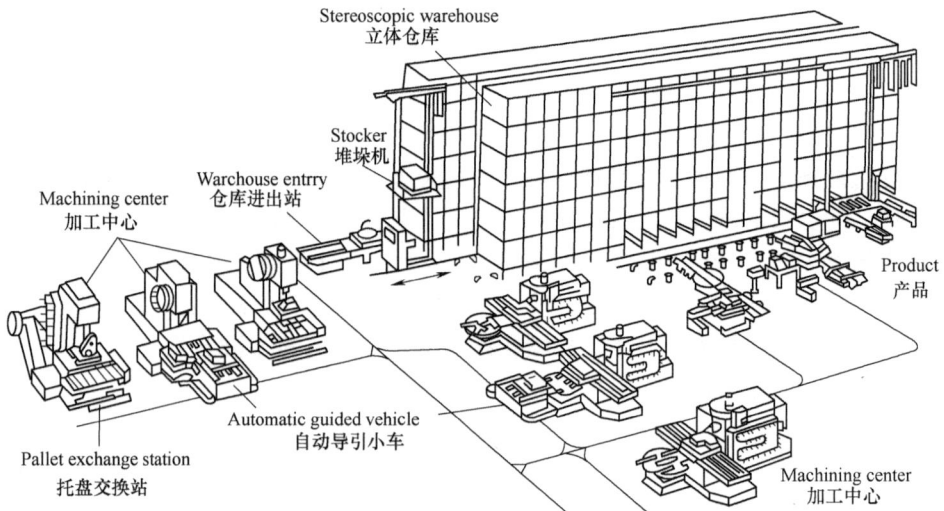

Fig. 4-8 Typical FMS layout 典型的 FMS 布局图

It should be pointed that humans and computer play major roles in an FMS. The amount of human labor is much less than with a manually operated manufacturing system, of course. However, humans still play a vital role in the operation of an FMS. Human tasks include:

1) Equipment troubleshooting, maintenance, and repair.

2) Tool changing and setup.

3) Loading and unloading the workpieces.

4) Data input.

5) Changing of parts programs.

6) Development of programs.

Flexible manufacturing system equipment, like all manufacturing equipment, must be monitored for bugs, malfunctions, and breakdowns. When a problem is discovered, a human trouble-

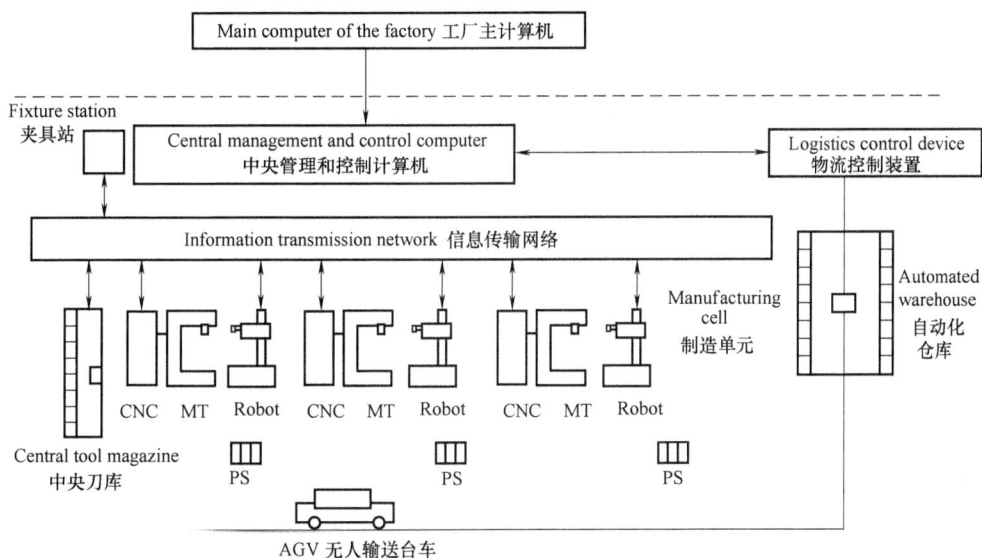

Fig. 4-7 FMS structural block FMS 结构框图

3）自动化仓库。它将毛坯、半成品和成品等进行自动调用或存储。

4）自动引导小车。它用来运输工件、刀具、夹具等，行走于各机床之间、机床与自动化仓库之间、机床与中央刀具库之间。自动引导小车可以是有轨的或无轨的。

5）制造单元。它由多台不同类型的 CNC 机床 MT 及工业机器人组成。其中 CNC 机床也包括加工中心 MC 或 FMC。

6）中央刀具库。它是刀具的集中存储区。

7）夹具站。它用于实现对夹具的调整、维护及其存储。

8）信息传输网络。它是 FMS 中的通信系统。

9）随行工作台。它用于实现从无人输送台车到制造单元之间的传送缓冲功能。

图 4-8 所示为一个典型的 FMS 布局图。

应该指出，人和计算机在 FMS 中起着重要作用。当然，人的工作量远小于手工操作的制造系统。然而，人仍然在 FMS 的运作中起着至关重要的作用。人的任务包括：

1）设备故障排除、维护和修理。

2）换刀和调整。

3）装卸零件。

4）数据输入。

5）零件程序变更。

6）程序开发。

和所有的制造设备一样，柔性制造系统设备也必须对运行中出现的毛病、失灵和停机等故障进行监控。当发现问题时，检修人员必须确定其来源和提出纠正措施。人还要按照规定的措施修理有故障的设备。即使所有系统都正常运行，定期维护也是必要的。

操作者还要调整机器，更换工具，必要时还得对系统进行重构。FMS 的刀具运储能力增加了，但仍需人去换刀和调刀。FMS 中的装卸也是如此。一旦原材料被装入自动物料

shooter must identify its source and prescribe corrective measures. Humans also undertake the prescribed measures to repair the malfunctioning equipment. Even when all systems are properly functioning, periodic maintenance is necessary.

Human operators also set up machines, change tools, and reconfigure systems as necessary. The tool handling capability of an FMS increases, but does not eliminate human involvement in tool changing and setup. The same is true of loading and unloading the FMS. Once raw material has been loaded onto the automated materials handling system, it is moved through the system in the prescribed manner. However, the original loading onto the materials handling system is still usually done by human operators, as is the unloading of finished products.

Humans are also needed for interaction with the computer. Humans develop part programs that control the FMS via computers. They also change the program as necessary when reconfiguring the FMS to produce another type of parts. Humans play less labor-intensive roles in an FMS, but the roles are still critical.

2. Types of FMS and Its Scope of Application

In manufacturing there have always been tradeoffs between production rates and flexibility. It can be seen from the spectrum shown in Fig. 4-9 that at one end of the spectrum are transfer lines capable of high production rate, but low flexibility, at the other end of the spectrum are independent CNC machines that offer maximum flexibility, but are capable only of low production rates. Flexible manufacturing falls in the middle of the spectrum. There has always been need in manufacturing for a system that could produce higher volume and production runs than could independent machines, while still maintaining flexibility.

Transfer lines are capable of producing large volumes of parts at high production rates. The line takes a great deal of setup, but can turn out identical parts in large quantities. Its chief shortcoming is that even minor design changes in a part can cause the entire line to be shut down and reconfigured. This is a critical weakness because it means that transfer lines cannot produce different parts, even parts from within the same family, without costly and time-consuming shutdown and reconfiguration.

CNC machines have been used to produce small volumes of parts that differ slightly in design. Such machines are ideal for this purpose because they can be quickly reprogrammed to accommodate minor or even major design changes. However, as independent machines they cannot produce parts in large volumes or at high production rates.

FMS can handle higher volumes and production rates than independent CNC machines. They cannot quite match such machines for flexibility, but they come close. What is particularly significant about the middle ground capabilities of flexible manufacturing is that most manufacturing situations require medium production rates to produce medium volumes with enough flexibility to quickly reconfigure to produce another parts or product. Flexible manufacturing fills this long-standing void in manufacturing.

It can be seen from Fig. 4-9 that flexible manufacturing can be divided into three different levels:

运储系统后，它会按预定的方式在系统中移动。然而，最初把原材料装到物料运储系统通常仍是由人工完成的，把成品从物料运储系统上卸下来也是如此。

还需要人与计算机交互。通过计算机控制 FMS 的零件程序是由人开发的。当需要重新调整 FMS 去生产另一种类型的零件时，也需要人更改程序。虽然人在 FMS 中的劳动强度小了，但其作用还是极其重要的。

2. FMS 的类型和适用范围

在制造业中，总是存在生产率和柔性的权衡问题。由图 4-9 所示的应用范围可以看出，其一端是具有高生产率和低柔性的流水线，另一端是具有高柔性、低生产率的单独的数控机床。柔性制造落在了中间区域。在制造中，总是需要一种既能比单机生产率高而又具有柔性的系统。

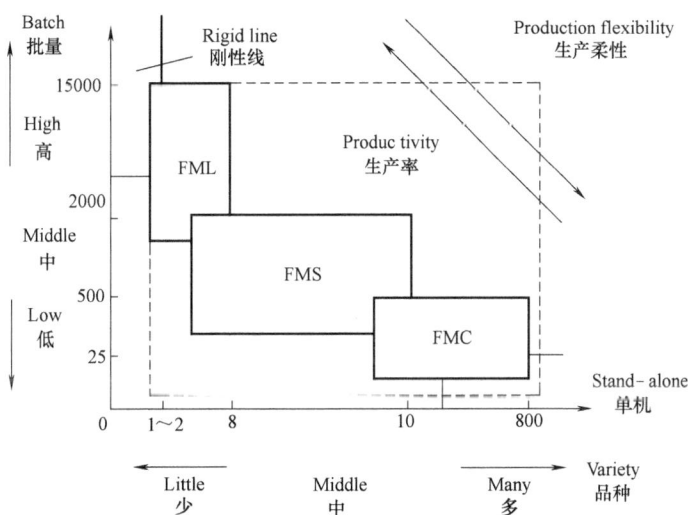

Fig. 4-9　Application scope of FMS　FMS 的应用范围

流水线能够以高生产率生产大量零件。该生产线耗费了大量的调整时间，但可以大批量生产相同的零件。其主要缺点是，即使零件上有微小的结构变化，也可能会导致全线停车并重新配置。这是一个致命弱点，因为这意味着如果没有高成本和耗时的停业和重新配置，流水线就不能生产不同的零件，即使是同一零件族的零件也不能生产。

数控机床一直被用于加工结构上略有不同的小批量零件。因此，这种机床是理想的，因为它们可以迅速重新编程以适应较小甚至重大的设计变化。然而作为单机，它们不能生产大批量的零件或不具有高的生产率。

FMS 比单独的数控机床具有更高的生产率。它们虽然在柔性上不能完全与数控机床匹敌，但它们接近。对于柔性制造中间区域的能力，特别重要的是大多数制造业要求以中等生产率生产中等数量的产品，并具有足够的柔性，以便快速重新配置去生产别的零件或产品。柔性制造填补了制造业长期存在的空白。

由图 4-9 看出，柔性制造可分为 3 个不同的层次：

（1）柔性制造单元（Flexible Manufacturing Cell，FMC）　如图 4-10 所示，柔性制造单元由卧式加工中心、环形工件交换工作台、工件托盘及托盘交换装置（Automatic Pallet Chang-

(1) Flexible manufacturing cell (FMC) Fig. 4-10 shows a schematic diagram of an FMC. It is composed of a horizontal machining center, a circular workpiece exchange worktable, pallet and automatic pallet changer (APC). The circular worktable is an independent general component; the pallets with workpieces are driven by the circular chain and move along the guideway of circular worktable.

(2) Flexible manufacturing system (FMS) The larger FMS has more than two FMCs or several CNC machine tools, and uses a material handling system to connect all the machine tools. The control and management functions of FMS are more powerful than that of FMC, but it has higher requirement for data management and communication network. FMS is suitable for the medium and small volume production (1000-30000 pieces per year) with many varieties of products (10-50 varieties).

(3) Flexible manufacturing line (FML) The difference between traditional rigid production line and FML is that the FML can machine a small number of different parts simultaneously or sequentially. The CNC modular machine tools are most often used while the general CNC machine tools are used in the FML. This kind of production line is equivalent to the automatic transfer line with numerical control. FML is suitable for the medium and large volume production (5, 000~200, 000 pieces per year) with 2-10 varieties.

4.3.3 Machining System in FMS

Machining system is the most basic part of FMS, it is mainly composed of CNC machine tools, machining centers and other processing equipment (some are equipped with the workpiece cleaning device, on-line detection and other auxiliary equipment). The structure of machining system and the number, specification, type of the machine tools in the system depend not only on the shape, size, and accuracy requirements of the workpiece, but also depend on the production lot and the degree of automation.

1. Commonly Used Configuration Form in the Machining System

(1) Interchangeable configuration (see Fig. 4-11a) Machine tools are distributed in parallel, the functions of machine tools can be substituted each other. If there is malfunction in a machine tool, the system can still maintain normal work.

(2) Complementary configuration (see Fig. 4-11b) Machine tools are distributed in series. The functions of machine tools are complementary each other, each machine has its specific machining tasks. This configuration has high productivity, but low reliability.

(3) Hybrid configuration (see Fig. 4-11c) In FMS, some machine tools presents in interchangeable configuration, and some are arranged in complementary form in order to give full play to their respective advantages.

2. Auxiliary Devices of Machining System

Auxiliary devices of machining system include machine tool fixture, pallet, pallet exchanging device, etc.

(1) Machine tool fixture The fixture in the FMS has two important development trends: One

er，APC）组成。环形工作台是一个独立的通用部件，装有工件的托盘在环形工作台的导轨上由环形链条驱动进行回转。

（2）柔性制造系统（FMS）　较大的 FMS 由两个以上的 FMC 或多台 CNC 机床组成，并用一个物料输送系统将机床联系起来。FMS 的控制与管理功能比 FMC 强大，对数据管理和通信网络要求高。FMS 适用于多品种（10～50 个品种）、中小批量（1000～30000 件/年）的生产。

（3）柔性制造线（Flexible Manufacturing Line，FML）　它与传统的刚性生产线的不同之处在于能同时或依次加工少量不同的零件。其加工设备在采用通用数控机床的同时，更多地采用数控组合机床。这种生产线相当于数控化的自动生产线。FTL 适合于多品种（2～10 个品种）、中大批量（5000～200000 件/年）生产。

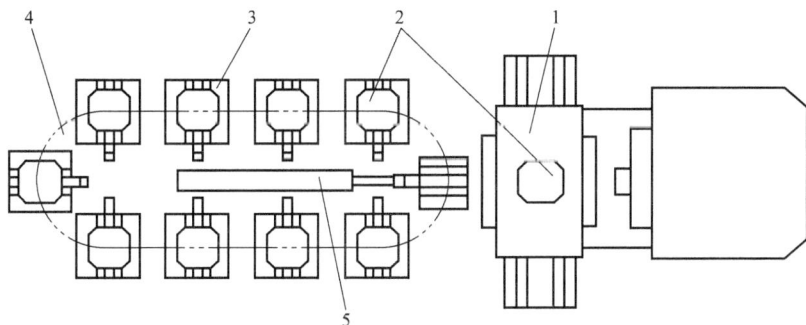

Fig. 4-10　Flexible manufacturing cell 柔性制造单元
1—Machining center 加工中心　2—Pallet 托盘　3—Pallet station 托盘站
4—Circular worktable 环形工作台　5—Workpiece exchange worktable 工件交换台

4.3.3　FMS 的加工系统

加工系统是 FMS 最基本的组成部分，它主要由数控机床、加工中心等加工设备（有的还带有工件清洗、在线检测等辅助设备）构成。加工系统的结构形式以及所配备的机床数量、规格、类型，取决于工件的形状、尺寸和精度要求，同时也取决于生产的批量及加工自动化程度。

1. 加工系统常用配置形式

（1）互替式配置（图 4-11 a）　机床布局呈并联关系，各机床功能可以互相代替。若某台机床出现故障，系统仍能维持正常的工作。

（2）互补式配置（图 4-11 b）　机床布局呈串联关系。机床功能是互相补充的，各自完成特定的加工任务。这种配置形式具有较高的生产率，但可靠性低。

（3）混合式配置（图 4-11 c）　即在 FMS 中，有些机床按互替形式布置，有些则按互补形式布置，以发挥各自的优点。

2. 加工系统的辅助装置

加工系统的辅助装置包括机床夹具、托盘、托盘交换装置等。

（1）机床夹具　用于 FMS 的夹具有两个重要的发展趋势：一是大量使用组合夹具，可针对不同的服务对象快速拼装出所需的夹具，提高夹具的重复利用率；二是开发柔性夹具，使一个夹具能为多个加工对象服务。

is the use of a large number of modular fixtures so as to assemble required fixture rapidly according to the different service objects and to improve the reused rate of the fixture; the other is the development of flexible fixture so that one fixture is able to serve several objects to be machined.

(2) Pallet In FMS, the pallet is the carrier of the workpiece and fixture. When the workpiece is machined in the machine tool, the pallet will become the worktable which supports the workpiece to complete the machining task. The pallet supports the workpiece and fixture and conveys them from one machine tool to any of other machine tools in handling. In order to connect each machine tool into a whole system, all the pallets in the system must have the same structure form.

(3) Pallet exchanging device Pallet exchanging device is the most common loading/unloading device in the machining center. There are two types of pallet exchanging devices, rotary type and linear reciprocating type. Fig. 4-12 shows a rotary pallet exchange device with two working stations, which has two parallel guideways for guiding the movement of the pallet. After the machining is finished, the pallet changer removes the pallet with machined parts from the worktable of machine tool. Then it rotates 180°and sends the pallet with the workpiece to be machined to the machining position on the machine tool.

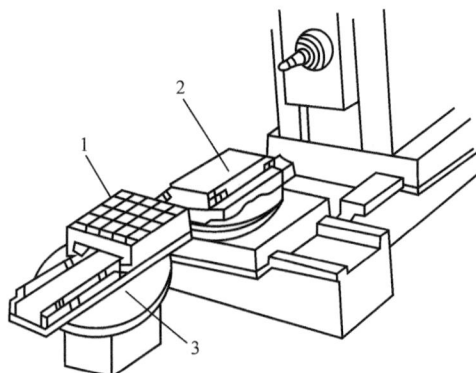

Fig. 4-12 Rotary pallet exchanging device 回转式托盘交换装置
1—Pallet 托盘   2—Pallet fastening device 托盘紧固装置   3—Rotary table 回转工作台

Fig. 4-13 is a reciprocating pallet exchanging device with several pallets. It consists of a pallet storage and a pallet exchanging device. After the machining is finished, the worktable moves transversely to the unloading position and sends the pallet with machined workpieces to the vacant of a pallet storage. Then, the worktable moves transversely to the loading position, pallet exchanging device moves the pallet with the workpiece to be machined onto the worktable.

4.3.4 Material Handling System in FMS

The automated material handling system is fundamental component that helps mold a group of independent DNC machines into a comprehensive FMS. The system must be capable of accepting workpieces mounted on pallets and moving them from workstation to workstation as needed. It must also be able to place workpieces on hold as they wait to be processed at a given workstation.

The material handling system must be able to unload a workpiece at one station and load anoth-

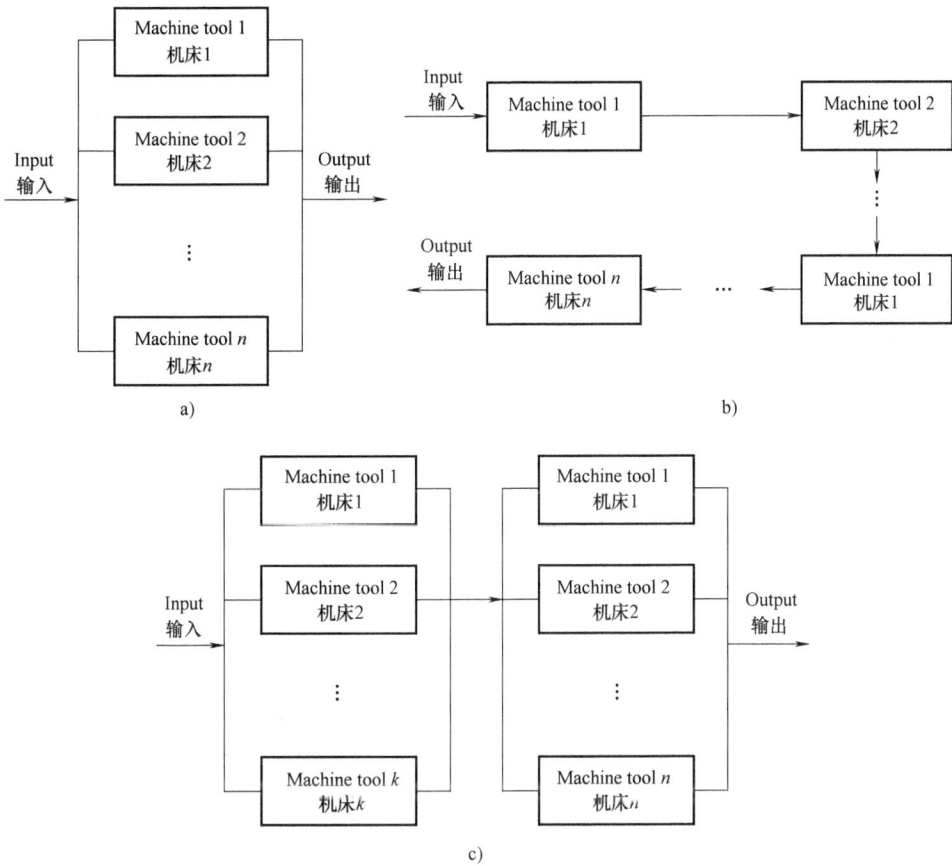

Fig. 4-11　FMS machine configuration form　FMS 机床配置形式

a）Interchangeable type 互替式配置　b）Complementary type 互补式配置　c）Hybrid type 混合式配置

（2）托盘　在 FMS 中，托盘是工件和夹具的一个承载体。当工件在机床上加工时，托盘成为机床工作台，支撑着工件完成加工任务。当工件输送时，托盘又承载着工件和夹具在机床之间进行传送。为使各台机床连接成为一个系统整体，系统中的所有托盘必须采用同一种结构形式。

（3）托盘交换装置　托盘交换装置是加工中心最为常见的上下料装置。托盘交换装置有回转式和直线往复式两种。图 4-12 所示为回转式托盘交换装置，其上有两条平行导轨以供托盘移动导向之用。当机床加工完毕后，托盘交换装置从机床工作台上移出已加工工件的托盘，然后旋转 180°，将装有待加工工件的托盘送到机床的加工位置。

图 4-13 所示为一个多托盘的往复式托盘交换装置。它由一个托盘库和一个托盘交换装置组成。当机床加工完毕后，工作台横向移动到卸料位置，将已加工的工件托盘移至托盘库的空位上。然后工作台横移至装料位置，托盘交换装置再将待加工的工件托盘移至工作台上。

4.3.4　FMS 的物料运储系统

自动物料运储系统是把一组独立的 DNC 机床集成为一个综合 FMS 的基本部件。该系统

er for transport to the next station. It must accommodate computer control and be completely compatible with other components in the flexible manufacturing system. The materials handling system for an FMS must be able to withstand the rigorous shop environment. Some FMSes are configured with automated guided vehicles (AGVs) as a principal means of materials handling.

Material handling system is generally composed of the following parts:

(1) Loading/unloading station of workpieces The loading/ unloading station of workpieces is located at the entrance of the FMS, the blanks and the machined workpieces are usually completed by hand.

(2) Pallet buffer station Pallet buffer station is generally set near the machine tool and can store several workpiece-pallet combinations. If the machine tool signals that it is ready to accept the workpiece, the system will send the workpiece from the pallet buffer station to the machine tool for machining through the pallet exchanging device.

(3) Automated warehouse Generally, the automatic warehouse adopts the multi-layer spatial structure, and is mainly composed of shelf, stacker and computer control system. Fig. 4-14 shows the layout of an automated warehouse.

The shelf is the main structure of the warehouse. There is a laneway or sometimes several laneways between the shelves if necessary. In general, the entrance and the exit are arranged at one end of the laneway. Each laneway has its own exclusive stacker which is responsible for the access of materials. The main tasks of computer control and management system are as follows:

1) Register and identification of material information. Automatic identification of materials is the key to the operation of automated warehouse. First of all, the container is coded, and then the bar code is attached to the appropriate part of the container. When the container is stored, the bar code reader scans the bar code automatically, and the information relevant to parts in the container is automatically input into the computer.

2) Automatic material access. The computer can control the movement of the stacker in the laneway in light of the information of the bar code, and automatically retrieve the storage address of the material to be stored or withdrawn. When the stacker arrives at the designated place, it stops moving, and pushes the workpiece to be stored into the storage cage or takes the required materials from the storage cage.

3) Warehouse management. The management of the computer control system can manage the materials, accounts, storage location and other material information of whole warehouse, and print various reports regularly or irregularly.

(4) Material handling device Material handling device is directly responsible for the transportation of the workpieces, tools and other materials, including the conveying and handling of materials between machine tools, automatic warehouse and pallet buffer station, and pallet buffer station and machine tool. In the FMS, the common material handling devices are the conveying belt, AGV and the transfer robot. Conveying belts are virtually useless in modern material handling systems. With the maturity of FMS control technology, more and more AGVs are used. Because of its flexibility and unique visual and tactile ability, the transfer robot has been used more and more widely in

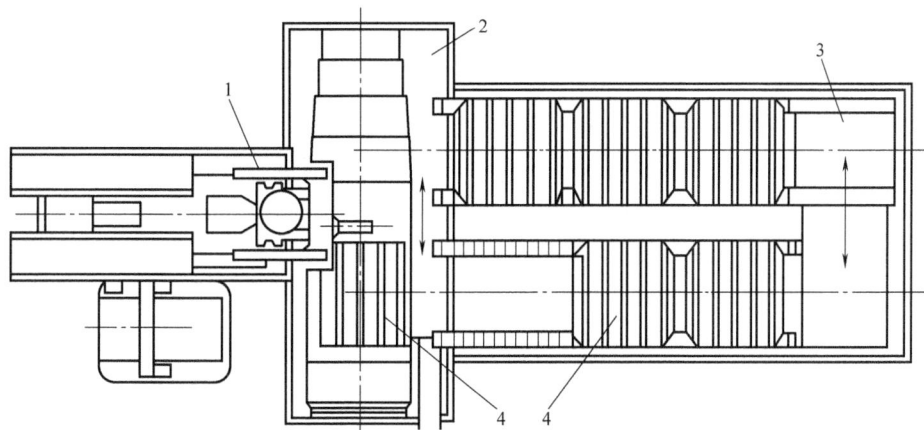

Fig. 4-13　Reciprocating pallet exchanging device 往复式托盘交换装置

1—Machining center 加工中心　2—Worktable 工作台　3—Pallet storage 托盘库　4—pallet 托盘

必须能够接纳安装在托盘上的工件，并根据需要把工件从一个工作站运送到另一个工作站。当工件在指定的工作站等待加工时，它也必须能够缓存工件。

物料运储系统必须能够在一个工作站卸下工件并装上另一个运输到下一个工作站。它必须适应计算机控制，并与柔性制造系统中的其他部件完全兼容。FMS 中的物料运储系统必须能够承受严酷的车间环境。有些 FMS 还配备有自动导引小车（Automatic Guided Vehicle，AGV）作为主要的物料运输工具。

物料运储系统一般由以下几个部分组成：

（1）工件装卸站　工件装卸站设在 FMS 的入口处，通常由人工完成对毛坯和已加工工件的装卸。

（2）托盘缓冲站　托盘缓冲站一般设置在机床附近，可存储若干个工件/托盘组合体。若机床发出信号已准备好接受工件信号时，系统通过托盘交换装置将工件从托盘缓冲站送到机床上进行加工。

（3）自动化仓库　自动化仓库一般采用多层立体结构形式，主要由货架、堆垛机和计算机控制管理系统等部分组成。图 4-14 为自动化仓库的布局示意图。

货架是仓库的主体结构。货架之间根据需要留有一条或多条巷道。一般情况下入库口和出库口都布置在巷道的某一端。每个巷道都有自己专有的堆垛机，负责物料的存取。计算机控制与管理系统的主要任务如下：

1）物料信息的登录和识别。物料自动识别是自动化仓库运行的关键。首先对货箱进行编码，然后将条形码贴在货箱的适当部位。当货箱入库时，条形码阅读器自动扫描条形码，将货箱零件的有关信息自动录入计算机。

2）物料自动存取。计算机可根据条形码信息控制堆垛机在巷道内移动，自动检索要存放或提取的物料的存储地址。堆垛机到达指定地点后便停止移动，把要存放的工件推入存储笼内，或从存储笼内取出所需的物料。

3）仓库管理。计算机控制管理系统可对全仓库进行物资、账目、存放位置以及其他物料信息进行管理，定期或不定期地打印各种报表。

FMS in recent years.

4.3.5 Tool Managing System in FMS

1. Elements of Tool Management System

The main functions of the tool management system are to be responsible for the transportation, storage and management of cutting tools, to provide the required cutting tools for machining unit timely, to monitor and manage the use of cutting tools, to take away the cutting tool which has reached the tool life or out-of-service cutter in time so as to decrease farthest the tool cost under the normal production conditions. The function and flexibility of tool management system have direct influence on the flexibility and productivity of FMS.

The automatic tool management system of FMS has two forms. One is to equip the tool magazine with a certain capacity on the machining center. Its shortcoming is that the limited tool storage capacity in each machining center. The other is to install an independent central tool magazine, which can provide tool exchange services for several machining centers by means of the tool changing robot or tool conveying vehicle. The second form is the development direction of tool management system.

The typical automatic tool management system of FMS is composed of the tool magazine system, tool presetting and tool loading/unloading station, tool exchanging device and computer of tool workstation.

2. Tool Changer

The tool exchange in FMS is usually realized by the tool changing robot or the tool conveying vehicle. The tool conveying vehicle is the same as the AGV, only there is a loading toolrest placed on the AGV. The toolrest can accommodate 5-20 cutting tool. The tool conveying vehicle is responsible for handling and changing cutting tools among tool loading/unloading station, each machine tool and the central tool magazine. There is also a small robot attached to the tool conveying vehicle, as shown in Fig. 4-15. When the vehicle reaches a goal, the robot will change the cutting tool.

3. Tool Monitoring and Management

(1) Monitoring of tools The main purpose of tool monitoring is to know the changes in performance of the on-line cutting tools due to the wear and torn. At present, the monitoring is carried out mainly from the tool life, tool wear, tool torn and other form of tool failures. After the cutting tool is installed in the machine tool, the administrator can inquire about the usage of cutting tool through the computer, and decide the replacement plan of the current tool.

The tool wear is gradually varied, the tool failure (including the tool tipping, cracking, etc.) is random, and the change of cutting force caused by them is also different. The difference in machining conditions (such as the workpiece material, cutting tool material and cutting variables) and the difference in cutting state (continuous cutting and intermittent cutting) as well as poor cutting conditions result in the complication in tool monitoring. It is required that the tool monitoring system be able to determine its characteristic quantity and discriminant criterion according to the damage

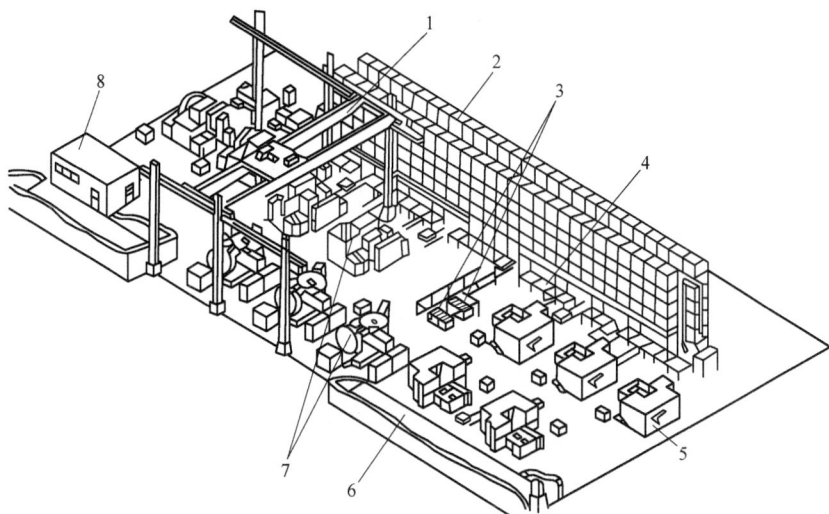

Fig. 4-14　Schematic diagram of automatic warehouse 自动化仓库的布局示意图

1—Automatic conveying crane 自动输送起重机　2—Automatic classifying shelf 自动分类货架

3—Pallet transceiver station 托盘收发站　4—Workpiece installation preparation station 工件安装准备站

5—CNC lathe 数控车床　6—Tool presetting place 刀具预调处　7—Machining center 加工中心

8—Inspection room 检查室

（4）物料运输装置　物料运输装置直接担负着工件、刀具及其他物料的运输，包括物料在加工机床之间、自动仓库与托盘存储站之间，以及托盘存储站与机床之间的输送与搬运。FMS 中常见的物料运输装置有传送带、自动运输小车和搬运机器人等。传送带在现代物料运输系统中已几乎不用。随着 FMS 控制技术的成熟，采用自动导向无轨小车的也越来越多。搬运机器人由于其工作灵活性强且具备独有的视觉和触觉能力，近年来在 FMS 中的应用越来越广。

4.3.5　柔性制造系统的刀具管理系统

1. 刀具管理系统的组成

刀具管理系统的主要职能是负责刀具的运输、存储和管理，适时地向加工单元提供所需的刀具，监控管理刀具的使用，及时取走已达到刀具寿命或报废的刀具，在保证正常生产条件下，最大限度地降低刀具成本。刀具管理系统的功能和柔性直接影响到整个 FMS 的柔性和生产率。

FMS 的刀具自动管理系统主要有两种形式：一种是在加工中心配置一定容量的刀库。其缺点是每台加工中心的刀库容量有限；另一种是设置独立的中央刀库，采用换刀机器人或刀具输送小车，为若干台加工中心进行刀具交换服务。第二种形式是刀具管理系统发展的方向。

典型的 FMS 刀具自动管理系统由刀库系统、刀具预调及刀具装卸站、刀具交换装置以及刀具工作站计算机组成。

2. 刀具交换装置

FMS 中的刀具交换通常由换刀机器人或刀具运输小车来实现。刀具运输小车与工件输送自动小车相同，只是在 AGV 运载小车上放置了一个装载刀架。该刀架可容纳 5~20 把刀

form and to send a signal immediately when the damage occurs (or will occur soon) so as to take corresponding measures quickly to avoid the accident. The simplest way to detect tool wear is to record the actual cutting time of each tool and to compare it with the tool life. Once the tool life is reached, the tool changing signal is sent. The simplest way to detect the tool breakage is to move each tool before cutting or after finishing cutting near to the detecting device to check whether there is any damage. These two methods have been widely used.

(2) Tool information management system The tool information in FMS is divided into dynamic information and static information. The dynamic information refers to the tool parameters which are constantly changing in the use, such as tool life, working diameter and length and other geometric parameters involved in cutting. These information varies with the continuation of the machining process, and directly reflects the tool service time, tool wear, and the effect on the machining accuracy and surface quality of the workpiece. The static information refers to some changeless information in the machining process, such as the tool code, type, attributes, geometry and some structural parameters. In order to facilitate the input, retrieval, modification and output control of the tool, the tool information is centrally managed in the form of a database in FMS.

4.3.6 Control System in FMS

The control system in FMS is actually an information flow system to realize the control, coordination, scheduling, monitoring and management of material flow process in the FMS machining process. It consists of computer, industrial control unit, programmable controller, communication network, database and the corresponding control and management software. It is the nerve center and the core part of a FMS.

1. The Architecture of FMS Control System

Since FMS is a complex automation integration system, the architecture and performance of the control system have a direct influence on the flexibility, reliability and automation of the entire FMS.

In order to avoid the centralized control by using a computer, currently, almost all of the FMS use the hierarchical control structure with multi-level computers to share the load of the host computer so as to improve the reliability of the control system. At the same time, it is also convenient for the design and maintenance of the control system.

The control system of FMS generally uses three-level hierarchical control structure, including system management and control, process coordination and monitoring, and equipment control. The reference model is shown in Fig. 4-16.

(1) System management and control level It is also known as the unit control level. It performs the production tasks assigned by superior, makes system production plan, allocates tasks timely to various workstations, working places, monitors system operation, coordinates the work and mutual support between each department and FMS.

(2) Process coordination and monitoring level It is also known as the workstation control level. It coordinates mainly the workpiece flow in the system and completes the connection between the equipment, the monitoring of system running status, the distribution of machining program, the col-

具，刀具运输小车在刀具装卸站、各加工机床与中央刀具库之间搬运与交换刀具，也有在刀具运载小车上附设一个小型机器人，如图 4-15 所示。当小车到达一个目标时，由附设的机器人进行刀具交换。

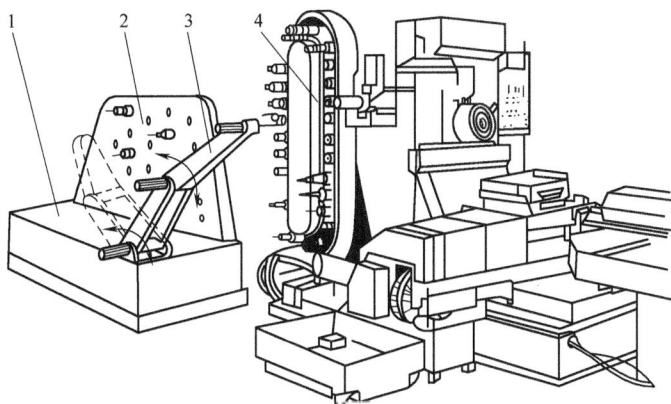

Fig. 4-15　AGV for handling cutting tools 运送刀具的 AGV
1—AGV　2—Loading toolrest 装载刀架　3—Robot 机器人　4—Tool magazine 刀库

3. 刀具的监控与管理

（1）刀具的监控　刀具监控的目的主要是及时了解在线刀具因磨损、破损而发生的性能变化。目前，监控主要从刀具寿命、刀具磨损、刀具破损以及其他形式的刀具故障等方面进行。当刀具装入机床后，管理员可通过计算机查询刀具的使用情况，并决定当前刀具的更换计划。

刀具的磨损是逐渐变化的，刀具破损（包括崩刃、破裂等）则是随机的，它们引起切削力的变化情况也不同。加工条件（如工件材质、刀具材料及切削用量等）的不同，切削状态（连续切削和断续切削）的不同，还有切削环境恶劣，使得刀具监控复杂化。刀具监控系统要能根据刀具的破损形式确定其特征量和判别基准，在破损发生（或即将发生）时能立即发出信号以便迅速采取相应措施，避免产生事故。刀具磨损最简单的检测方法是记录每把刀具的实际切削时间，并与刀具寿命进行比较。达到刀具寿命时就发出换刀信号。刀具破损最简单的检测方法是将每把刀具在切削加工开始前或切削加工结束后移近固定的检测装置，以检测是否破损。这两种方法已得到广泛应用。

（2）刀具的信息管理系统　FMS 中的刀具信息分为动态信息和静态信息两大部分。动态信息是指在使用过程中不断变化的一些刀具参数，如刀具寿命、工作直径、工作长度以及参与切削加工的其他几何参数。这些信息随加工过程的延续不断发生变化，直接反映了刀具使用时间的长短、磨损量的大小、对工件加工精度和表面质量的影响。静态信息是指一些加工过程中固定不变的信息，如刀具的编码、类型、属性、几何形状以及一些结构参数等。为了便于刀具的输入、检索、修改和输出控制，FMS 以数据库形式对刀具信息进行集中管理。

4.3.6　柔性制造系统的控制系统

FMS 的控制系统实际上是实现 FMS 加工过程中的物料流动过程的控制、协调、调度、监测和管理的信息流系统。它由计算机、工业控制机、可编程控制器、通信网络、数据库和相应的控制与管理软件组成，是 FMS 的神经中枢和核心部分。

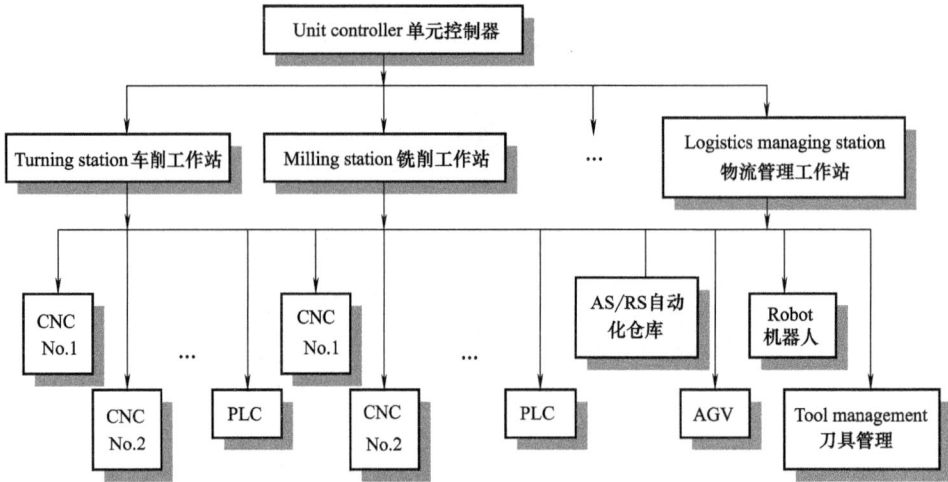

Fig. 4-16 Hierarchical control structure of FMS FMS 递阶控制结构

lection of working conditions and equipment operation data as well as the report to the higher controller. The field operators complete the real-time control and on-site scheduling over the entire system mainly through the interface of this level.

（3）Equipment control level   It consists of CNC device and PLC device which belong to the CNC machine tool, robot, AGV, automatic warehouse and other equipment. It controls directly the automatic operation cycle of various processing equipment and material system, receives and performs the control instructions from higher-level system, and feeds the field data and control information back to the higher-level system.

In the three-level hierarchical control structure mentioned above, the information flow of each level is two-way flow. That is, it can give the control instructions, allocate control tasks, and monitor the operation process of lower level downward; it can also feed operating state back to the higher level and report on-site production data upward. However, the control computers in each level have the differences in real-time control and processed information amount. The lower the level is, the stronger the real-time control, but the less the processed information; the higher the level is, the more the processed information, but the lower the real-time requirement. For example, the planning time of equipment-level is usually in the range of several milliseconds to several minutes; the planning time of the workstation level can be from a few minutes to several hours; and the planning period of the top level can be up to several hours or even several weeks.

2. Functions of FMS Control System

Some typical functions of an FMS control system are illustrated in Fig. 4-17.

The scheduler function involves planning how to product the current volume of orders in the FMS, considering the current status of machine tools, work-in-process, tooling, fixtures, and so on. The scheduling can be done automatically or can be assisted by an operator. Most FMS control systems combine automatic and manual scheduling; the system generates an initial schedule that can be changed manually by the operator. The dispatcher function involves carrying out the schedule and

1．FMS 控制系统的体系结构

由于 FMS 是一个复杂的自动化集成体，其控制系统的体系结构和性能直接影响整个 FMS 的柔性、可靠性和自动化程度。

为了避免用一台计算机过于集中地控制，目前几乎所有的 FMS 都采用了多级计算机递阶控制结构，由此来分担主控计算机的负荷，提高控制系统的可靠性，同时也便于控制系统的设计和维护。

FMS 控制系统一般采用三级递阶控制结构，包括系统管理与控制级、过程协调与监控级、设备控制级。其参考模型如图 4-16 所示。

（1）系统管理与控制级 系统管理与控制级又称单元控制级。它执行上级下达的生产任务，制订系统生产作业计划，实时分配作业任务给各个工作站、点，监控系统的运行，协调各部门与 FMS 的工作及相互支援等。

（2）过程协调与监控级 过程协调与监控级又称工作站控制级，主要协调工件在系统中的流动，完成各设备间的交接、系统运行状态的监控、加工程序的分配、工况和设备运行数据的采集以及向上级控制器报告等。现场操作人员主要通过这一级界面完成对整个系统的实时运控与现场调度。

（3）设备控制级 设备控制级由数控机床、机器人、AGV、自动化仓库等设备的 CNC 装置和 PLC 装置组成。它直接控制各类加工设备和物料系统的自动工作循环，接受和执行上级系统的控制指令，并向上级系统反馈现场数据和控制信息。

在上述三级递阶控制结构中，每层的信息流都是双向流动的，即向下可下达控制指令，分配控制任务，监控下层的作业过程；向上可反馈运行状态，报告现场生产数据。然而，各层控制计算机在控制的实时性和处理信息量方面则有所不同：越往底层，其控制的实时性越强，而处理的信息量则越少；越到上层，其处理的信息量越大，而对实时性要求则越小。例如，设备层控制规划时间通常在几毫秒到几分钟范围内；在工作站层，其规划时间可以从几分钟到几小时；而最上层其规划期可高达几小时甚至几周。

2．FMS 控制系统的功能

某 FMS 控制系统的一些典型功能如图 4-17 所示。

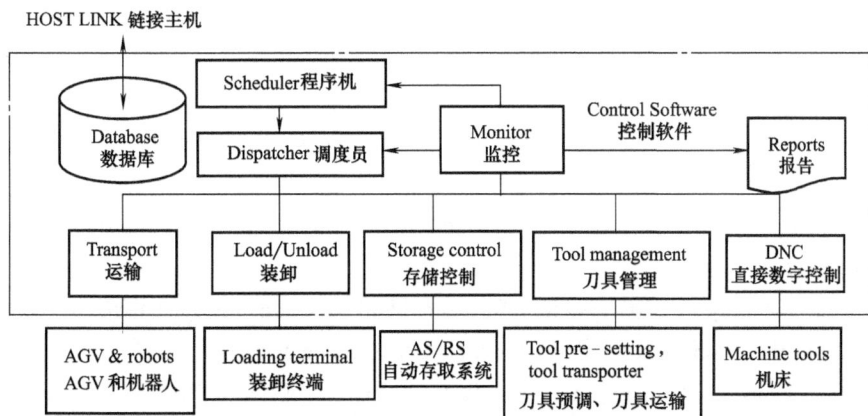

Fig 4-17 Diagram of functions of an FMS control system FMS 控制系统的功能图

coordinating the activities on the shop floor, that is, deciding when and where to transport a pallet, when to start a process on a machining center and so on.

The monitor function is concerned with monitoring work progress, machine status, alarm messages and so on, and providing input to the scheduler and dispatcher as well as generating various production report and alarm messages. A transport control module manages the transportation of parts and palettes within the system. Having an AGV system with multiple vehicle, the routing control logic can become rather sophisticated and a critical part of the FMS control software. A load/unload module with a terminal at the loading area shows the operators which parts are to introduce to the system and enables him or her to update the status of the control system when parts are ready for collection at the loading area. A storage control module keeps an account of which parts are stored in the AS/RS as well as their exact location. The tool management module keeps an account of all relevant tool data and the actual location of tools in the FMS.

Tool management can be rather comprehensive since the number of tools normally exceeds the number of parts in the system. And furthermore, the module must control the preparation and flow of tools. The DNC function provides interfaces between the FMS control program and machine tools and devices on the shop floor. The DNC capabilities of the shop floor equipment are essential to an FMS; a "full" DNC communication protocol enabling remote control of the machines is required.

Review Questions and Problems

4-1　What is the meaning of manufacturing automation?

4-2　What are the differences between the manufacturing automation in narrow sense and the manufacturing automation in broad sense?

4-3　What is the goal of manufacturing automation?

4-4　What are the characteristics of rigid automation? What problems does the flexible automation mainly solve? What are their typical representatives?

4-5　What is the meaning of NC technology? State briefly the components of CNC machine tool.

4-6　What are the differences between modern CNC machine tool and general CNC machine tool?

4-7　What is the 5-axis linkage CNC machine tool? What advantages does the 5-axis CNC machine tool have in machining?

4-8　What is the parallel kinematic machine tool? Compared with the traditional machine tool, what advantages does it have?

4-9　What is the open CNC system? Why should we develop the open CNC system?

4-10　What the main features does the open CNC system have?

4-11　How many kinds of architectures does the PC-based open CNC system have? What are they? What are the characteristics of each?

4-12　What is a flexible manufacturing system? What parts is it composed of?

程序机的功能在于考虑机床、在制品、工具、夹具等的当前状态，对如何在 FMS 中生产当前的订单进行规划。调度任务可以自动完成，或由操作者协助完成。大多数 FMS 控制系统将自动调度和手动调度相结合；系统生成一个初始的调度计划，然后再由操作员手工修改。调度员的作用在于执行调度计划并协调车间内的各个环节，即决定何时运送托盘、运到何处、何时开始在加工中心上加工等。

监控功能关注的是对工作过程、机器状态、报警信息等进行监控，并输入给程序机和调度员，以及生成各种各样的生产报告和报警信息。运输控制模块管理零件和托盘在系统内的运行。如果采用具有多个运输小车的 AGV 系统，路径控制逻辑可能成为 FMS 控制软件的相当复杂和极其重要的部分。具有装载区终端的装卸模块向操作者显示哪些零件要送入系统，并当零件在装载区收集好时，使操作者能够更新控制系统的状态。存储控制模块保存着 AS/RS 中的零件以及它们的确切位置。刀具管理模块保存所有相关的刀具数据和 FMS 中刀具存放的实际位置。

由于系统中刀具的数量通常超过零件的数量，刀具管理相当全面。此外，刀具管理模块必须控制刀具的制备和流动。DNC 功能为 FMS 控制程序与车间机床和装置之间提供了接口。车间设备的 DNC 能力对于一个 FMS 是必不可少的；一个能对机床远程控制的全 DNC 通信协议也是需要的。

复习题与习题

4-1 制造自动化的含义是什么？

4-2 "广义制造"中的制造自动化与"狭义制造"中的制造自动化有何区别？

4-3 制造自动化的目标是什么？

4-4 刚性自动化的特点是什么？柔性自动化主要解决什么问题？其典型代表是什么？

4-5 数控技术的含义是什么？简述数控机床的构成。

4-6 现代数控机床与普通数控机床有哪些区别？

4-7 何谓五轴联动数控机床？五轴联动数控机床在零件加工方面具有哪些优势？

4-8 何谓并联运动机床？它与传统机床相比具有哪些优点？

4-9 什么是开放式数控系统？为什么要开发开放式数控系统？

4-10 开放式 CNC 系统具有哪些主要特征？

4-11 基于 PC 的开放式 CNC 有哪几种体系结构？各有什么特点？

4-12 何谓柔性制造系统？它由哪几部分组成？

4-13 为什么说人在柔性制造系统中仍起着重要作用？人的主要任务是什么？

4-14 FMS 有哪几种类型？各适用于何种场合？

4-15 FMS 加工系统常用的配置形式有哪几种？分析互替式与互补式机床配置形式的特点。

4-16 物料运储系统起什么作用？它一般由哪几个部分组成？

4-17 计算机控制与管理系统的主要任务有哪些？

4-18 FMS 中常见的物料运输装置有哪些？

4-19 刀具管理系统的主要职能是什么？典型的 FMS 刀具自动管理系统由哪几个部分组成？

4-20 FMS 的控制系统由哪几个部分组成？一般采用哪三级递阶控制结构？

4-13   Why do the people still play an important role in flexible manufacturing system? What are the main tasks of the people?

4-14   What are the types of FMS? Which situation does each adapt to?

4-15   What are the configuration forms commonly used in the FMS machining system? Try to analyze respectively the characteristics of the configuration form with reciprocal machine tool and the configuration form with complementary machine tool.

4-16   What is the function of the material handling system? How many parts does it usually consist of?

4-17   What are the main tasks of computer control and management system?

4-18   What are the common materials handling devices in FMS?

4-19   What are the main functions of the tool management system? What the components does a typical automatic tool management system consist of?

4-20   What parts does a FMS control system consist of? Which three-level hierarchical control structures are generally used?

Chapter 5   Information Management Technology in Manufacturing Enterprise

第5章　制造企业的信息管理技术

5.1 Introduction

5.1.1 Manufacturing Information and Its Characteristics

Manufacturing systems have three main flows, that is material flow, energy flow and information flow, where the information flow has been the most active driving factor. The competition among enterprises has changed from the competition of production scale and capital to the competition of ability to acquire information quickly and to use information. The main performances are to realize the informatization of enterprise by means of information technology (IT) and Internet technology and to realize the advanced manufacturing strategy of enterprise with information technology support, thus making the best use of material, knowledge, capital, information and other resources.

Manufacturing information in manufacturing process is divided into design information and manufacturing information. Manufacturing information has the following characteristics:

(1) Polymorphism In addition to general structured information, there is a lot of unstructured information in manufacturing system, such as graphic information, solid models, NC programs, expert knowledge, design experience and other data types, which makes manufacturing information more polymorphic.

(2) Structure complexity The structure of a lot of information such as graphic information, solid models, NC programs, process files is quite complex, and very difficult to store and process in the form of model table. This proposes higher requirement to relational database.

(3) Distributivity Manufacturing information distributed in each application unit of manufacturing system. And the differences in the time of establishing database, the system environment, and the application purpose make the database structure and application environment have obvious heterogeneity. This distribution and heterogeneity makes it difficult to ensure the consistency, security and reliability of data, as well as information conversion and communication.

(4) Real-time Real-time database technology should be used when the bottom manufacturing system should consider the collection, analysis and management of real-time processing and monitoring information.

(5) Integration During the manufacturing process, there is frequent data exchange among the various information subsystems. In order to share information, reduce the redundancy of information, information integration requirements must be considered in the design of database.

Both product design and manufacturing process design are all information processing processes. The product design process is to map function and performance requirements to product design information, and the manufacturing process is to map product design information to manufacturing process control information. Design information determines manufacturing information, and is the input of manufacturing information. The manufacturing information will restrict the design information, and the manufacturability must be considered in the product design. Design information and manufacturing information have interactivity and unity, the ultimate goal of design and manufacture is to produce products that meet the needs of customers.

5.1　概述

5.1.1　制造信息及其特征

制造是一个包括产品设计和文档、材料的选择、规划、生产、质量保证、管理和营销的过程。商品制造系统有三大主流，即物质流、能量流和信息流，其中信息流已成为最活跃的驱动因素。企业间的竞争已从生产规模和资本的竞争转向快速获取信息和运用信息能力的竞争。其主要表现是，利用信息技术（IT）和互联网技术实现企业的信息化，在信息技术支持下实现企业的先进制造战略，从而使物质、知识、资本以及信息等资源得到最佳利用。

制造过程中的信息分为设计信息和制造信息。这些信息有以下特点：

（1）多态性　在制造系统中，除一般的结构化信息外，还有大量的非结构化信息，如图形信息、实体模型、数控程序、专家知识、设计经验等数据类型，因此使制造信息呈现明显的多态性。

（2）结构复杂性　制造系统中的许多信息，如图形、实体模型、数控程序、工艺文件等信息结构十分复杂，很难采用或参照模型表的形式进行存储和处理，这对关系型数据库提出了更高的要求。

（3）分布性　制造信息分布在制造系统的各个应用单元中，并且由于数据库建立的时间差异、系统环境的差异、应用目的的差异，数据库的结构和应用环境具有明显的异构性。这种分布和异构的情况，给保证数据的一致性、安全性和可靠性以及信息转换和通信带来了较大的难度。

（4）实时性　例如底层制造系统要考虑实时加工和监控信息的收集、分析和管理，这就要求采用实时数据库技术。

（5）集成性　在生产过程中，各个信息分系统之间频繁地进行着数据交换。为了实现信息的共享，减少信息的冗余，在数据库设计时必须考虑信息的集成要求。

产品设计和制造过程设计都是信息的处理过程。产品设计过程是把功能和性能要求映射为产品设计信息，制造过程则是把产品设计信息映射为制造过程控制信息。设计信息决定制造信息，是产生制造信息的输入。制造信息又会制约设计信息，产品设计时必须考虑可制造性。设计信息和制造信息之间具有交互性和统一性，设计和制造的最终目的是生产出能够满足客户需求的产品。

5.1.2　制造业信息化的内涵

在传统制造企业中，信息的产生、传递、复制和存储的主要形式是图样、文件、报表和各种会议。信息的传递不仅缓慢，而且经常中断，从而导致管理层次多、机构重叠、相互推诿、工作效率低下。

制造业信息化就是将 IT 技术、自动化技术、现代管理技术与制造技术相结合，带动产品设计方法和工具的创新、企业管理模式的创新、企业间协作关系的创新，实现产品设计制造和企业管理的信息化、生产过程控制的智能化、制造装备的数控化、咨询服务的网络化。

5. 1. 2　Connotation of Manufacturing Informatization

In traditional manufacturing enterprises, the main forms of information production, information transmission, information replication, and information storage are the blueprint, document, report forms and meetings. The transmission of information is not only slow, but also often interrupted, thus leading to many management levels, organizational overlapping, mutual evasion and low work efficiency.

Manufacturing informatization is to combine IT technology with automation technology, modern management technology and manufacturing technology, and to promote the innovations in the product design methods and tools, the enterprise management model, and the cooperation relationship among enterprises so as to realize the informatization of product design & manufacturing and enterprise management, intelligentization of production process control, numerical control of manufacturing equipment, networking of consulting services. The connotation of manufacturing informatization is mainly reflected as follows:

1) Information on product design, process planning, NC machining, coordinate measurement, flexible manufacturing and rapid prototyping, etc.

2) Marketing and management of informatization, mainly including management information system (MIS), manufacturing resource planning (MRP II), product data management (PDM) and enterprise resources planning (ERP), etc.

3) Manufacturing simulation and virtual manufacturing.

4) Network manufacturing and e-manufacturing based on Internet and LAN.

5) Intelligent manufacturing.

6) Virtual enterprise and supply chain, enterprise dynamic alliance.

7) Manufacturing engineering database and decision support system.

Manufacturing informatization involves product development, production and the value chain of marketing process. It has changed the simple money, goods transaction relationship among manufacturers, suppliers and customers. Suppliers can participate in the manufacture and transportation of product through supply chain management (SCM), and customers can participate in the design and manufacture of purchased products through customer relationship management (CRM) and product lifecycle management (PLM), thus having the enterprise solve different problems for customers in the application, maintenance and waste disposal of products.

5. 1. 3　Connotation of Enterprise Informatization

Enterprise informatization means that applying advanced management ideas and methods, and taking computer and network technology as a means, the existing production, business, design, manufacturing and management are integrated to provide timely, accurate and effective data information for enterprise decision-making so as to respond quickly to customer requests. The essence of enterprise informatization is to strengthen the core competitiveness of enterprise.

Enterprise informatization can be divided into four main business areas and their information systems, namely, ERP system, SCM system, CRM system and PLM system. The organic combination of these four information systems constitutes the enterprise information system. Aiming at a kind

制造业信息化的内涵主要体现在：

1）信息化的产品设计、工艺设计、数控加工、坐标测量、柔性制造以及快速成形等。

2）信息化的营销和管理，主要有管理信息系统、制造资源规划、产品数据管理和企业资源计划等。

3）制造仿真和虚拟制造。

4）基于互联网和局域网的网络制造、电子化制造。

5）智能制造。

6）虚拟企业和供应链、企业动态联盟。

7）制造工程数据库及决策支持系统。

制造业信息化涉及产品开发、生产和营销过程价值链，它改变了制造商、供应商和客户之间单纯的钱、货交易关系。供应链管理（SCM）使得供应商可以参与产品的制造和运输，客户关系管理（CRM）和产品生命周期管理（PLM）使得客户能够参与所购买产品的设计和制造过程，使企业为客户解决产品的使用、维护和废弃处理中的各种问题。

5.1.3 企业信息化的内涵

企业信息化是指运用先进的管理思想和方法，以计算机和网络技术为手段，整合企业现有的生产、经营、设计、制造和管理，为企业的决策提供及时、准确和有效的数据信息，以便对客户要求做出快速反应。企业信息化的本质是加强企业的核心竞争力。

企业信息化可分为4个主要的业务领域及其信息系统，即企业资源计划系统、供应链管理系统、客户关系管理系统和产品生命周期管理系统，这4种信息系统的有机结合构成了企业信息化体系。企业可根据自身情况，面向某类特定的业务问题，选用一种或几种系统来构建自己的企业信息化框架体系。

大多数企业有4个主要运作职能领域，即市场营销和销售（M/S）、供应链管理（SCM）、会计和财务（A/F）、人力资源（HR）。每个领域由各种较窄的业务功能组成，这些功能是特定于该操作职能领域的活动。运作职能领域和业务功能的例子见表5-1。

Tab. 5-1 Examples of functional areas of operation and their business functions

运作职能领域和业务功能的例子

Functional area of operation 运作职能领域	Marketing and sales 营销和销售	Supply chain management 供应链管理	Accounting and finance 会计与财务	Human resources 人力资源
Business functions 业务功能	Marketing a product 营销产品	Purchasing goods and raw materials 采购货物和原材料	Financial accounting of payments from customers and to suppliers 客户和供应商付款的财务核算	Recruiting and hiring 招聘和录用
	Taking sales orders 获取销售订单	Receiving goods and raw materials 收到货物和原材料	Cost allocation and control 成本分配与控制	Training 培训

（续）

Functional area of operation 运作职能领域	Marketing and sales 营销和销售	Supply chain management 供应链管理	Accounting and finance 会计与财务	Human resources 人力资源
Business functions 业务功能	Customer support 顾客支持	Transportation and logistics 运输和物流	Planning and budgeting 计划和预算	Payroll 工资
	Customer relationship management 顾客关系管理	Scheduling production runs 调度生产运行	Cash-flow management 现金流管理	Benefits 效益
	Sales forecasting 销售预测	Manufacturing goods 制造货物	Government compliance 顺应政府	
	Advertising 广告	Plant maintenance 设备维护		

of specific business problem, according to its own situation, the enterprise can choose one or more systems to build its own enterprise information framework system.

Most companies have four main functional areas of operation: marketing and sales (M/S), supply chain management (SCM), accounting and finance (A/F), and human resources (HR). Each area is composed of a variety of narrower business functions, which are activities specific to that functional area of operation. Examples of the business functions of each area are shown in Tab. 5-1.

Historically, businesses have had organizational structures that separated the functional areas. Although what happens in one functional area is not closely related to what happens in others, functional areas are interdependent, each requiring data from the others. The better a company can integrate the activities of each functional area, the more successful it will be in today's highly competitive environment. Integration contributes to improvements in communication and workflow. Each area's information system depends on data from other functional areas.

5.2 Enterprise Resource Planning

5.2.1 Concept and Development of （enterprise resource planning ERP）

In the late 1980s and early 1990s, with the development of the market all over the world and the application of information management means, the enterprises have gradually formed the scale development and entered international development space. Any enterprise should bear the pressure of competition from international enterprises. The time when the enterprise sold what it produced has gone forever. In this market context, ERP was born at the historic moment on the basis of manufacturing resource planning (MRP Ⅱ). In 2001, American Production and Inventory Control Society (APICS) defined the ERP as "a means to implement efficient planning and control of achieving, making, purchasing all required resources, and considering customer's orders in manufac-

从历史上看，企业的组织结构分离各项职能领域。虽然在一个职能领域发生的事情与其他职能领域发生的事情不密切相关，然而职能领域是相互依存的，每一个职能都需要其他职能的数据。在今天高度竞争的环境中，如果一个企业可以更好地整合各职能的活动，它将更为成功。整合有助于改善通信和工作流程。每个领域的信息系统都依赖于其他职能领域的数据。

5.2 企业资源计划

5.2.1 企业资源计划（ERP）的概念和发展

20世纪80年代末和90年代初，随着世界各国市场的开发和信息化管理手段的应用，企业逐步形成规模化发展并进入了国际化发展空间。任何企业都要承受来自国际化企业的竞争压力。以前企业生产什么就卖什么的年代一去不复返了。在这种市场背景下，企业资源计划在制造资源规划的基础上应运而生。美国生产与库存管理协会将企业资源计划定义为"一种进行高效规划与控制全部所需资源的获取、制作、采购的方法，并在制造、分配或服务过程中考虑顾客的订单"。

企业资源计划可以描述为：一个覆盖整个企业范围的管理工具，用于平衡需求和供给，包含连接客户和供应商成为一个完整的供应链的能力，采用经过验证的业务流程进行决策，并提供在销售、营销、制造、运营、物流、采购、财务、新产品开发和人力资源领域的跨职能整合，从而使人们以高水平的客户服务和生产力运行他们的业务，同时降低成本和库存；而且为提供有效的电子商务提供了基础。

ERP系统是一个业务管理系统，包括集成的综合软件套装，成功实施时，可用于管理和整合组织内的所有业务职能。这些套装通常包括财务和成本会计、销售与分销、物料管理、人力资源、生产计划和计算机集成制造、供应链、客户信息等一套成熟的业务应用和工具。这些软件包可使组织中的所有供应链流程之间（内部和外部）的信息流动。此外，ERP系统可以作为一种工具，通过帮助减少循环时间来改善供应链网络的业绩水平。然而，它一直被应用于资本密集型产业，如制造业、建筑业、航空航天业和国防业。最近，ERP系统已扩展到制造业之外，并引入金融、医疗保健、连锁酒店、教育、保险、零售和电信部门。

ERP允许企业整合各部门信息，它已经从人力资源管理应用程序发展到跨越IT管理的工具。对于许多用户来说，ERP是一个"做一切"的系统，可以做从销售订单输入到客户服务的所有事情。它试图依靠组织的制造环境整合供应商和客户。例如，一个采购信息进入订单模块将订单传递给制造应用，然后制造应用将物料请求发送给供应链模块，供应链模块从供应商处获取必要的零件，并使用物流模块将它们送到工厂。

图5-1所示为ERP系统模型，在每个模块中都包含有一些最普通的功能。然而，在不同的软件供应商提供的ERP系统中的模块名称和数量可能会有所不同。通过允许模块之间共享和传递信息，自由地集中信息到一个所有模块都可以访问的单一数据库，典型的系统可

turing, distribution, or service".

Enterprise resource planning can be described as follows: An enterprise-wide set of management tools that balances demand and supply, containing the ability to link customers and suppliers into a complete supply chain, employing proven business processes for decision-making, and providing high degrees of cross-functional integration among sales, marketing, manufacturing, operations, logistics, purchasing, finance, new product development, and human resources, thereby enabling people to run their business with high levels of customer service and productivity, and simultaneously lower costs and inventories; and providing the foundation for effective e-commerce.

ERP system is a business management system that comprises integrated sets of comprehensive software, which can be used, when successfully implemented, to manage and integrate all the business functions within an organization. These sets usually include a set of mature business applications and tools for financial and cost accounting, sales and distribution, materials management, human resource, production planning and computer integrated manufacturing, supply chain, and customer information. These packages have the ability to facilitate the flow of information between all supply chain processes (internal and external) in an organization. Furthermore, an ERP system can be used as a tool to help improve the performance level of a supply chain network by helping to reduce cycle times. However, it has traditionally been applied in capital-intensive industries, such as manufacturing, construction, aerospace and defence and so on. Recently, ERP systems have been expanded beyond manufacturing and introduced to the finance, health care, hotel chains, education, insurance, retail and telecommunications sectors.

ERP allows companies to integrate various departmental information. It has evolved from a human resource management application to a tool that spans IT management. For many users, an ERP is a "do it all" system that performs everything from entry of sales orders to customer service. It attempts to integrate the suppliers and customers with the manufacturing environment of the organization. For example, a purchase entered in the order module passes the order to a manufacturing application, which in turn sends a materials request to the supply-chain module, which gets the necessary parts from suppliers and uses a logistics module to get them to the factory.

An overview of ERP systems including some of the most popular functions within each module is shown in Fig. 5-1. However, the names and numbers of modules in an ERP system provided by various software vendors may differ. A typical system integrates all these functions by allowing its modules to share and transfer information by freely centralizing information in a single database accessible by all modules.

Although ERP systems have certain advantages, such as low operating cost and improving customer service, they have some disadvantages due to the tight integration of application modules and data. Huge storage needs, networking requirements and training overheads are frequently mentioned ERP problems. However, the scale of business process re-engineering (BPR) and customization tasks involved in the software implementation process are the major reasons for ERP dissatisfaction. Its high cost prevents small businesses from setting up an ERP system, the privacy concern within an ERP system and lack of trained people may affect ERP's efficiency. Implementation of an ERP

以集成所有这些功能。

ERP 系统虽然具有一定的优势,如运营成本低和提高客户服务,但由于应用模块和数据的紧密集成,ERP 也存在一些缺点。巨大的存储需求、网络需求和培训开销是经常被提到的 ERP 问题。然而,大规模的业务流程再造(BPR)和定制任务所涉及的软件实施过程则是对 ERP 不满意的主要理由。ERP 的高成本限制了小企业建立 ERP 系统,在 ERP 系统中的隐私问题和缺乏训练有素的人可能会影响 ERP 的效率。实施一个 ERP 项目是很费劲的,定制是昂贵和费时的。ERP 系统的各种缺陷,如功能性和技术性等,如图 5-2 所示。

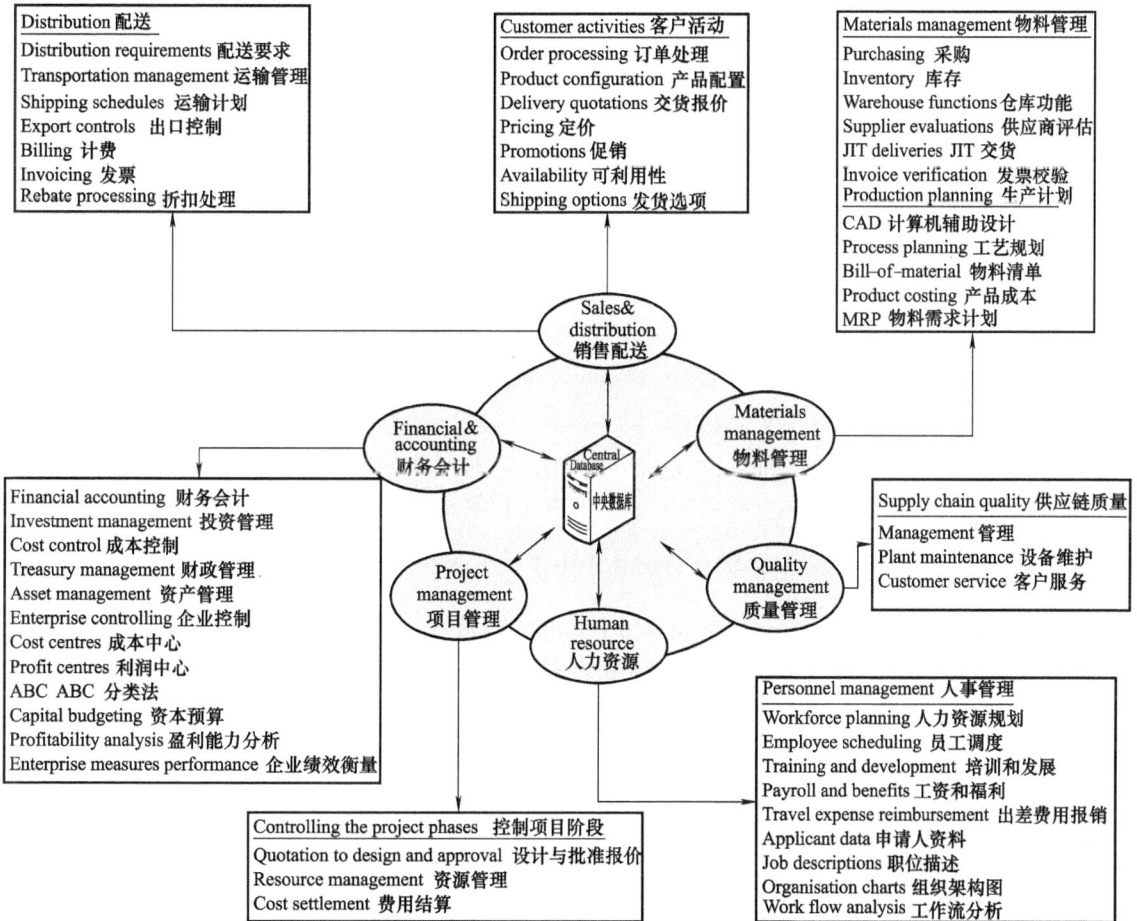

Distribution 配送
Distribution requirements 配送要求
Transportation management 运输管理
Shipping schedules 运输计划
Export controls 出口控制
Billing 计费
Invoicing 发票
Rebate processing 折扣处理

Customer activities 客户活动
Order processing 订单处理
Product configuration 产品配置
Delivery quotations 交货报价
Pricing 定价
Promotions 促销
Availability 可利用性
Shipping options 发货选项

Materials management 物料管理
Purchasing 采购
Inventory 库存
Warehouse functions 仓库功能
Supplier evaluations 供应商评估
JIT deliveries JIT 交货
Invoice verification 发票校验
Production planning 生产计划
CAD 计算机辅助设计
Process planning 工艺规划
Bill-of-material 物料清单
Product costing 产品成本
MRP 物料需求计划

Sales& distribution 销售配送
Financial& accounting 财务会计
Central Database 中央数据库
Materials management 物料管理

Financial accounting 财务会计
Investment management 投资管理
Cost control 成本控制
Treasury management 财政管理
Asset management 资产管理
Enterprise controlling 企业控制
Cost centres 成本中心
Profit centres 利润中心
ABC ABC 分类法
Capital budgeting 资本预算
Profitability analysis 盈利能力分析
Enterprise measures performance 企业绩效衡量

Project management 项目管理
Human resource 人力资源
Quality management 质量管理

Supply chain quality 供应链质量
Management 管理
Plant maintenance 设备维护
Customer service 客户服务

Personnel management 人事管理
Workforce planning 人力资源规划
Employee scheduling 员工调度
Training and development 培训和发展
Payroll and benefits 工资和福利
Travel expense reimbursement 出差费用报销
Applicant data 申请人资料
Job descriptions 职位描述
Organisation charts 组织架构图
Work flow analysis 工作流分析

Controlling the project phases 控制项目阶段
Quotation to design and approval 设计与批准报价
Resource management 资源管理
Cost settlement 费用结算

Fig. 5-1 ERP system modules ERP 系统模型

5.2.2 ERP 软件与系统的意义与效益

ERP 系统具有以下优点:

1)ERP 使全球一体化更容易。生产数据可以跨国界进行集成。

2)ERP 集成人和数据,同时不需要更新和修护许多独立的计算机系统。

3)ERP 允许管理人员实施管理操作,而不仅仅是监视它们。ERP 系统已经拥有所有的

project is painful, and customization is costly and time-consuming. The various shortcomings of the ERP systems, such as functionality and technicality, are shown in Fig. 5-2.

5.2.2 Significance and Benefits of ERP Software and Systems

ERP systems offer the following benefits:

1) ERP allows easier global integration. Data can be integrated across international borders.

2) ERP integrates people and data while eliminating the need to update and repair many separate computer systems.

3) ERP allows management to actually manage operations, not just monitor them. The ERP system already has all the data, allowing the manager to focus on improving processes. This focus enhances management of the company as a whole, and makes the organization more adaptable when change is required.

An ERP system can dramatically reduce costs and improve operational efficiency. For example, EZ-FLO International, Inc. , a family-run plumbing products manufacturer and distributor, had been experiencing double-digit growth and needed a scalable business platform to support its global growth. The company selected SAP Business All-in-One customized for wholesale distribution. As a result of the integrated process capabilities, EZ-FLO has been able to greatly improve its inventory management processes and has eliminated its annual inventory count, which used to take 100 employees two days to complete. In addition, the company has reduced manufacturing lead times in its domestic plants by two weeks. These process improvements led to improved customer service, and a 20% increase in the number of new customers—with a 12% growth in sales per customer.

5.2.3 Evolution of Enterprise Resource Planning

1. Step One—Material Requirements Planning (MRP)

ERP began life in the 1960s as material requirements planning (MRP), an outgrowth of early efforts in bill of material processing. MRP's inventors were looking for a better method of ordering material and components, and they found it in this technique. The logic of material requirements planning asks the following questions: What are we going to make? What does it take to make it? What do we have? What do we have to get?

This is called the universal manufacturing equation. Its logic applies wherever things are being produced whether they are jet aircraft, tin cans, machine tools, chemicals, cosmetics…or thanksgiving dinner.

MRP simulates the universal manufacturing equation. It uses the master schedule (What are we going to make?), the bill of material (What does it take to make it?), and inventory records (What do we have?) to determine future requirements (What do we have to get?). For a visual depiction of this and the subsequent evolutionary steps, are illustrated in Fig. 5-3.

2. Step Two—Closed-Loop MRP

Early users soon found that MRP contained capabilities far greater than merely giving better signals for reordering. They learned this technique could help to keep order due dates valid after the

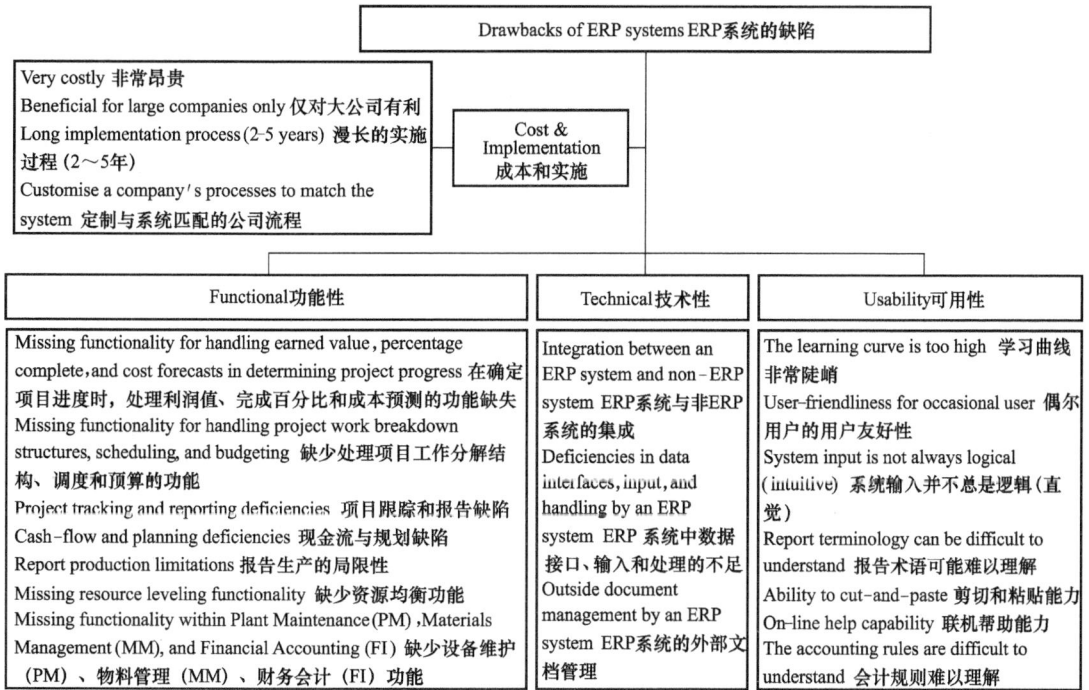

Drawbacks of ERP systems ERP系统的缺陷		

Cost & Implementation 成本和实施
- Very costly 非常昂贵
- Beneficial for large companies only 仅对大公司有利
- Long implementation process (2-5 years) 漫长的实施过程 (2～5年)
- Customise a company's processes to match the system 定制与系统匹配的公司流程

Functional 功能性	Technical 技术性	Usability 可用性
Missing functionality for handling earned value, percentage complete, and cost forecasts in determining project progress 在确定项目进度时，处理利润值、完成百分比和成本预测的功能缺失 Missing functionality for handling project work breakdown structures, scheduling, and budgeting 缺少处理项目工作分解结构、调度和预算的功能 Project tracking and reporting deficiencies 项目跟踪和报告缺陷 Cash-flow and planning deficiencies 现金流与规划缺陷 Report production limitations 报告生产的局限性 Missing resource leveling functionality 缺少资源均衡功能 Missing functionality within Plant Maintenance (PM), Materials Management (MM), and Financial Accounting (FI) 缺少设备维护 (PM)、物料管理 (MM)、财务会计 (FI) 功能	Integration between an ERP system and non-ERP system ERP系统与非ERP系统的集成 Deficiencies in data interfaces, input, and handling by an ERP system ERP系统中数据接口、输入和处理的不足 Outside document management by an ERP system ERP系统的外部文档管理	The learning curve is too high 学习曲线非常陡峭 User-friendliness for occasional user 偶尔用户的用户友好性 System input is not always logical (intuitive) 系统输入并不总是逻辑(直觉) Report terminology can be difficult to understand 报告术语可能难以理解 Ability to cut-and-paste 剪切和粘贴能力 On-line help capability 联机帮助能力 The accounting rules are difficult to understand 会计规则难以理解

Fig. 5-2　Drawbacks of ERP systems ERP 系统的缺陷

数据，允许管理人员专注于改进流程。这将提高公司的整体管理水平，使组织在需要变化时有更强的适应性。

ERP 系统可以大大降低成本，提高运营效率。例如，EZ-FLO 国际公司是一个家庭水暖产品制造商和销售商，需要一个可扩展的业务平台来支持其全球化增长。该公司选择 SAP 业务一体化定制的批发分销。得益于此系统的综合处理能力，EZ-FLO 已经大大优化了其库存管理流程，消除了它的年度盘点，以前这一工作需要 100 名员工 2 天才能完成。此外，该公司缩短了其国内工厂的生产提前期两周。这些流程的改进带来了客户服务的改进，并增加了 20% 的新客户——每个客户的销售额也增长了 12%。

5.2.3　ERP 的演变

1. 第一阶段——物流需求计划（MRP）

ERP 在 20 世纪 60 年代作为物料需求计划（MRP）开始出现，它是物料清单处理方面早期工作的产物。MRP 的发明者在寻找一个订购材料和部件的更好的方法，他们在寻找中发现了 MRP。物料需求计划的逻辑包括下述问题：要生产什么？用什么生产它？有什么？必须得到什么？这就是所谓的通用制造方程。它的逻辑适用于任何正在生产的东西，无论是喷气式飞机、锡罐、机床、化学品、化妆品，还是晚餐。

物料需求计划模拟通用制造方程。它使用主计划（要生产什么？）、物料清单（用什么生产它？）和库存记录（有什么？）确定未来的需求（必须得到什么？）。对于 MRP 和随后的进化阶段的可视化描述，如图 5-3 所示。

orders had been released to production or to suppliers. MRP could detect when the due date of an order (when it's scheduled to arrive) was out of phase with its need date (when it's required).

This was a breakthrough. The function of keeping order due dates valid and synchronized with these changes is known as priority planning.

So, did this breakthrough regarding priorities solve all the problems? Was this all that was needed? Hardly, the issue of priority is only half the battle. Another factor—capacity—represents an equally challenging problem (see Tab. 5-2).

Tab. 5-2　Priority vs. Capacity 优先级与能力

Priority 优先级	Capacity 能力
Which one? 哪一个?	Enough? 是否足够?
Sequence 顺序	Volume 数量
Scheduling 时序	Loading 负荷

Techniques for helping plan capacity requirements were tied in with material requirements planning. Further, tools were developed to support the planning of aggregate sales and production levels (sales & operations planning); the development of the specific build schedule (master scheduling); forecasting, sales planning, and customer-order promising (demand management); and high-level resource analysis (rough-cut capacity planning). Systems to aid in executing the plan were tied in: various plant scheduling techniques for the inside factory and supplier scheduling for the outside factory—the suppliers. These developments resulted in the second step in this evolution: closed-loop MRP (see Fig. 5-4).

Closed-loop MRP has a number of important characteristics:

1) It's a series of functions, not merely material requirements planning.

2) It contains tools to address both priority and capacity, and to support both planning and execution.

3) It has provisions for feedback from the execution functions back to the planning functions. Plans can then be altered when necessary, thereby keeping priorities valid as conditions change.

3. Step Three—Manufacturing Resource Planning (MRP Ⅱ)

Manufacturing resource planning or MRP Ⅱ (to distinguish it from material requirements planning, MRP) is a direct outgrowth and extension of closed-loop MRP. MRP Ⅱ is a method for the effective planning of all resources of a manufacturing company. Its functions cover the market operation plan, material supply, purchasing management, equipment management, cost management, workshop management, inventory management and other aspects. It implements the integrated managing system centered on business plan, sales plan, master production plan, material requirement plan, purchase plan, capacity plan up to production schedule, etc.

4. Step Four—Enterprise Resource Planning (ERP)

The fundamentals of ERP are the same as with MRP Ⅱ. However, thanks in large measure to enterprise software, ERP as a set of business processes is broader in scope, and more effective in dealing with multiple business units. Financial integration is even stronger. Supply chain tools, supporting business across company boundaries, are more robust. A graphical view of ERP is shown

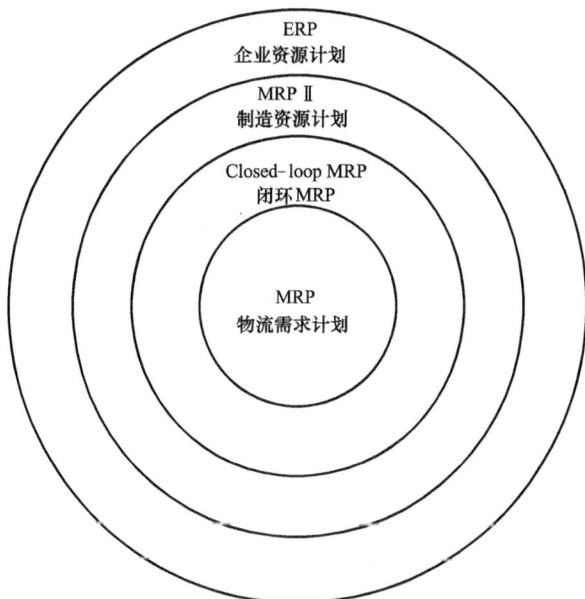

Fig. 5-3　Evolution Of ERP ERP 的演化

2. 第二阶段——闭环 MRP

早期的 MRP 用户很快就发现 MRP 的能力远比仅仅给出重新订购信息要好得多。在订单被发送到生产或供应商后，他们知道这种技术可以在到期日前帮助保持订单的有效性。MRP 可以检测到订单的到期日（预定到达的时间）与需求日期（需要时间）是不同的。这是一个突破。保持到期日前订单有效性的功能与这些变化同步被称为优先计划。

那么，这种突破优先考虑解决了所有的问题吗？这一切都是需要的吗？没有，优先权的问题只是问题的一半。另一个因素——能力——是一个同样具有挑战性的问题。（见表 5-2。）

能力需求计划的辅助技术与物料需求计划紧密相关。此外，支持总销售计划和生产水平（销售和运营规划）、专业的建设计划（主生产计划）、预测销售计划、客户订单承诺（需求管理）、高级资源分析（粗能力计划）等工具已经被开发，以协助执行系统和适用于内部工厂的各种调度技术和适用于外部工厂——供应商的供应商调度捆绑在一起。这些发展导致了这种演变的第二个阶段：闭环 MRP（见图 5-4）。

Fig. 5-4　Closed-loop MRP 闭环 MRP

in Fig. 5-5.

ERP is a direct outgrowth and extension of MRP, and includes all of MRP Ⅱ's capabilities. ERP is more powerful in that: ①applies a single set of resource planning tools across the entire enterprise; ②provides real-time integration of sales, operating, and financial data; ③connects resource planning approaches to the extended supply chain of customers and suppliers.

The primary purpose of implementing ERP is to run the business, in a rapidly changing and highly competitive environment, far better than before.

5. Step Five—The Future of ERP

ERP systems have now reached a level of maturity where both software vendors and users understand the technical, human resource and financial resources required for implementation and ongoing use. Hardware and software architectural platforms within and between firms will increasingly become commoditized with data modeling tools and translation software possessing the ability to move any amount of data in any format, and/or language, anywhere in near real-time. Portals to both internal and external business information will become commonplace. "Push" information based on user-defined interest areas will be integral to the enterprise system architecture.

Systems will become much more intelligent. Data mining and intelligence tools including expert systems, and advanced planning systems (with optimization) will increasingly be used to make/suggest business decisions. Simulation will become an increasingly important element of an integrated extended enterprise planning and execution system. Examples of major areas to receive the benefits of simulation include cost accounting, forecasting, capacity planning, order rate and response capacity planning, available to promise/capable to match, lead time, and supply network planning.

5. 2. 4　Implement of ERP

There are two different strategic approaches to ERP software implementation. In the first approach, an organization has to re-engineer the business process to accommodate the functionality of the ERP system, which means the profound changes in long-term management mode and the vacillation of the roles and duties of important persons. The other approach is customization of the software to fit the existing process, which will slow down the project, introduce dangerous bugs into the system and make upgrading the software to the ERP vendor's next release excruciatingly difficult, because the customizations will need to be torn apart and rewritten to fit with the new version.

Historically, ERP implementations have had to deal with the critical issue of changing the business process or modifying the software. Since each alternative has drawbacks, the solution can be a compromise between complete process redesign and massive software modification. However, many companies tend to take the advice of their ERP software vendor and focus more on process changes.

ERP implementation should involve the analysis of current business processes and the chance of re-engineering, rather than designing an application system that makes only the best of bad processes. Therefore, ERP implementation and BPR activities should be closely connected. In principle, it would be always better to carry out BPR in advance of ERP. Pragmatically, it may not be easy to do so because BPR is effort intensive and costs money and time. Also, carrying out BPR in advance

闭环 MRP 有一些重要的特征：

1）它具有一系列功能，而不仅仅是物料需求计划。

2）它包含用来解决优先级和能力的工具，并支持计划和执行。

3）它提供从执行功能到计划功能的反馈服务。因此，必要时可以改变计划，以在条件变化时保持优先事项的有效性。

3. 第三阶段——制造资源计划（MRP Ⅱ）

制造资源计划或称为 MRP Ⅱ，区别于物料需求计划 MRP，是闭环 MRP 的直接产物和延伸。制造资源计划（MRP Ⅱ）是一种有效规划制造公司所有资源的方法，其功能覆盖市场经营计划、物资供应、采购管理、设备管理、成本管理、车间管理、库存管理等方面，并执行以经营计划、销售计划、主生产计划、物料需求计划、采购计划、生产能力计划直至生产作业计划等为中心的一体化管理系统。

4. 第四阶段——企业资源计划（ERP）

ERP 的基本原理与 MRP Ⅱ一样。然而，在很大程度上要感谢企业软件的协助，ERP 的业务流程范围更广，可以更有效地处理多个业务单元，实现金融一体化，功能更强大，供应链工具也更加强大，可支持跨越公司边界的业务。图形化的 ERP 如图 5-5 所示。

Fig. 5-5　Enterprise resource planning 企业资源计划

企业资源计划是制造资源计划的直接产物和延伸，它囊括了 MRP Ⅱ 的所有能力。ERP 的功能更强大，这是因为：①ERP 拥有适用于整个企业的一组资源规划工具；②ERP 提供实时集成的销售、运营和财务数据；③ERP 把资源规划方法连接到包括客户和供应商的扩展供应链。

实施 ERP 的主要目的是在瞬息万变、竞争激烈的环境中比以前更好地经营企业。

5. 第五阶段——ERP 的未来

ERP 系统已经达到了成熟水平，软件供应商和用户都理解实施和持续使用 ERP 所需的技术、人力资源和财务资源。企业内和企业间的硬件和软件架构平台将日益商品化，数据建模工具和翻译软件可以接近实时地移动任何数量、任何格式和/或任何语言的数据。内部和外部业务信息的接入将成为家常便饭。企业系统架构将集成基于用户感兴趣的领域"推送"信息。

of ERP implies that the enterprises need to put resources into two successive projects. In addition, ERP packages offer many best business practices that might be worth including as a part of BPR. After the ERP implementation, one could get into continuous process re-engineering. Several enterprises may have different primary objectives in implementing ERP. They would probably fall in one of the following: standardization of objectives, BPR, elimination of organizational and technical bottlenecks, improvement in quality of information, replacement of out-of-date procedures and systems, integration of business processes, reduction in stand alone systems and interfaces, and covering areas previously neglected. The objectives and the corresponding expectations should be clearly documented.

Implementing an ERP system is an expensive and risky. In fact, 65% of executives believe that ERP systems have at least a moderate chance of hurting their businesses because of the potential for implementation problems. Numerous researchers have identified a variety of factors that can be considered to be procedures of an ERP implementation.

1. Strategic Aims

ERP implementations require that key people throughout the organization create a clear, compelling vision of how the company should operate in order to satisfy customers, empower employees, and facilitate suppliers for the next years. There must also be clear definitions of goals, expectations, and deliverables. The organization must carefully define why the ERP system is being implemented and what critical business needs the system will address.

2. Top Management Support

Successful implementations require strong leadership, commitment, and participation by top management. The implementation project should have an executive management planning committee that is committed to enterprise integration, understands ERP, fully supports the costs, demands payback, and champions the project.

3. Project Management

Successful ERP implementation requires that the organization engage in excellent project management. This includes a clear definition of objectives, development of both a work plan and a resource plan, and careful tracking of project progress. And the project plan should establish aggressive, but achievable, schedules that instill and maintain a sense of urgency. If management decides to implement a standardized ERP package without major modifications, this will minimize the need to customize the basic ERP code. This, in turn, will reduce project complexity and help keep the implementation on schedule.

4. Change Management

The existing organizational structure in most companies is not compatible with the structure, tools, and types of information provided by ERP systems. Therefore, implementing an ERP system may force the reengineering of key business processes and/or developing new business processes to support the organization's goals. The changes may significantly affect organizational structures, policies, processes, and employees. If people are not properly prepared for the imminent changes, then denial, resistance, and chaos will be predictable consequences of the changes created by the

系统将变得更加智能化。数据挖掘和智能工具包括专家系统和高级计划系统（优化）将越来越多地用于做出或建议业务决策。仿真模拟将成为集成、扩展的企业规划和执行系统的越来越重要的部分。应用仿真的主要领域包括成本核算、预测、能力规划、订单率和响应能力规划、承诺/能力匹配、交货时间和供应网络规划。

5.2.4　ERP 的实施

ERP 软件的实施有两种不同的战略方法。第一种方法是企业必须重新设计业务流程，以适应 ERP 系统的功能。这就意味着企业的长期经营方式的深刻变化和重要人物的角色及职责的动摇。另一种方法是定制软件以适应现有流程，这将减缓项目进度，在系统中引入危险的漏洞，而且使软件升级变得非常困难，因为定制需要打乱和改写以适应新的版本。

从历史上看，ERP 的实施必须应对业务流程的变化或修改软件的问题。由于每一种选择都有缺点，最终解决方案可以是重新设计业务流程和修改软件之间的妥协。然而，许多公司倾向于采纳他们的 ERP 软件供应商的意见，并更多地关注业务流程的变化。

ERP 的实施应包括当前业务流程与再造机会的分析，而不是设计一个应用系统使流程最好。因此 ERP 的实施与 BPR 活动应紧密相连。原则上在实施 BPR 后再实施 ERP 是比较好的。实际上，这可能不容易做到，因为 BPR 不但费力，而且花费大量的时间和金钱。同时，实施 ERP 前先实施 BPR 意味着企业需要投入两个连续的项目。此外，ERP 软件包提供了许多最佳业务实践，这可以作为 BPR 的一部分。ERP 实施后，一方面可以进入持续的流程再造，在实施 ERP 时一些企业的主要目标可能有不同，它们可能会是下面的一种：标准化的目标、业务流程再造、消除组织和技术的瓶颈、提高信息质量、更换过时的程序和系统、整合业务流程、减少独立系统和接口并覆盖以前被忽视的区域。应该清楚地记录企业的目标和相应的预期。

实施 ERP 系统是昂贵且危险的。事实上，65% 的高管认为，因为潜在的实施问题，ERP 系统至少有一定概率伤害到他们的企业。研究者已经确定了一些因素，它们被认为是实施 ERP 的关键点。

1. 战略目标

ERP 的实施需要整个组织的关键人物设立一个清晰的、引人注目的、公司在下一年应该如何运作以满足客户、员工、供应商的愿景。组织必须仔细定义为什么实施 ERP 系统，以及系统将解决哪些关键业务需求。

2. 高层管理的支持

成功实施 ERP 需要高层管理人员强有力的领导、承诺和参与。实施 ERP 项目应该有一个致力于企业整合、理解 ERP、完全支持成本、要求投资回报和捍卫项目执行的管理规划委员会。

3. 项目管理

成功的 ERP 实施要求组织实行优秀的项目管理。这包括明确的目标定义、工作计划和资源计划的制订以及详细的项目进度跟踪。项目计划应建立一个积极且可完成、使人具有紧迫感的日程表。如果管理层决定实施一个没有重大修改的标准化 ERP 软件包，这将最大限度地减少需要定制的 ERP 代码。反之，会降低项目的复杂性并帮助实施计划。

implementation. However, if proper change management techniques are utilized, the company should be prepared to embrace the opportunities provided by the new ERP system.

5. Implementation Team

The people of ERP implementation team should be entrusted with critical decision making responsibility. Management should constantly communicate with the team, but should also enable empowered, rapid decision making.

6. Education and Training

Education/training is probably the most important critical success factor. ERP implementation requires knowledge to enable people to solve problems. If the employees do not understand how a system works, they will invent their own processes using those parts of the system they are able to manipulate.

5.3 Supply Chain Management

Over the past dozen years, a wide spectrum of manufacturing and distribution companies have come to view the concept and practice of supply chain management (SCM) as perhaps their most important strategic discipline for corporate survival and competitive advantage. This is not to say that companies have been unmindful of the tremendous breakthroughs in globalization, information technologies, communications networking, e-commerce, and the Internet exploding all around them. It is widely recognized that these management practices and technology toolsets possess immense transformational power and that their ability to continuously innovate the very foundations of today's business structures have by no means reached it catharsis.

SCM is important because companies have come to recognize that their capacity to continuously reinvent competitive advantage depends less on internal capabilities and more on their ability to look outward to their networks of business partners in search of the resources to assemble the right blend of competencies that will resonate with their own organizations and core product and process strategies. Today, no corporate leader believes that organizations can survive and prosper isolated from their channels of suppliers and customers. In fact, perhaps the ultimate core competency an enterprise may possess is not to be found in a temporary advantage it may hold in a product or process, but rather in the ability to continuously assemble and implement market-winning capabilities arising from collaborative alliances with their supply chain partners.

Today, three major changes have enabled companies to actualize the power of supply chains to a degree impossible in the past. To begin with, today's technologies have enabled the convergence of SCM and computerized networking toolsets capable of linking all channel partners into a single trading community. Second, new SCM management concepts and practices have emerged that continually cross-fertilize technologies and their practical application. Finally, the requirements of operating in a global business environment have made working in supply chains a requirement. Simply, those companies that can master technology-enabled SCM are those businesses that are winning in today's highly competitive, global marketplace.

4. 变更管理

大多数公司现有的组织结构与 ERP 系统所提供的信息的结构、工具和类型是不兼容的。因此，实施 ERP 系统可能会迫使关键业务流程再造和/或开发新的业务流程以支持组织的目标。这些变化会显著影响组织结构、政策、流程和员工。如果人们对即将发生的变化没有做好充分的准备，那么拒绝、反抗和混乱将是实施 ERP 所带来的可预测的后果。然而，如果使用适当的变更管理技术，公司应该准备抓住新的 ERP 系统提供的机会。

5. 实施团队

ERP 实施团队的成员应该被委以关键决策的责任。管理层应不断与团队沟通，但也应启用授权和快速决策。

6. 教育和培训

教育和培训可能是最重要的关键成功因素。ERP 的实施需要能使人们解决问题的知识。如果员工不了解一个系统如何工作，他们将使用自己能操纵的那部分系统创造自己的流程。

5.3　供应链管理

在过去的十几年里，许多制造业和分销公司已经开始将供应链管理的概念和实践视为企业生存和获得竞争优势的最重要的战略方向。这并不是说公司不关注围绕全球化、信息技术的通信网络、电子商务和互联网爆炸等方面的巨大突破。这些管理措施和技术工具拥有巨大的变革力量，它们不断创新的能力形成了今天的企业结构的基础。

供应链管理是重要的，因为企业已经认识到它们不依赖于内部而更多地从企业外部业务合作伙伴网络中寻求资源与自己的组织和核心产品及流程战略的最佳组合，从而不断创造竞争优势的能力。今天，没有一个企业领导人认为企业可以独立于他们的渠道供应商和客户而生存和发展。事实上，企业拥有的最终核心竞争力可能不是拥有一种产品或流程的暂时优势，而是从供应链伙伴协作联盟获得的不断整合资源和赢得市场的能力。

今天，三个重大的变化使企业能够实现在过去不可能实现的供应链的力量。首先，今天的技术使供应链管理和计算机网络工具结合起来，能把所有的渠道合作伙伴链接成一个交易社区。其次，新的供应链管理的概念和实践出现，并不断融合技术及其实际应用。最后，全球商业环境运作呼唤供应链。简单地说，那些能够掌握供应链管理的公司能在竞争激烈的全球市场中获胜。

5.3.1　供应链管理的定义

1. 供应链

供应链不仅包括制造商和供应商，也包括运输、仓库、零售商甚至客户自己。在每个组织中，如制造商，供应链包括接受和满足客户请求所涉及的所有功能。这些功能包括但不限于运营、分销、财务和客户服务。

例如，当顾客走进沃尔玛商店购买洗涤剂，供应链开始于顾客对洗涤剂的需求。这个供应链的下一阶段是顾客所去的沃尔玛零售店。沃尔玛库存的货架上备有成品仓库或分销商使用第三方卡车供应的货物。分销商的库存由制造商提供（如 Procter & Gamble，下文称宝洁）。宝洁公司的生产工厂接受来自各种供应商的原材料，这些供应商可能还由下级供应商

5. 3. 1　Definition of Supply Chain Management

1. Supply Chain

A supply chain consists of all parties involved, directly or indirectly, in fulfilling a customer request. The supply chain includes not only the manufacturer and suppliers, but also transporters, warehouses, retailers, and even customers themselves. Within each organization, such as a manufacturer, the supply chain includes all functions involved in receiving and filling a customer request. These functions include, but are not limited to, operations, distribution, finance, and customer service.

Consider a customer walking into a Wal-Mart store to purchase detergent. The supply chain begins with the customer and his or her need for detergent. The next stage of this supply chain is the Wal-Mart retail store that the customer visits. Wal-Mart stocks its shelves using inventory that may have been supplied from a finished-goods warehouse or a distributor using trucks supplied by a third party. The distributor in turn is stocked by the manufacturer (say, Procter & Gamble [P&G] in this case). The P&G manufacturing plant receives raw materials from a variety of suppliers, who may themselves have been supplied by lower-tier suppliers. For example, packaging material may come from Pactiv Corporation (formerly Tenneco Packaging) while Pactiv receives raw materials to manufacture the packaging from other suppliers. This supply chain is illustrated in Fig. 5-6, with the arrows corresponding to the direction of product flow.

A supply chain is dynamic and involves the constant flow of information, product, and funds between different stages. In our example, Wal-Mart provides the product, as well as pricing and availability information, to the customer. The customer transfers funds to Wal-Mart. Wal-Mart conveys point-of-sales data as well as replenishment orders to the warehouse or distributor, who transfers the replenishment order via trucks back to the store. Wal-Mart transfers funds to the distributor after the replenishment. The distributor also provides pricing information and sends delivery schedules to Wal-Mart. Wal-Mart may send back packaging material to be recycled. Similar information, material, and fund flows take place across the entire supply chain.

The primary purpose of any supply chain is to satisfy customers' needs and, in the process, generate profit for itself. The term supply chain conjures up images of product or supply moving from suppliers to manufacturers to distributors to retailers to customers along a chain. This is certainly part of the supply chain, but it is also important to visualize information, funds, and product flows along both directions of this chain. The term supply chain may also imply that only one player is involved at each stage. In reality, a manufacturer may receive material from several suppliers and then supply several distributors. Thus, most supply chains are actually networks. It may be more accurate to use the term supply network or supply web to describe the structure of most supply chains, as shown in Fig. 5-7.

A typical supply chain may involve a variety of stages, including customers, retailers, wholesalers/distributors, manufacturers, component/raw material suppliers and so on.

Each stage in a supply chain is connected through the flow of products, information, and funds. These flows often occur in both directions and may be managed by one of the stages or an intermediary.

供应。例如，包装材料可能来自 Pactiv 公司，而 Pactiv 接收其他供应商的原材料制造包装材料。洗涤剂供应链的各个阶段如图 5-6 所示，箭头方向对应于产品的流动方向。

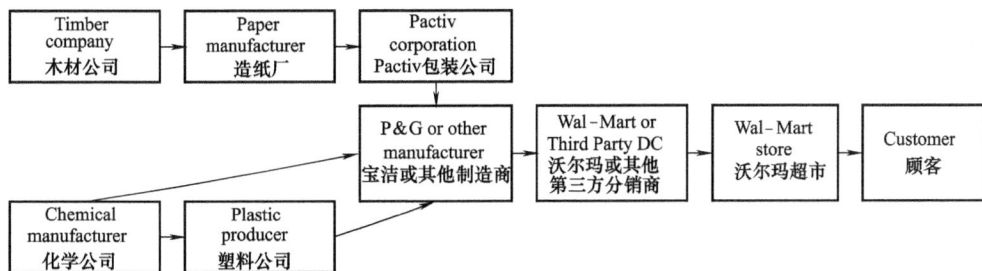

Fig. 5-6　Stages of a detergent supply chain 洗涤剂供应链的各个阶段

供应链是动态的，它涉及信息、产品和资金在不同阶段之间的不断流动。在上述的例子中，沃尔玛给顾客提供产品以及定价和可用性信息。顾客将资金转移到沃尔玛。沃尔玛向仓库或分销商传送销售数据和补货订单，分销商通过卡车将补货订单转移到仓库。沃尔玛将资金转移给经销商补货。分销商还提供定价信息，并将交货时间表发送到沃尔玛。沃尔玛可能会退回可回收的包装材料。类似的信息、材料和资金流发生在整个供应链中。

供应链的主要目的是满足客户的需求，并在这个过程中产生它的利润。从供应商、制造商、分销商、零售商到客户的产品或供应的链状结构被称为供应链。这当然是供应链的一部分，但沿着供应链的两个方向上的信息、资金和产品流动的可视化也是重要的。供应链这个术语让人联想到产品或供应沿着一条链从供应商到制造商，再到分销商，再到零售商，直到最终用户的样子。因此，大多数供应链实际上是网络结构。用"供应网络"或"供应网"来描述大多数供应链的结构可能更为准确，如图 5-7 所示。

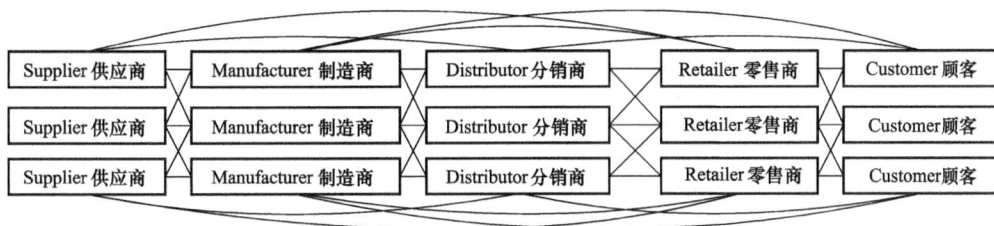

Fig. 5-7　Supply chain stages 供应链的阶段

一个典型的供应链可能涉及几个结点，包括顾客、零售商、批发商/分销商、制造商和零部件/原材料供应商等。供应链中的每个结点通过产品流、信息流和资金流联系起来。这些流经常是双向的，可以由一个结点或中介进行管理。

图 5-7 中的每个阶段不是必须出现在供应链中的。合理的供应链设计取决于客户的需求和所涉及的结点所扮演的角色。例如，为了满足客户需求，戴尔有两个供应链结构。对于需要定制个人计算机的团体客户和部分个体客户，戴尔按订单生产，也就是说，客户订单启动戴尔制造。在满足这些客户的供应链中，戴尔没有一个所谓的独立零售商、分销商或批发商。自 2007 年以来，戴尔也通过在美国的沃尔玛和中国的国美电器集团出售其个人电脑。

Each stage in Fig. 5-7 need not be present in a supply chain. The appropriate design of the supply chain depends on both the customer's needs and the roles played by the stages involved. For example, Dell has two supply chain structures that it uses to serve its customers. For its corporate clients and also some individuals who want a customized personal computer (PC), Dell builds to order; that is, a customer order initiates manufacturing at Dell. For these customers, Dell does not have a separate retailer, distributor, or wholesaler in the supply chain. Since 2007, Dell has also sold its PCs through Wal-Mart in the United States and the GOME Group, China's largest electronics retailer. Both Wal-Mart and the GOME Group carry Dell machines in inventory. This supply chain thus contains an extra stage (the retailer) compared to the direct sales model also used by Dell. In the case of other retail stores, the supply chain may also contain a wholesaler or distributor between the store and the manufacturer.

2. Definition of Logistics Management

Over the centuries, logistics has been associated with the planning and coordination of the physical storage and movement of raw materials, components, and finished goods. There are many definitions of logistics. The Council of Supply Chain Management Professionals (CSCMP) defines logistics as "that part of SCM that plans, implements, and controls the efficient, effective forward and reverse flow and storage of goods, services, and related information between the point of origin and the point of consumption in order to meet customers' requirements". In a similar vein, the Association for Operations Management (APICS) defines logistics as "the art and science of obtaining, producing, and distributing material and product in the proper place and in proper quantities".

A very simple, yet comprehensive defines logistics as consisting of the Seven Rs: that is, having the right product, in the right quantity and the right condition, at the right place, at the right time, for the right customer, at the right price.

Understanding the organizational boundaries and functional relationships of logistics can be seen by separating it into two separate, yet closely integrated spheres of processes, as illustrated in Fig. 5-8. The materials management sphere is concerned with the incoming flow of materials, components, and finished products into the enterprise. This sphere comprises the flow of materials and components as they move from purchasing through inbound transportation, receipt, warehousing and production, and presentation of finished goods to the delivery channel system. The sphere of physical distribution is concerned with the outbound flow of goods from the place of production to the customer. Functions in this sphere encompass warehouse management, transportation, value-added processing, and customer order administration.

Finally, logistics management is a connective function, which coordinates and optimizes all logistics activities as well as links logistics activities with other business functions, including marketing, sales, manufacturing, finance, and information technology.

3. Definition of Supply Chain Management

Although it can be said that logistics remains at the core of what supply chains actually do, the concept of SCM encompasses much more than simply the transfer of products and services through the supply pipeline. SCM is about a company integrating its process capabilities with those of its

沃尔玛和国美电器集团都在库存戴尔的机器。因此，相对于戴尔的直接销售模式，供应链中包含了一个额外的阶段（零售商）。在其他零售店的状况下，供应链也可能包含商品库存和制造商之间的批发商或分销。

2. 物流管理的定义

几个世纪以来，物流一直与原材料、零部件和成品的物理储存以及移动的规划和协调联系在一起。物流有许多定义。供应链管理专业协会（CSCMP）定义"物流是供应链过程的一部分，是对货物、服务及相关信息从起源地到消费地的有效率、有效益的正向和反向流动和储存进行计划、执行和控制，以满足顾客要求"。与此类似，运营管理协会（APICS）将物流定义为"在正确的地点，以正确的数量获得、生产和分配材料和产品的艺术和科学"。

关于物流的非常简单且全面的定义是7R，即将恰当的产品以恰当的数量和恰当的质量、恰当的价格，在恰当的时间，送到恰当的场所、恰当的顾客手中。

通过了解物流的组织边界和功能关系，可以看出，把物流功能分为两个单独的但紧密集成的过程领域，就可以理解物流的组织边界和功能关系，如图5-8所示。其中，材料管理涵盖企业的材料、零件、产成品的输入流，范围包括由采购而输入企业的材料和部件的运输、收货、仓储和生产以及产成品的配送系统。物质流通管理涉及货物从生产地到客户的输出流，范围包括仓库管理、运输、增值处理和客户订单管理。最后，物流管理是一个连接功能，它协调和优化所有物流活动以及其他业务部门，包括市场营销、销售、生产、财务和信息技术。

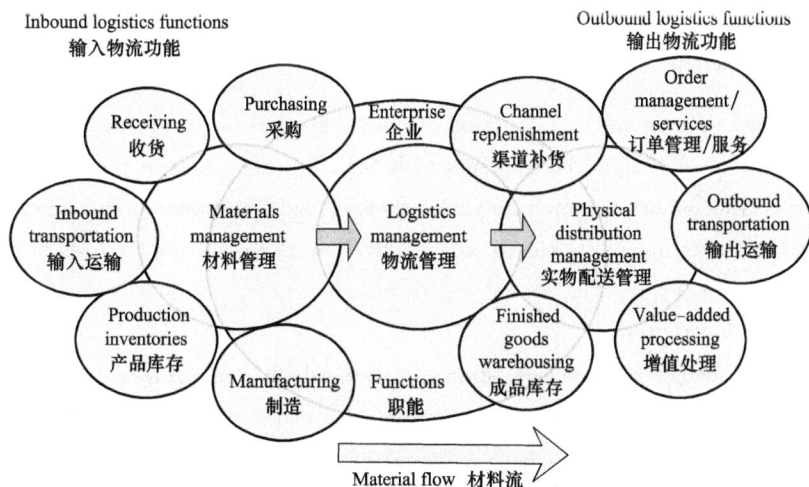

Fig. 5-8　Logistics management functions 物流管理职能

3. 供应链管理的定义

虽然可以说物流仍然是供应链的核心，但供应链管理的概念不仅仅是通过供应渠道来传递产品和服务。供应链管理是公司在战略层面上将自身的业务能力与其供应商和客户整合。综合供应链由许多贸易伙伴通过一个包含多层次的能力和各种类型关系驱动的协作网络同时参与。供应链管理在公司共同体中利用彼此优势建立高级供应和交付流程，为客户提供全面的价值，能够激发企业的协力优势。

suppliers and customers on strategic level. Integrative supply chains consist of many trading partners participating simultaneously in a collaborative network containing multiple levels of competencies and driven by various types of relationships. SCM enables companies to activate the synergy to be found when a community of firms utilizes the strengths of each other to build superlative supply and delivery processes that provide total customer value.

SCM can be viewed from several perspectives. Like most management philosophies, definitions of SCM must take into account a wide spectrum of applications incorporating both strategic and tactical objectives. For example, APICS defines SCM as the design, planning, execution, control, and monitoring of supply chain activities with the objective of creating net value, building a competitive infrastructure, leveraging worldwide logistics, synchronizing supply with demand, and measuring performance globally.

In their book *Designing and Managing the Supply Chain*, Simchi Levi and Kaminsky define SCM as a set of approaches utilized to efficiently integrate suppliers, manufacturers, warehouses, and stores, so that merchandise is produced and distributed at the right quantities, to the right locations, and at the right time, in order to minimize system wide costs while satisfying service level requirements.

Finally, the CSCMP defines SCM very broadly as encompassing the planning and management of all activities involved in sourcing and procurement, conversion, and all logistics management activities. Importantly, it also includes coordination and collaboration with channel partners, which can be suppliers, intermediaries, third-party service providers, and customers. In essence, SCM integrates supply and demand management within and across companies.

5.3.2 Supply Chain Competencies

The supply and delivery network-building attributes of SCM have revolutionized the role of the supply chain and infused channel constituents with radically new ways of providing total customer value. Instead of a focus just on purchasing, warehouse, and transportation management, today's SCM practices require entire supply chains to collectively work to activate an array of competencies, as illustrated in Fig. 5-9.

1. Customer Management

In the past, customer management strategies focused on optimizing economies of scale and scope by pushing standardized goods and services into the marketplace regardless of actual customer wants and needs. Today, companies can no longer compete by pursuing strategies built solely on volume and throughput, but instead must migrate to a supply chain perspective where the collective competencies and resources of their channel ecosystems can be leveraged in the pursuit of unique avenues of customer value and superior service. As customers demand to be more involved in product/service design, pricing, and configuration of their own buying solutions, companies have come to understand that creating customer value rests on establishing enriching customer relationships.

In today's era of the customer-centric marketplace, businesses cannot hope to survive without the activation of an array of techniques that have come to coalesce around CRM. The goal of CRM is to provide complete visibility to all aspects of the customer, from facilitating the service process, to

SCM 可以从几个角度看。与大多数管理哲学一样，SCM 的定义必须考虑到战略和战术目标在内的广泛应用。例如，APICS 把供应链管理定义为供应链活动的设计、规划、执行、控制和监督，其目的是创造网络价值、建设有竞争力的基础设施、充分利用全球物流、供应与需求同步与系统绩效评价。

Simchi-Levi 和 Kaminsky 在他们的《供应链设计与管理》著作中指出，供应链管理是为了满足服务水平要求，降低系统成本，利用一套高效整合供应商、制造商、仓库和商店的方法，生产和配送正确数量的商品，并在正确的时间送到正确的地点。

CSCMP 对供应链管理的定义非常广泛：供应链管理涵盖了涉及采购、外包和转化等过程的全部计划和管理活动以及全部物流管理活动。更重要的是，它包括了涉及提供商、中间商、第三方服务提供商和顾客、渠道伙伴之间的协调与合作。供应链管理在本质上是企业内部和企业之间的供给和需求管理的集成。

5.3.2 供应链的竞争力

供应链管理的供应和配送网络建设特性已经彻底改变了供应链的角色，而且注入了能够提供全面客户价值的全新方式的渠道因素。今天的供应链管理要求整个供应链协同工作从而激发一系列的能力，而不是专注于采购、仓库和运输管理，如图 5-9 所示。

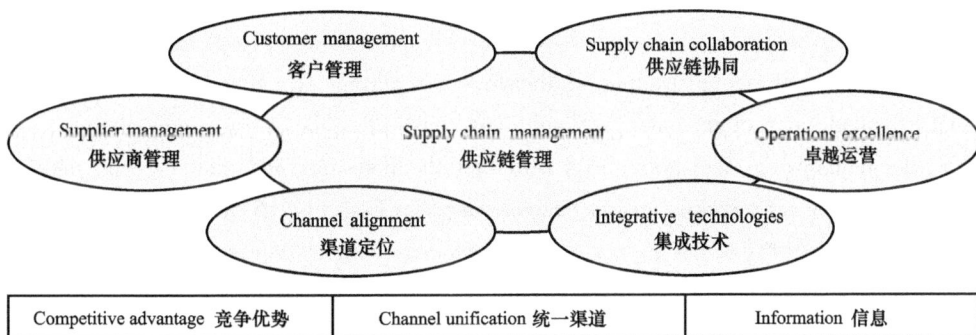

Fig. 5-9　Supply chain management competencies 供应链管理竞争力

1. 客户管理

在过去，客户管理战略的重点是推动标准化的商品和服务进入市场，优化经济规模和范围，而不管客户的实际想法和需求。今天，企业可以不再追求建立在单纯产量上的竞争策略，而是必须从供应链的视角，利用渠道生态系统的集体能力和资源，寻求实现客户价值和卓越服务的独特途径。公司已经认识到，客户的需求将更多地参与到产品和服务的设计、定价和他们自己的购买方案配置中，创造顾客价值的基础是建立丰富的客户关系。

在当今以客户为中心的市场，企业必须利用一系列围绕 CRM 的整合技术才能继续生存。CRM 的目标是为客户提供全方位的可见性，从促进服务流程到收集有关客户购买历史的数据到优化购物体验。

2. 供应商管理

有效的外包管理功能是供应链的核心竞争力。然而，管理采购不仅仅是购买产品和服

collecting data concerning customer buying history, to optimizing the buying experience.

2. Supplier Management

Effective management of the sourcing function resides at the very core of competitive supply chains. Managing procurement, however, is more than just the acquisition of products and services. In fact, for several decades, businesses have known that the relationship between buyers and sellers, and not just the price and quality of the goods, determined the real value-add component of purchasing. With the application of today's enabling technologies, this viewpoint has spawned a new concept and set of business practices termed supplier relationship management (SRM).

The mission of SRM is to activate the real-time synchronization of inventory and service requirements of buyers with the supply capabilities of channel partners. The goal is to actualize a customized, unique customer buying experience while simultaneously pursuing cost reduction and continuous improvement performance objectives. SRM seeks to fuse supplier management functions—information systems, logistics, resources, skills, cost management, and improvement—found across the entire supply chain into an efficient, seamless process driven by relationships founded on trust, shared risk, and mutual benefit.

3. Channel Alignment

The geography of any supply chain is composed of its supply and delivery nodes and the links connecting them. In the past, this network was characterized as a series of trading dyads. In this model, channel partners created trading relationships one partner at a time without consideration of the actual extended chain of customers and suppliers constituting the entire supply chain ecosystem. In reality, a company like Wal-Mart deals with literally thousands of suppliers, and their supply chain resembles more a networked grid of business partners (see Fig. 5-10). Maintaining a strategy of trading partner dyads as the supply chain expands risks decay of cost management objectives, leveraging resource synergies, and maintaining overall marketplace competitiveness.

4. Supply Chain Collaboration

The keystone of SCM can, perhaps, be found in the willingness of supply network partners to engage in and constantly enhance collaborative relationships with each other. Collaboration can be defined as an activity pursued jointly by two or more entities to achieve a common objective. It can mean anything from exchanging raw data by the most basic means, to the periodic sharing of information through technology-based tools, to the structuring of real-time architectures capable of leveraging highly interdependent infrastructures in the pursuit of complex, tightly integrated functions ensuring planning, execution, and information synchronization.

The intensity of the collaborative content can vary. It can be internally driven and focused on the achievement of local objectives. On the other hand, it could seek to use technology to deepen interchannel operations linkages, drive shared processes and codevelopment, and even foster a common competitive vision for the whole channel. The value of collaboration is gauged by how effectively firms are leveraging the competencies of the distributed knowledge of the channel base, reducing redundant functions and wastes, sharing a common vision of the supply chain, and constructing the technical and social architectures, thereby enabling whole channel networks to achieve marketplace leadership.

务。事实上，几十年来，企业认识到买卖双方之间的关系决定了购买的真正价值，而不仅仅是商品的价格和质量。随着新技术的应用，这一观点已催生了被称为供应商关系管理（SRM）的新概念和商业行为。

SRM 的任务是依靠渠道合作伙伴的供应能力使库存和顾客的服务要求实现即时同步。其目标是实现定制的、独特的客户购买体验，同时追求降低成本和持续提高绩效。SRM 旨在融合信息系统、物流、资源、能力、成本管理和改善供应商管理功能，使整个供应链成为建立在信任、风险共担、利益共享关系基础上的高效、无缝的流程。

3. 渠道定位

任何供应链在地理上都是由供应和交货节点以及连接他们的链条组成的。在过去，这个网络具有一系列交易关系的特点。在这种模式下，渠道合作伙伴仅和一个合作伙伴创建交易关系，而不考虑客户和供应商链条的扩展实际上形成了整个供应链生态系统。在现实中，例如，沃尔玛要处理数以千计的供应商公司，其和他们的供应链更像一个商业伙伴网络（图 5-10）。保持贸易伙伴关系战略使供应链扩展成本管理目标的风险衰减，可以充分利用资源的协同效应，并保持整体的市场竞争力。

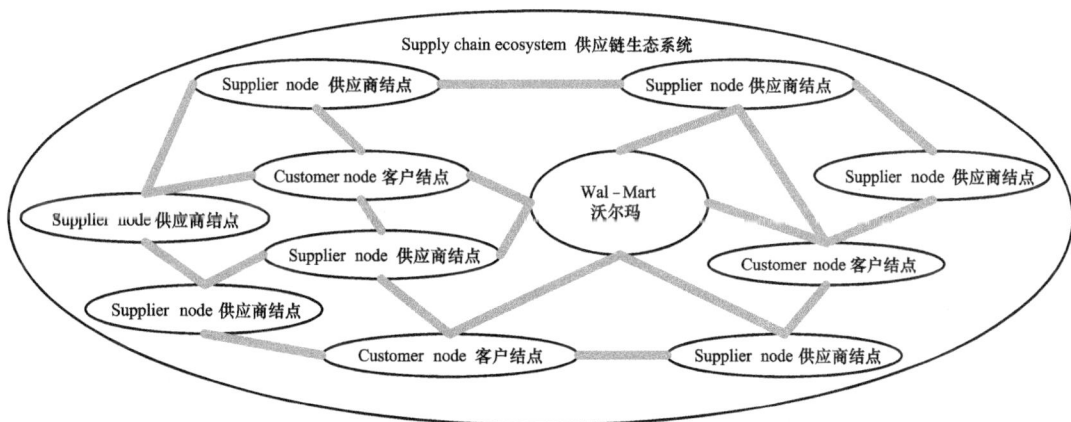

Fig. 5-10 Supply chain as a network grid 供应链网络

4. 供应链协同

供应网络合作伙伴的参与意愿和不断加强相互协同关系是供应链管理的基石。协同可以概括为两个或更多的实体共同追求并实现共同的目标的一种活动。这意味着追求复杂的、功能紧密集成的确保计划、执行和信息同步的所有事务，从用最基本的方式交换原始数据，到通过技术工具周期性共享信息，到利用高度相互依赖的基础设施的实时结构组合。

协作内容的强度可以变化。它可以内部驱动，专注于实现本地目标。另一方面，它可能会寻求利用技术深化渠道间的业务联系，推动共享的流程和共同发展，甚至形成全渠道共同竞争体。协同的价值通过公司有效地利用渠道基础的分布式知识的能力，减少冗余功能和浪费，共享供应链共同体，构建技术和社会架构，使整个渠道网络成为市场领导来体现。

5. 卓越运营

所有组织都寻求优化生产，同时降低成本。在供应链管理中，这个目标变得更加重要。理想的情况下，卓越运营迫使在渠道网络中的每一个公司不但优化自身的绩效，而且通过扩

5. Operations Excellence

All organizations seek to optimize productive functions while removing costs, and this objective becomes even more critical in SCM. Ideally, operations excellence compels every firm in the channel network to optimize both their own performance and, by extension, the performance of the entire supply chain. Simply, as the competency of each individual channel node is increasingly integrated with other participants, collectively the supply chain will have access to a range of processes and benefits individual companies would be incapable of achieving acting on their own. By the collaborative nature of SCM, supply network participants are compelled to look beyond the performance of their own organizations to the supply chain in its entirety.

6. Integrative Technologies

It has been argued earlier that it is virtually impossible to think of SCM without the power of the enabling technologies that have shaped and driven its development into a management science. In today's highly competitive global marketplace, having the best product or service is simply not enough: Now, having the best information has determined market leaders and followers. Supply channel transparency requires a single version of data encompassing demand, logistics, demand-capability alignment, production and processes, delivery, and supplier intelligence across the firms comprising the end-to-end supply chain. Generating a single view of the supply chain requires information technologies that enable collection, processing, access, and manipulation of complex views of data necessary for determining optimal supply chain design and execution configurations.

Technology applications to SCM can be divided into three broad areas. To begin with, the most complex technology toolsets can be found in the utilization of major business systems, such as ERP, that seek to cover the entire enterprise. On the next level can be found point technology solutions, such as advanced planning systems (APS) and transportation management systems (TMS), which seek to optimize specific functions or provide for cross-channel visibility. In the third and final level of technology tools can be found execution solutions, such as electronic data interchange (EDI), the Internet, or radio frequency identification (RFID).

5.3.3 Current SCM

Today, SCM has progresses to Stage 5 in its development. Through the application of integrative information technologies like the Internet, SCM has evolved into a powerful strategy capable of generating digital sources of competitive value based on the real-time convergence of networks of suppliers and customers into collaborative supply chain systems. Actualizing technology-integrated SCM is a three-step process. Companies begin first with the integration of supply channel facing functions within the enterprise through technology solutions like ERP. The next step would be to integrate across trading partners channel operations functions such as transportation, channel inventories, and forecasting. Finally, the highest level would be achieved by utilizing the power of technologies such as the Internet and e-business to synchronize the entire supply network into a single, scalable "virtual" enterprise capable of optimizing core competencies and resources from anywhere at any time in the supply chain to meet market opportunities.

展优化整个供应链的绩效。简单地说，由于每个单独的渠道节点的能力与其他参与者越来越集成，总的供应链将获得一系列的流程优化和利益，这是单个公司靠自己的行动无法实现的。对于供应链整体而言，供应链管理的协同性迫使供应网络参与者超越自身的组织绩效。

6. 集成技术

曾有人认为，如果没有使其发展成为管理科学的有利技术的力量，就几乎不可能想到供应链管理。在当今竞争激烈的全球市场上，拥有最好的产品或服务是不够的；现在，谁拥有最好的信息决定了谁是市场领导者和追随者。供应渠道的透明需要单一版本的数据，包含跨公司组成的端到端供应链的所有数据，涵盖需求、物流、需求能力调整、生产加工、配送和供应商的情报。形成单一供应链视图要求信息技术能够收集、处理、访问和操作所需数据的复杂视图，以确定最佳的供应链设计和执行配置。

供应链管理的技术应用可分为三大领域。首先是专业系统中的最复杂的技术方案，如ERP，试图覆盖整个企业。第二个层面是点的技术解决方案，如高级计划系统（APS）和运输管理系统（TMS），目标是优化特定的功能或提供跨渠道的可视化。第三个层面即最后一级是执行解决方案，如电子数据交换（EDI）、互联网或无线电频率识别（RFID）。

5.3.3　当前的供应链管理

随着互联网等综合信息技术的应用，供应链管理已经发展成为一个强大的战略，能够基于实时汇聚的供应商和客户的协同供应链系统，产生有竞争价值的数字资源。实施技术集成供应链可以分三步走：首先，公司利用像ERP这样的技术解决方案，实施面向功能的供应渠道集成；其次是跨贸易伙伴的渠道运作功能整合，如运输、渠道库存和预测；最后，利用如互联网和电子商务的技术力量，协同整个供应网络成单一的、可扩展的"虚拟"企业，在供应链中优化核心能力和资源，从而可以在任何时间任何地点满足市场机会，实现供应链的最高水平。

信息集成技术的应用给供应链管理带来了巨大的变化，这些变化的要点如下：

1. 产品与工艺设计

由于产品生命周期的不断缩短和开发成本持续飙升，公司应迅速利用技术手段使客户和设计过程联系起来，促进协作，组成跨公司设计团队，整合有形与无形资产和能力增加响应市场的速度和提高从时间获得的利润。一方面，利用过去传统的产品数据管理系统和设计数据交换方式成本高，手续烦琐，效率低下；另一方面，现在的互联网技术可以使贸易伙伴之间的联系具有互操作性、成本低和实时联系的特点。

2. 电子市场和交易

传统上，买家和供应商关注以长期关系、冗长合同的谈判、长的提前期和固定利润为特征的自营渠道。今天，像互联网这样的技术正在彻底改变这种环境。现在，公司可以在各种各样的互联网市场购买和销售，范围包括独立和私下的交易到拍卖网站。

3. 协同规划

从历史上看，企业不愿分享有关预测、销售需求、供应需求和新产品介绍的关键规划信息。今天，许多组织越来越多地将非核心业务外包给网络合作伙伴，能够实时转移规划信息到"虚拟"供应链已成为一种必然。集成技术成为了跨商业网络转移产品和规划信息实现联合决策所必需的双向协作的关键。

The application of integrative information technologies brought about the enormous changes to SCM, the high points of these changes are as follows:

1. Product and Process Design

As product life cycles continue to decline and development costs soar, firms have been quick to utilize technology enablers to link-in customers to the design process, promote collaborative, cross company design teams, and integrate physical and intellectual assets and competencies in an effort to increase speed-to-market and time-to-profit. In the past, efforts utilizing traditional product data management systems and exchange of design data had been expensive, cumbersome, and inefficient. Internet technologies, on the other hand, now provide interoperable, low cost, real-time linkages between trading partners.

2. E-Marketplaces and Exchanges

Buyers and suppliers have traditionally been concerned with proprietary channels characterized by long-term relationships, negotiation over lengthy contracts, long-lead times, and fixed margins. Today, technologies like the Internet are completely reshaping this environment. Companies can now buy and sell across a wide variety of Internet-enabled marketplaces ranging from independent and private exchanges to auction sites.

3. Collaborative Planning

Historically, enterprises were averse to sharing critical planning information concerning forecasts, sales demand, supply requirements, and new product introduction. Today, as many organizations increasingly outsource non-core functions to network partners, the ability to transfer planning information real-time to what is rapidly becoming a "virtual" supply chain has become a necessity. Today, integrative technologies provide the backbone to transfer product and planning information across the business network to achieve the two-way collaboration necessary for joint decision-making.

4. Fulfillment Management

The collapse of the dot-com era at the beginning of the century revealed one of the great weaknesses of e-business. Customers may have access to product information and can place orders at the speed of light, but actual fulfillment is still a complex affair that occurs in the physical world of materials handling and transportation. Solving this crucial problem requires the highest level of supply chain collaboration and takes the form of substituting as much as possible information for inventory. Some of the methods incorporate traditional tools, such as product postponement, while others utilize Web-based network functions providing logistics partners with the capability to consolidate and ship inventories from anywhere in the supply network and generate the physical infrastructures to traverse the "last mile" to the customer.

5.3.4　Examples of Supply Chains

Wal-Mart has been a leader at using supply chain design, planning, and operation to achieve success. From its beginning, the company invested heavily in transportation and information infrastructure to facilitate the effective flow of goods and information. Wal-Mart designed its supply chain with clusters of stores around distribution centers to facilitate frequent replenishment at its retail

4. 执行管理

21世纪初，互联网时代的崩溃揭示了电子商务的一大弱点。客户可以访问产品信息，快速下订单，但发生在材料处理和运输的物理世界的实际履行仍是一项复杂的事务。解决这个关键问题需要最高水平的供应链合作并采取方法替代众多的库存信息。一些方法包含传统的工具如延迟制造，同时，其他企业使用基于 Web 的网络功能，能够在供应网络上的任何地方合并和运输库存，实现在物理设施上通过"最后一公里"，将产品送达客户。

5.3.4 供应链案例

沃尔玛一直致力于利用供应链的设计、规划和运作以实现商业成功。公司从一开始就大量投资于交通和信息基础设施，借以促进货物和信息的有效流动。沃尔玛在设计供应链的同时，将零售商店放在了配送中心周围，使零售商店可以以经济的方式频繁补货。频繁补货使商店比竞争对手更有效地匹配供应和需求。沃尔玛一直是以信息共享和与供应商协同来降低成本、提高产品可用性的领导者。在其 2010 年的年报中，该公司报告的销售收入约为 4080 亿美元，净收入超过 143 亿美元。这对一个在 1980 年销售额只有 10 亿美元的公司来说是个非常好的结果。销售额增长意味着年复合增长率超过 20%。

戴尔是一个在其供应链的设计、规划和运营上取得了巨大成功的例子，但随后不得不适应其供应链，以应对技术和客户期望的变化。从 1993 年到 2006 年，通过构建供应链，并以合理的成本迅速为客户提供定制的个人计算机，戴尔实现了收入和利润的空前增长。到 2006 年年底，戴尔的净收入超过 35 亿美元，收入超过 560 亿美元。这一成功是基于两个关键的供应链功能：快速支持和低成本定制。首先是戴尔决定直接向终端客户销售，绕过分销商和零售商。戴尔的供应链的第二个关键方面是在几个地方集中生产和库存，在接到顾客订单后再进行最终装配。结果，戴尔能够提供大量的个人计算机配置，同时保持低水平的零件库存。

5.4 制造执行系统

为了在生产中实现有效的价值创造，需要能完全满足这些新需求的设备。现有市场上的 ERP 系统主要是行政管理和会计系统。所需要的新系统必须包括能进行规划、记录和控制并实时做出反应的功能。针对这些需求提出了制造执行系统（MES）的概念。由于 MES 涉及多个领域，每个部门都站在自己的立场解读此概念。此外，还有各种软件供应商提供他们自己的系统作为 MES 系统。

5.4.1 MES 的定义

1990 年，美国先进制造研究机构的报告中首次提出"制造执行系统"这一概念，并将制造执行系统定义为"位于上层的计划管理系统与底层的工业控制之间的面向车间层的管理信息系统"。该系统集成了车间中生产调度、工艺管理、质量管理、设备维护、过程控制等相互独立的系统，使这些系统之间的数据实现完全共享，完全解决了信息孤岛状态下的数据重叠和数据矛盾的问题。同时，制造执行系统可以收集生产过程中大量的实时数据，对实时事件进行及时处理的同时与计划层和生产控制层保持双向通信能力，从上下两层接收相应

stores in a cost-effective manner. Frequent replenishment allows stores to match supply and demand more effectively than the competition. Wal-Mart has been a leader in sharing information and collaborating with suppliers to bring down costs and improve product availability. The results are impressive. In its 2010 annual report, the company reported a net income of more than $ 14. 3 billion on revenues of about $ 408 billion. These are dramatic results for a company that reached annual sales of only $ 1 billion in 1980. The growth in sales represents an annual compounded growth rate of more than 20%.

Dell is an example of a company that enjoyed tremendous success based on its supply chain design, planning, and operation but then had to adapt its supply chain in response to shifts in technology and customer expectations. Between 1993 and 2006, Dell experienced unprecedented growth of both revenue and profits by structuring a supply chain that provided customers with customized PCs quickly and at reasonable cost. By 2006, Dell had a net income of more than $ 3. 5 billion on revenues of just over $ 56 billion. This success was based on two key supply chain features that supported rapid, low-cost customization. The first was Dell's decision to sell directly to the end customer, bypassing distributors and retailers. The second key aspect of Dell's supply chain was the centralization of manufacturing and inventories in a few locations where final assembly was postponed until the customer order arrived. As a result, Dell was able to provide a large variety of PC configurations while keeping low levels of component inventories.

5. 4   Manufacturing Executive System

In order to achieve effective value creation in production, equipment is needed to meet these new demands completely. Existing enterprise resource planning (ERP) systems established on the market are largely administrative and accounting systems. The new systems needed must include functions for planning, logging, and control that not only act but also react in real time. For these systems, the concept of a manufacturing execution system (MES) has arisen. Since MES is a multi-faceted area, each sector interprets the concept from its own standpoint. In addition, there are various software suppliers who are offering their systems as MES.

5. 4. 1   Definition of MES

The concept "manufacturing execution system (MES)" was put forward for the first time in the report of advanced manufacturing research institutions of America in 1990, and is defined as "workshop-oriented management information system between the project control system in the upper and the industrial control at the bottom". The system integrated the production scheduling, process management, quality management, equipment maintenance, process control and other mutually independent systems in the workshop. It makes the data among these systems achieve full sharing and solves completely the problem of data overlapping and data conflict under the condition of information island. Besides, MES can collect a large amount of real-time data in the production, and handle the

数据并反馈处理结果和生产指令。

MES 发展成为柔性生产、网络化生产的战略工具。所有的生产管理任务都归结到一个集成平台。因此 MES 不是一个松散的软件组件（混合体），而是允许使用个别功能模块且能够被公司其他软件系统使用的集成系统，例如采用面向服务的架构方式。

作为一个数据库，MES 需要一个完整的、一致的数据模型，包含生产路线及所有资源和生产数据（或者说产品定义数据）。因此，MES 必须和 PLM 系统紧密结合，携手工作。主数据包含在 MES 中，例如要生产的物品，通过主数据管理也能被其他 IT 系统使用。

5.4.2 制造执行系统的功能结构以及与其他信息系统的关系

制造执行系统在计划管理层与底层控制之间架起了一座桥梁，填补了两者之间的空隙，如图 5-11 所示。

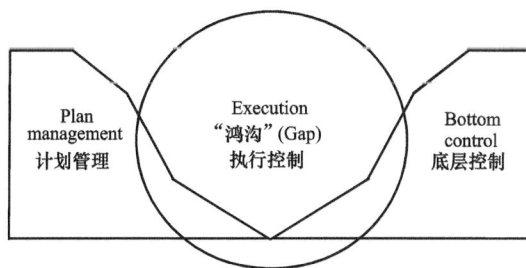

Fig. 5-11 Bridge function of MES 制造执行系统的桥梁作用

AMR 于 20 世纪 90 年代提出如图 5-12 所示的企业集成模型，它清楚地描述了制造执行系统在企业系统中的位置。

计划层强调企业的计划性，它以客户订单和市场需求为计划源头，充分利用企业内的各种资源，降低库存，提高企业效益。执行层强调计划的执行和控制，通过制造执行系统把企业资源计划与企业的生产现场控制有机地集成起来。控制层强调设备的控制，包括分布式控制系统、可编程控制器、分布式数控等。过程控制系统（process control system，PCS）结合了包括顺序、运动和过程控制在内的多种控制平台，并具备更强的信息处理能力。这种技术提供了开放的工业标准、增强的区域功能以及公共的开发平台。

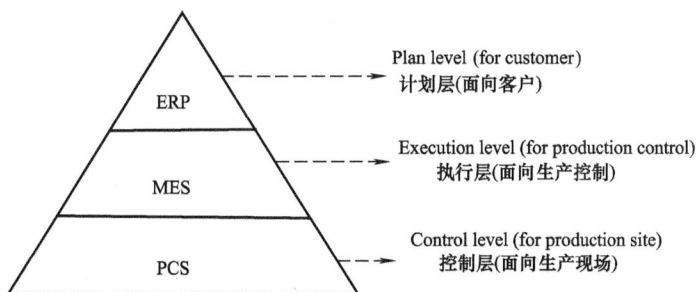

Fig. 5-12 3-level enterprise integration model 三层企业集成模型

以上三层在企业经营与生产过程中既相互独立又紧密联系，并在部分功能上存在着信息重叠的现象。

real-time event in time and maintain two-way communication capability from the production plan level and production control level at the same time, receive the corresponding data from the upper and lower levels and feed the processing result and production instructions back.

The MES is developing into a strategic instrument for flexible, networked production. All production management tasks are summarized in an integrated platform. The MES therefore is not a loose collection of software components (patchwork) but rather an integrated system that allows the modular use of individual functions and makes these functions available to other software systems in the company, for example, by means of a service-oriented architecture.

As a database, the MES requires a complete and consistent data model that contains both a map of production with all its resources and the product data (or rather, the data for product definition). Therefore, the MES must be closely integrated to the PLM system and work hand in hand with it. The master data contained in the MES, for example, on the articles to be produced, are managed through master data management and are also made available to other IT systems.

5.4.2 Functional Structure of MES and Relationship with Other Information Systems

MES plays the role of bridge between the plan management and the bottom control, and fills the gap between them, as shown in Fig. 5-11.

MES proposed enterprise integration model in 1990s, as shown in Fig. 5-12, which described clearly the position of the MES in enterprise system.

Plan level emphasizes the planning of the enterprise. Taking customer order and market demand as the source of the plan, it makes full use of different resources in the enterprise to reduce inventory and improve the benefits of enterprise. Execution level emphasizes on the implementation and control of the plan. It integrates organically the EPS with the production control through MES. Control layer emphasizes the control of the equipment. It includes distributed control system, programmable logic controller, distributed numerical control system, etc. Process control system (PCS) combines multiple control platforms including sequence control, motion control and process control and has stronger information processing ability. This technology provides open industrial standards, enhanced regional functions, and public development platforms.

The three levels mentioned above are independent and closely related in the enterprise management and production process, and have the information overlapping phenomenon in part of functions.

MES is based on the manufacturing process, it shares and interacts information inevitably with other manufacturing management systems. These systems include SCM, ERP, SSM, P/PM and bottom production control and other management systems. MES plays the role of the information concentrator to connect above information systems. Fig. 5-13 shows the relationship between MES and other management systems of enterprise.

5.4.3 Functions of MES

The 11 function groups of a MES are as follows:

制造执行系统是面向制造过程的，它必然与其他的制造管理系统共享和交互信息。这些系统包括供应链管理、企业资源计划管理、销售和客户服务管理、产品及产品工艺管理以及底层生产控制等管理系统。制造执行系统起到连接以上各信息系统的信息集线器的作用。图 5-13 所示为制造执行系统与企业其他管理系统之间的关系。

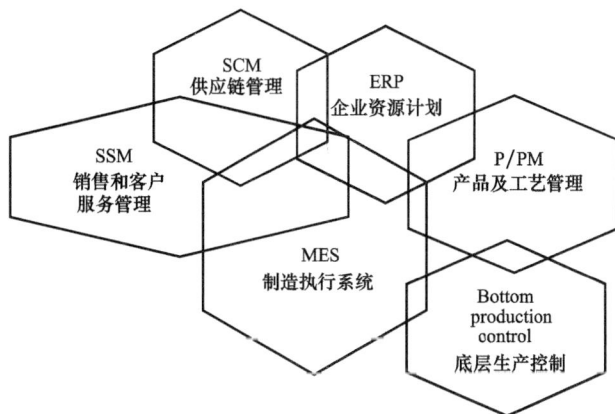

Fig. 5-13　Relationship between MES and other management systems of enterprise
制造执行系统与企业其他管理系统之间的关系

5.4.3　MES 的功能

MES 具有 11 个功能：

1）详细的工作计划。根据可用资源，相关的基础条件（设置时间、处理时间等）确定最优序列计划。

2）具有状态维护的资源管理。管理和监控相关资源（员工、机器、工具等）。

3）生产单元控制。根据订单、批次等及时控制产品流，如果有必要，也直接调整计划。

4）信息控制。员工可以在正确的时间和地点查阅所有与生产过程相关的信息（CAD、设计、测试规范、环境规范要求、安全说明等），并可以使用该系统记录偏差。

5）操作数据记录。自动或人工记录所有与产品生产相关的操作数据。

6）员工管理。记录员工的工作时间并能够在员工缺勤、度假等情况下进行编辑。

7）质量管理。实时分析生产相关测量数据，以保证产品质量，及时发现问题和薄弱环节。

8）过程管理。监控实际生产过程，包括报警管理功能。

9）维护管理。记录使用的操作材料和使用时间，以启动定期和预防性维护任务。该系统还支持维护的执行。

10）产品的可追溯性。记录整个生产链中所有与生产相关的数据以确保生产的每一件产品都可追溯。

11）性能分析。如果可行的话，快速生成从制造尺寸到停机时间、中断时间、件数等生产的关键管理数据，以便于简单评估生产率和发现问题，并用各种图表形式提供给用户。

1）Fine planning of workflow. This group envisions optimal sequence planning with regard to the relevant basic conditions (setup times, processing time, etc.) based on the resources available.

2）Resource management with status maintenance. Management and monitoring of the relevant resources (staff, machines, tools, etc.).

3）Production unit control. Control of the flow of production units based on orders, batches, etc. Events during ongoing production are responded to immediately, and if necessary, the plan is adjusted.

4）Information control. All information relevant to the production process (CAD, designs, test specifications, environmental compliance requirements, safety instructions, etc.) is made accessible to the staff at the right time and right place. Staff can use the system to record deviations.

5）Operating data logging. Automatic or manual logging of all production-related operating data linked with the production unit.

6）Staff management. Recording of staff working hours and potential to edit in case of absence, holiday, etc.

7）Quality management. Analyses of production-related measurement data in real time in order to safeguard product quality and be able to identify problems and weak points in good time.

8）Process management. Monitoring of the actual production process, including alarm management functions.

9）Maintenance management. Recording the use of operating material and hours of use in order to initiate periodic and preventive maintenance tasks. The system also supports the execution of maintenance.

10）Lot traceability. Recording of all production-related data across the entire production chain to ensure that every product manufactured is traceable.

11）Performance analysis. From the manufactured sizes to down time, disruptions, piece counters, etc., managerial key figures are produced promptly, in real time, if feasible, in order to allow for simple assessment of production efficiency, detection of problems, etc. Display in various diagram formats is made available to the user.

5.4.4   Benefits of a MES

1. Integrated Data Transparency

What is lacking in today's production systems is the punctual recording of production data and their usually isolated evaluation using table calculation. Generally, there is no integrated overall picture of all data for an assessment of the overall situation. A MES is the instrument for integrated data recording and performance monitoring in real time and, on the other hand, for more long-term analyses.

2. Reducing Time Usage

Savings are obvious when processing times for production orders are reduced. The time usage for processing an order can be divided into the following time areas: administrative processing, operative order planning, setup, production, interim storage, and final storage.

5.4.4　MES 的优势

1. 综合数据透明化

在现在的生产系统中，缺乏的是准时记录生产数据，通常用表格计算来进行孤立的评估。一般来说，对整体形势进行评估时，并没有完整地了解所有的数据。MES 一方面是记录集成数据和实时监控性能的工具，另一方面也是进行较长期分析的工具。

2. 减少使用时间

当生产订单的处理时间减少时，节省时间是显而易见的。处理订单的使用时间可分为以下时间区域：行政处理、运作顺序规划、计划、生产、临时储存和最后存储。

3. 减少管理费用

实施 MES，其目标就是要消除或大大减少不直接创造价值的活动。在管理合理化领域的分析得出的结论是，仅用传统方法至少可以节省 20% 的管理费用。如果把这些分析与MES 结合起来，节约效果必定会上升到 30% 以上。

4. 改善对客户的服务

在今天的市场上，可靠的交货日期和订单进度的信息是必不可少的。有了 MES，这些要求可以得到适当的满足。虽然这些好处不能直接测量，但生产公司形象的改善对于订单的增加是很有助益的。

5. 提高质量

通过集成的 MES 提供持续不断的、自动的测量和监控机器的工艺参数支持，确保工艺能力。如果正确使用这个工具，将可能大大减少废品和返工，也会因此降低与此相关的成本。

6. 预警系统，实时成本控制

通过对生产过程中各影响参数的实时控制，可以立即确定不可接受的偏差并采取相应的措施。在超出计划费用的原因方面，实时成本控制起到一个早期预警系统的作用。

7. 提高员工生产力

集成 MES 给机械工人提供了有序生产、错误尽可能少的实时信息。虽然这种效益不可能量化，但还是可以清楚看到的。

5.4.5　MES 的案例

Acker 公司于 1949 年在德国成立。在 20 世纪 60 年代初，Acker 是工业用布的领先制造商之一。Acker 产品一般提供给其他公司（如汽车供应商），然后经进一步处理以形成最终客户的产品（通常没有消费），例如汽车网、行李舱盖和胶布、防晒面料。

除了增加透明度和可重复性的流程及维护过程外，MES 的任务还包括以下内容：主数据管理、订单监控和确认数据、操作数据采集（ODA/MDA）、精细计划与控制、产品追踪、维护（TPM）、生产性能记录、物料管理。

MES 系统实现采用循序渐进的方法。第一步，已经存在的孤立系统被扩展集成到 MES 互连中。这种迁移迫切需要保留现有的数据记录，即使最终实现 MES 将完全替代一些孤立的系统。因为具有非常特殊的性质，实验室的任务是用它自己的数据库在自己的软件模块中实现。该模块可以通过一个共同的数据库完全集成到最终的 MES 中。第二步，为所有任务

3. Reducing Administration Expenses

The aim is to check to what extent indirect value-creation activities can be eliminated or reduced considerably through the implementation of a MES.

Analyses in the field of administration rationalization led to the conclusion that with conventional methods alone, savings of at least 20% can be made. If you couple the analyses with the effects of a MES, the savings effect certainly will rise to over 30%. Indirect value-creation activities are eliminated or at least reduced considerably.

4. Improved Customer Service

Reliable delivery dates and information about order progress are essential on the market today. With a MES, these requirements can be suitably covered. Although the benefit of these options cannot be measured directly, the image of a production company improves, and order increases are more probable.

5. Improved Quality

Within the scope of the strategic 6 Sigma initiative, a MES becomes a significant tool. The ensured process capability is largely supported through continual, automatic measurement and monitoring of process parameters on machines in an integrated MES. If the tools mentioned are used correctly, it should be possible to reduce rejects and reworks considerably and thus also reduce the costs associated with this source of loss.

6. Early Warning System, Real-Time Cost Control

Unacceptable deviations are recognized immediately by the real-time control of all influencing parameters in a production process, and measures can be taken accordingly. Primarily, real-time cost control acts as an early warning system in connection with the reasons for exceeding planned costs.

7. Increasing Employee Productivity

An integrated MES provides machine workers electronically with real-time information needed for orderly production with error as few as possible. It is also impossible to quantify this benefit, but its benefits can be seen clearly.

5.4.5 Examples of MES

The company, Acker, was founded in 1949 in Germany. Acker is one of the leading manufacturers of technical fabrics. Acker products generally are supplied to other companies (e. g. , automotive suppliers) and then processed further to form products for the end customer (usually not yet the consumer). Examples from the product range are car nets, luggage compartment covers, and fabrics for adhesive plasters and sun screens.

In addition to increasing transparency and reproducibility of processes and safeguarding processes, the following tasks are also covered by MES: master data administration, order monitoring and confirmation data, operational data acquisition (ODA/MDA), fine planning and control, product tracing, maintenance (TPM) , recording production performance, material management

For realization, a step-by-step process is selected. In the first step, already existing isolated solutions are extended to integration in the MES interconnections. This migration was urgently nee-

区域（即部门）创建统一的软件前端（基于 . NET Framework 2.0），并借助于自己的数据库提供专业的特定业务流程。在第二步中，提供了一个完整的 MES 的核心标准产品。这个标准产品呈现出所需要的信息管理、数据采集以及与控制有关联的过程。因此，标准产品的任务包括跟踪和追踪、全员生产维护（TPM）和资源管理。个别的 MES 终端使用标准产品的 Web 技术来实现要求的信息管理。因此，所有的信息和查询可通过 Web 可视化界面显示在每一台计算机的前端。

下一步，Acker 计划整合提前分析的各个业务流程到 MES 系统。数据库的数量从 5 个减少到 2 个（MES 标准数据库和面向客户的数据库）。通过以客户为导向的软件开发和标准软件混合使用，Acker 有一个针对其业务优化调整的、面向未来的系统。由于使用了高度标准化的组件，该系统的维护成本是可控的。

复习题与习题

5-1　简述制造的定义与目的。

5-2　简述企业的主要运作职能领域。

5-3　什么是 ERP？

5-4　ERP 系统存在哪些缺点？

5-5　总结 ERP 的演化过程。

5-6　实施 ERP 的程序是什么？

5-7　什么是供应链管理？

5-8　供应链管理的竞争力包含哪几方面？

5-9　简述 MES 的定义及其主要功能。

5-10　何谓制造执行系统？

5-11　试论述制造执行系统与企业其他管理系统之间的关系。

ded to retain the existing data records, even if ultimately realization involved replacing some isolated solutions completely with the MES. The tasks of the laboratory are realized in its own software module with its own database owing to its very special nature. This module can be integrated completely into the final MES via a common database.

In the second step, uniform software front ends (based on the. NET framework 2. 0) were created for all task areas (i. e. , departments) that offer professional customer-specific business processes with the aid of their own databases.

Together with the second step, a standard product for the core of the complete MES was introduced. This standard product assumes, among other things, the required information management, data acquisition, and process linking to controls. Tasks of the standard product also include tracking and tracing, total productive maintenance (TPM) tasks, and resource management. The individual MES terminals use the Web technology of the standard product to implement the required information management. Thus all information and queries are available via Web visualization in the front end on every computer.

In further stages in the process, it is planned to integrate the individual business processes analyzed in advance into the MES system. The number of databases then is reduced from five to two (the MES standard database and a customer-oriented database). Through the mixture of customer-oriented software development and standard software used, Acker has an optimally adjusted, future-oriented system for its operations. The maintenance costs of the system are manageable owing to the high degree of standard components.

Review Questions and Problems

5-1　Briefly describe the definition and purpose of manufacturing.

5-2　Briefly describe the main functions of the enterprise.

5-3　What is ERP?

5-4　What are the disadvantages of ERP?

5-5　Summarize the evolution process of ERP development stages.

5-6　What are the procedures for implementing ERP?

5-7　What is supply chain management?

5-8　What are the competencies of supply chain?

5-9　Briefly describe the definition and main functions of MES.

5-10　What is manufacturing execution system (MES)?

5-11　Narrate the relationship between MES and other management systems of enterprise.

Chapter 6   Advanced Manufacturing Mode

第6章　先进制造模式

6.1　Concept of Advanced Manufacturing Mode

6.1.1　Concept and Evolution of Manufacturing Mode

1. Implication of Manufacturing Mode

Manufacturing mode refers to the pattern and operation mode of enterprise system, operation, management, production organization and technical system. Manufacturing mode can be understood as "a typical way to carry out production in manufacturing systems".

The difference between manufacturing mode and management are as follows: Manufacturing model is the concentrated embodiment of some characteristics of the manufacturing system, and also the crystallization of the integration of all management methods with engineering technology in manufacturing enterprises. Management is a subject, and a function of the business community. Manufacturing mode is a kind of state that represents the management way and technical pattern of manufacturing enterprises, but the management is a process for all organizations.

2. Evolution of Manufacturing Mode

Looking back the history, the development of the production mode of manufacturing industry has gone through four main stages:

(1) Manual and job production　The invention of Watt's steam engine had revolutionized the manufacturing industry, led to the Industrial Revolution the emergence of workshop type manufacturing factory. From handicraft to machine operation, from workshop to batch production, productivity had been greatly improved, thus opened a prelude to the modern industrial production.

(2) Large volume production　From the middle of the 19th century to the middle of the 20 th century, because of the "interchangeability" and large volume production proposed by E. Whitney, "scientific management" of F. W. Taylor, and the automatic assembly line initiated by H. Ford for auto assembly made the manufacturing industry start the conversion of the first production mode. This mode had promoted the industrialization process, provided a large number of economic products for society, promoted the development of market economy and become the production mode followed by countries.

(3) Flexible automatic production　The first CNC milling machine in the world produced successfully by Massachusetts Institute of Technology in 1952 opened the prelude to flexible automation. In 1958, the boring and milling machining center with automatic tool changer was developed successfully. On the basis of NC technology, the first industrial robot, automatic warehouse and AGV are developed successfully in 1962. In 1966, a CNC system appeared with which several NC machine tools could be controlled by means of a large-sized general computer. In 1968, the first automated manufacturing system named as flexible manufacturing system was built by British firm Molins, Ltd. and American Cincinnati Corporation. Various microcomputer numerical control systems, flexible manufacturing cell, flexible production lines and automatic factories emerged in 1970s.

Contrary to rigid automation which is characterized by process decentralization, fixed production

6.1　先进制造模式的概念

6.1.1　制造模式的概念与演化

1. 制造模式的概念

制造模式（Manufacturing Mode）是指企业体制、经营、管理、生产组织和技术系统的形态和运作的模式。制造模式可以理解为"制造系统实现生产的典型方式"。

制造模式与管理的区别是：制造模式是制造系统某些特性的集中体现，也是制造企业所有管理方法与工程技术融合的结晶。管理是一门学科，也是企业界的一项职能。制造模式是表征制造企业管理方式和技术形态的一种状态，而管理是面向一切组织的一种过程。

2. 制造模式的演化

回顾历史，制造业的生产方式的发展大致经历了 4 个主要阶段：

（1）手工与单件生产方式　瓦特蒸汽机的发明，促使制造业取得了革命性的变化，引发了工业革命，出现了工场式的制造厂。从手工业到机器作业，从作坊到批量生产，生产率有了较大提高，揭开了近代工业化大生产的序幕。

（2）大批量生产方式　从 19 世纪中叶到 20 世纪中叶，伊莱·惠特尼（Eli Whitney）提出了"互换性"与大批量生产，泰勒（F. W. Taylor）实施了"科学管理"，福特开创了汽车装配自动流水生产线，从而使制造业开始了第一次生产方式的转换。这种模式推动了工业化进程，为社会提供了大量的经济产品，促进了市场经济的高度发展，成为各国仿效的生产方式。

（3）柔性自动化生产方式　1952 年，美国麻省理工学院（MIT）试制成功世界上第一台数控铣床，揭开了柔性自动化的序幕。1958 年，自动换刀镗、铣加工中心研制成功。1962 年，在数控技术的基础上，第一台工业机器人、自动化仓库和自动导引小车研制成功。1966 年，出现了用一台大型通用计算机集中控制多台数控机床的 CNC 系统。1968 年，英国莫林公司和美国辛辛那提公司建造了第一条由计算机集中控制的自动化制造系统，定名为柔性制造系统。20 世纪 70 年代，出现了各种微型机数控系统、柔性制造单元、柔性生产线和自动化工厂。

与刚性自动化的工序分散、固定生产节拍和流水生产的特征相反，柔性自动化的共同特征是：工序相对集中、没有固定的节拍、物料的非顺序输送；将高效率和高柔性融于一体，生产成本低；具有较强的灵活性和适应性。

（4）高效、敏捷与集成经营生产方式　近些年来，在日本和美国，有关制造模式的新概念层出不穷，如精益生产、敏捷制造、智能制造、并行工程等，这些新方法的出现彻底动摇了原有的管理理论和生产方式。

6.1.2　先进制造模式的内涵与类型

1. 先进制造模式的内涵

从广义上讲，先进制造模式（Advanced Manufacturing Mode，AMM）是用于制造系统的具有相似特点的一类先进方式方法的总称。它以获取生产有效性为首要目标，以制造资源快速有效集成为基本原则，以人-组织-技术相互结合为实施途径，使制造系统获得精益、敏

tempo and transfer production, the common features of flexible automation are relative concentration of process, without fixed tempo, without sequential transportation of materials; integration high efficiency and flexibility, low production cost; with stronger flexibility and adaptability.

(4) Efficient, agile and integrated production management In recent years, new concepts about manufacturing mode have emerged in an endless stream in Japan and America, such as lean production, agile manufacturing, intelligent manufacturing, concurrent engineering and so on. The emergence of these new methods completely shook the original management theories and production modes.

6.1.2 Connotation and Types of Advanced Manufacturing Mode

1. Connotation of Advanced Manufacturing Mode

In broad sense, advanced manufacturing mode (AMM) is the general term for a class of advanced methods that have similar characteristics to the manufacturing system. Its primary goal is to achieve productive effectiveness. Taking the rapid and effective integration of manufacturing resources as the basic principle, the combination of human-organization-technology as the implementation approach, it makes the manufacturing system acquire lean, agile, high quality and efficiency and other features so as to adapt the new requirement of market change for time, quality, cost, service and environment.

AMM and AMT are two different concepts in manufacturing system. There was no clear distinction in the past, because they are very closely related, and AMM was classified as the system management technology of AMT. In fact, AMT is the basis for implementing AMM. AMT stresses on the exertion of function, and forms the technology group; AMM stresses on the embodiment of philosophy, management, and the synergism of environment and strategy.

2. Types of Advanced Manufacturing Mode

The manufacturing mode has a distinctive times. The main purpose of the Ford's large volume production in industrial age is to provide cheap products. The main purpose of flexible production, lean production and agile manufacturing in the information age is to meet the diverse needs of customers. The future development trend is the green manufacturing mode in the knowledge age, and its main purpose is to promote environmental protection and reduce energy consumption in the life cycle of the product. While various advanced manufacturing technologies formed with the traditional manufacturing technologies gradually developing, permeating and envolving to the modern high technologies, a series of advanced manufacturing modes emerged.

(1) Flexible manufacturing mode This mode is proposed for the first time by the British firm Molins, Ltd., and got wide application in the late of 1970s. This mode depends mainly on the high flexible manufacturing equipment mainly based on CNC machine tools to carry out the production with many varieties and small quantities so as to enhance the flexibility and resilience of the manufacturing industry, shorten production cycle, and improve equipment utilization and employee labor productivity.

(2) Computer-integrated manufacturing (CIM) mode The prominent feature of CIM is to emphasize the integrity of the manufacturing process. It integrates the requirement analysis, sale and service into the manufacturing system and fully orients to the market and users. Computer aided

捷、优质与高效的特征，以适应市场变化对时间、质量、成本、服务和环境的新要求。

AMM 与先进制造技术（AMT）应是制造系统中两个不同的概念。过去之所以没有明确区别，是因为二者具有十分密切的相关性，并把 AMM 归为 AMT 的系统管理技术。事实上，AMT 是实现 AMM 的基础。AMT 强调功能的发挥，形成了技术群；AMM 强调制造哲理的体现，偏重于管理，强调环境、战略的协同。

2. 先进制造模式的类型

制造模式具有鲜明的时代性。工业化时代的福特大批量生产模式以提供廉价的产品为主要目的；信息化时代的柔性生产模式、精益生产模式、敏捷制造模式等以快速满足客户的多样化需求为主要目的；未来发展趋势是知识化时代的绿色制造生产模式，其主要目的是在产品的整个生命周期中促进环境保护，减少能源消耗。在传统制造技术逐步向现代高新技术发展、渗透和演变，形成各种先进制造技术的同时，出现了一系列先进制造模式。

（1）柔性制造模式　这种模式由英国的莫林斯（Molins）公司首次提出，于 20 世纪 70 年代末得到推广应用。该模式主要依靠具有高度柔性的以 CNC 机床为主的制造设备来实现多品种小批量的生产，以增强制造业的灵活性和应变能力，缩短产品生产周期，提高设备利用率和员工劳动生产率。

（2）计算机集成制造（CIM）模式　CIM 的突出特点是强调制造过程的整体性。它将需求分析、销售和服务等都纳入制造系统范畴，充分面向市场和用户；计算机辅助手段提高了产品研制和生产能力，加速了产品更新换代；物流集成提高了制造过程的柔性、设备利用率和生产率；信息集成促进了经营决策与生产管理的科学化等。

（3）智能制造模式　该模式是在制造生产的各个环节中，应用智能制造技术和系统，以一种高度柔性和高度集成的方式，通过计算机模拟专家的智能活动，进行分析、判断、推理、构思和决策，以便取代或延伸制造过程中人的部分脑力劳动，使人类专家的制造智能得以进一步地完善、继承和发展。

（4）精益生产模式　该模式是由美国 MIT 于 1990 年在总结日本丰田汽车生产经验时提出的。其基本特点是实施"精简"的对策和"精益求精"的管理思想，消除制造企业因采用大量生产方式所造成的过于臃肿和浪费。该模式要求产品优质，且充分考虑人的因素，采用灵活的小组工作方式和强调合作的并行工作方式；采用适度的自动化技术，使制造企业的资源能够得到合理配置与充分利用。

（5）敏捷制造模式　该模式将柔性制造的先进技术、熟练掌握生产技能和高素质的劳动力、企业内部和企业之间的灵活管理三者集成在一起，利用信息技术对千变万化的市场机遇做出快速响应，最大限度地满足客户的要求。

（6）虚拟制造模式　该模式是利用制造过程的计算机仿真来实现产品的设计和研制。在产品投入制造之前，先在虚拟制造环境中以软产品原型代替传统的硬样品进行试验，并对其性能进行预测和评估，从而大大缩短产品设计与制造周期，降低产品开发成本，提高其快速响应市场变化的能力。

（7）极端制造模式　极端制造是指在极端环境下制造极端尺度或极高功能的器件和系统。当前，极端制造集中表现在微细制造、超精密制造、巨系统制造等。例如，制造航天飞行器、超常规动力装备、制造微纳电子器件、微纳光机电系统等极小尺度和极高精度的产品。

means improve the product development ability and the production capacity, and speed up the upgrading of products. Logistics integration improves the flexibility of manufacturing processes, the equipment utilization and productivity. Information integration promotes the scientific operating decisions and the scientific production management.

(3) Intelligent manufacturing mode   This mode applies intelligent manufacturing technology and system in every link of the manufacturing process, and in a highly flexible and highly integrated manner performs analysis, judgment, reasoning, conception and decision making by using computer to simulate expert's intelligence activities so that part of human mental labour in the manufacturing process can be replaced or extended, and the manufacturing intelligence of the human experts can be further improved, inherited and developed.

(4) Lean production mode   This mode was proposed by MIT of America in 1990 when summed up the production experience of TOYOTA. Its basic characteristics to implement the countermeasure of "streamlining" and the "excelsior" management principle and to eliminate the overstaffed and waste of manufacturing enterprises due to the adoption of a large volume of production. The mode requires high quality and full consideration of human factors, and adopts a flexible "team work" and emphasize concurrent work in cooperation, and adopts appropriate automation techniques to make the resources of the manufacturing enterprise get rational configuration and full utilization.

(5) Agile manufacturing mode   The mode integrates the advanced technologies of flexible manufacturing, the workforces with high quality and production skills with the flexible management inside the enterprise and between the enterprises, and makes the quick response to the ever-changing market opportunities by making use of information technology so as to furthest meet the requirements of customers.

(6) Virtual manufacturing mode   The mode means using the computer simulation of manufacturing process to achieve the product design and development. Before the product is put into production, the soft prototype instead of traditional hard prototype is first tested in the virtual manufacturing environment, and its performance is predicted and evaluated so as to greatly shorten the product design and manufacturing cycle, reduce the cost of product development and improve its ability to respond quickly to market changes.

(7) Extreme manufacturing mode   It refers to manufacture the device and system with extreme dimension or extreme function in extreme circumstances. Currently, extreme manufacturing focuses on micro-manufacturing, ultraprecision manufacturing, giant system manufacturing and so on, such as aerospace craft, transnormal power equipment, micro/nano electronic device, micro/nano opto mechatronics system and other products with minimal scale and sky-high precision.

(8) Green manufacturing mode   Green manufacturing is the eco-manufacturing technology which utilizes synthetically the achievements in biotechnology, "green chemistry", information technology and environmental science and does not produce or produce little wastes and pollutants in the manufacturing process. Increasingly stringent environmental and resource constraints make green manufacturing more and more important. Green manufacturing is a manufacturing mode to realize sustainable development of manufacturing industry.

（8）绿色制造模式　绿色制造是指综合运用生物技术、"绿色化学"、信息技术和环境科学等方面的成果，使制造过程中没有或极少产生废料和污染物的生态型制造技术。日趋严格的环境与资源约束，使绿色制造显得越来越重要。绿色制造是实现制造业可持续发展的制造模式。

6.2　计算机集成制造系统

20 世纪 70 年代以来，随着市场的全球化，市场竞争不断加剧，给企业带来了巨大的压力，迫使企业纷纷寻求有效方法，加速推出高性能、高可靠性、低成本的产品，以期更有力地参与市场竞争。与此同时，随着计算机在设计、制造、管理等领域的广泛应用，相继出现了许多单一目标的计算机辅助自动化技术，如 CAD、CAPP、MRP Ⅱ 等。由于缺少整体规划，这些单元技术的应用是相对独立的，彼此之间的数据不能共享，往往还会产生诸如数据不一致之类的矛盾和冲突，出现所谓的"自动化孤岛"现象，从而降低系统运行的整体效率，甚至造成资源浪费。只有把这些单项应用通过计算机网络和系统集成技术连接成一个整体，才能消除内部信息和数据的矛盾和冗余，由此出现了计算机集成制造系统（CIMS）。

6.2.1　CIMS 的内涵

计算机集成制造（Computer Integrated Manufacturing，CIM）的概念最早由美国的约瑟夫·哈林顿（Joseph Harrington）博士于 1973 年在《计算机集成制造》一书中首先提出。他强调了两个观点：①系统观点——企业各个生产环节是不可分割的，需要统一安排与组织；②信息化观点——产品制造过程实质上是信息采集、传递、加工处理的过程。CIM 是一种先进的哲理，其内涵是借助计算机，将企业中各种与制造有关的技术系统集成起来，进而提高企业适应市场竞争的能力。但由于受当时条件的限制，CIM 思想未能立即引起足够的注意，20世纪 80 年代后，才逐渐被制造领域重视并采用。

至今 CIM 和 CIMS 还没有一个公认的定义，不同的国家在不同时期对 CIMS 有各自的认识和理解。1991 年日本能源协会认为：CIMS 是以信息为媒介，用计算机把企业活动中的多种业务领域及其职能集成起来，追求整体效益的新型生产系统。1992 年，国际标准化组织（ISO）认为：CIM 是将企业所有的人员、功能、信息和组织诸方面集成为一个整体的生产方式。

6.2.2　CIMS 的基本组成

从系统功能考虑，CIMS 通常由 4 个功能分系统和 2 个支撑分系统组成，如图 6-1 所示。每个分系统都有其特有的结构、功能和目标。

1. 管理信息系统（Management Information System，MIS）

MIS 是 CIMS 的神经中枢，指挥与控制着其他各个部分有条不紊地工作。MIS 通常是以MRPII 为核心，包括预测、经营决策、各级生产计划、生产技术准备、销售、供应、财务、成本、工具等各项管理信息功能。图 6-2 所示为 CIMS 经营管理信息分系统模型。它集生产经营与管理于一体，各个功能模块可在统一的数据环境下工作，以实现管理信息的集成，从而缩短产品生产周期、减少库存、降低流动资金、提高企业应变能力。

6.2　Computer Integrated Manufacturing System

Since the 1970s, with markets globalization the market competition was continuously increased led to great pressure to enterprises. Enterprises were forced to seek effective method to accelerate the launch of products with high performance, high reliability and low cost, aiming to make enterprises more effectively participate in market competition. Meanwhile, with the widespread application of the computer in design, manufacturing and management, computer-aided automation technology was used for many single goals, such as CAD, CAPP, MRP Ⅱ, etc. Due to lacking overall plan, the application of these technologies was relatively independent and the data cannot be shared. It causes contradictions and conflicts such as data inconsistent, leading to the called "Island of Automation". It reduces the efficiency of the system operation, even causes resources waste. Only when single applications are connected to form a whole system by computer network and system integration technology, the contradiction and redundancy between inner information and data would be eliminated. So, computer integrated manufacturing system (CIMS) appeared to solve these issues.

6.2.1　Connotation of CIMS

The concept of computer integrated manufacturing (CIM) was first suggested by Ph. D. Joseph Harrington in his book *Computer Integrated Manufacturing* in 1973. He emphasized two viewpoints, namely: ①system viewpoint: Each production link was indivisible and needed to unified arrangement and organization; ② Information viewpoint: Product manufacturing process was essentially the process of information collection, transmission and processing. CIM is a kind of advanced philosophy. The connotation of CIM is that the technology related to the manufacturing system could be integrated by using computer, thereby improving the ability of enterprises to adapt to market competition. However, due to limit of conditions at that time, the idea of CIM could not immediately pay enough attention. In the 1980s, CIM got more attention and was used for manufacturing field.

Until now, there is no generally accepted definition about CIM and CIMS. Different countries had different knowledge and understanding for CIMS in different periods. In 1991, the Japan Association of Energy considered that: CIMS was a new production system in pursuit of overall efficiency which integrated a variety of business areas and functions in business activities through computer taking information as medium. In 1992, International Organization for Standardization (ISO) considered that CIM was a mode of production which can take companies of all staff, function, information and organization as a whole.

6.2.2　Basic Components of CIMS

Taking into consideration the system function, CIMS is usually composed of four function subsystem and two support subsystems, as shown in Fig. 6-1. Each subsystem has its unique structure, function and goal.

1. Management Information System (MIS)

MIS is the nerve center of CIMS. It commands and controls other parts work methodically.

Engineering design
automation system
工程设计自动化系统

Management
information system
管理信息系统

Computer aided
quality system
计算机辅助质量
控制系统

Manufacturing automation system
制造自动化系统

Database system
数据库系统

Computer network system
计算机网络系统

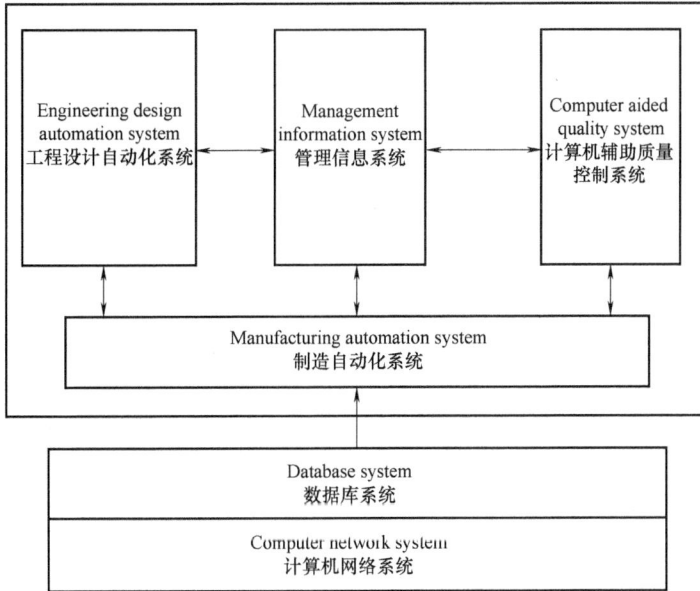

Fig. 6-1　Basic component of CIMS CIMS 的基本构成

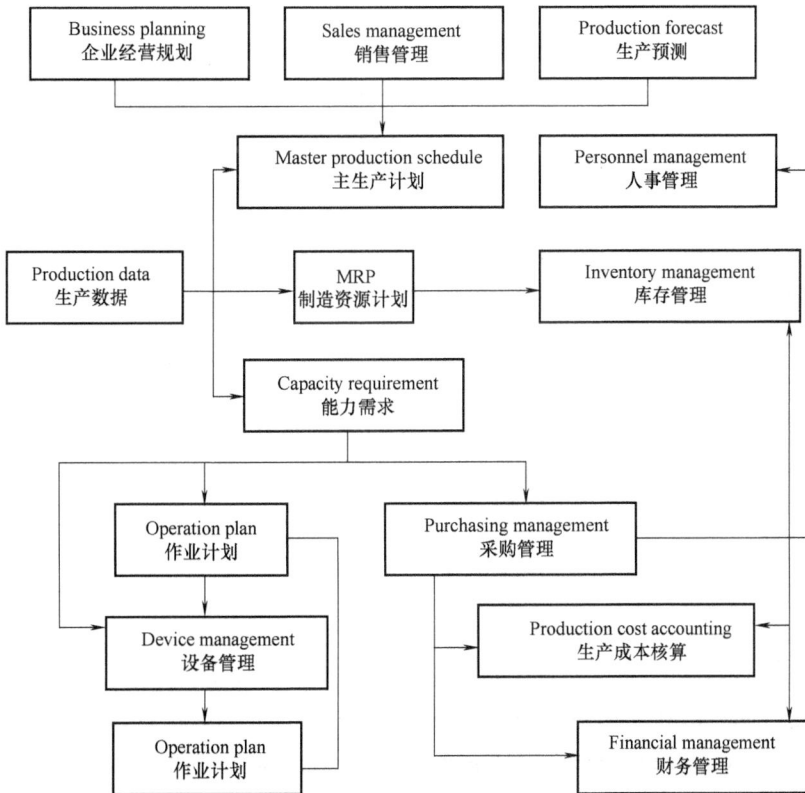

Business planning
企业经营规划

Sales management
销售管理

Production forecast
生产预测

Master production schedule
主生产计划

Personnel management
人事管理

Production data
生产数据

MRP
制造资源计划

Inventory management
库存管理

Capacity requirement
能力需求

Operation plan
作业计划

Purchasing management
采购管理

Device management
设备管理

Production cost accounting
生产成本核算

Operation plan
作业计划

Financial management
财务管理

Fig. 6-2　The model of CIMS business and management information subsystem
CIMS 经营管理信息分系统模型

The MIS which MRPII is generally the core of MIS includes forecast, management decision-making, production planning, production technical preparation, sales, supply, finance, cost, tools and so on. Fig. 6-2 shows a model of CIMS business and management information subsystem. It integrates production and marketing management as a whole, in which each functional module can operate in uniform data environment to realize the integrated management information, thus to shorten the production cycle, reduce inventory, reduce cash flow and improve the enterprise's responsive capacity to change.

2. Engineering Design Automation System (EDAS)

EDAS means in essence that the computer technology during the product development process is applied to make product development activities more efficient, higher quality, and more automatic. The product development activities include a series of preparation work such as concept design, analysis of engineering and structure, detailed design, process design, numerical control programming and manufacturing preparation, i. e. referred to CAD, CAPP, CAM in usual.

3. Manufacturing Automation System (MAS)

MAS is the integration point of information flow and material flow of CIMS, and is the gathering area of final economic benefits. It is usually made up of CNC machine tools, machining centers, FMC and FMS. Under the computer control and schedule, MAS machines work blanks one by one into qualified parts through execute NC codes and completes the tasks assigned by design and management department. Meanwhile, MAS transmits the information to the relevant departments in time or after initial processing to schedule and control them in time.

The goal of MAS is to realize the flexible automation of multi variety and small batch production; to achieve high quality, low cost, short cycle and high efficiency in production; to improve the market competitiveness of enterprises; to create a comfortable and safe working environment for operators.

4. Computer-Aided Quality System (CAQS)

In the fierce market competition, the quality is the key to enterprises survival. To gain a larger share of the market, enterprises must meet customer requirements in product performance, price, delivery time and after-sales service. So, enterprises need a set of complete quality assurance system. The quality assurance system in CIMS covers all phases of the product life cycle, manly including four subsystems.

(1) The quality planning subsystem　It is used to improve the quality objectives, to establish the quality standards and technical standards, to plan the attainable technical route, to estimate an attainable desired result, and to develop detection plan and detection procedure according to production programs and quality requirements.

(2) The quality detection management subsystem　It manages quality detection data for materials, bought-in components and outsourced parts. It adopts the automatic or manual testing to detect products and parts, and samples the quality data to check and preprocess. It establishes the finished product file and improves the quality of service.

(3) The quality analysis and evaluation subsystem　It analyzes product design quality, quality of bought-in components, quality of procedure control point, and quality cost, and assesses impact

2. 工程设计自动化系统（Engineering Design Automation System，EDAS）

EDAS 实质上是指在产品开发过程中运用计算机技术，使产品开发活动更高效、更优质、更自动地进行。产品开发活动包括产品的概念设计、工程与结构分析、详细设计、工艺设计以及数控编程等设计和制造准备阶段的一系列工作，即通常所说的 CAD、CAPP、CAM 三大部分。

3. 制造自动化系统（Manufacturing Automation System，MAS）

MAS 是 CIMS 的信息流和物料流的结合点和最终产生经济效益的聚集地，通常由 CNC 机床、加工中心、FMC 和 FMS 等组成。MAS 在计算机的控制与调度下，按照 NC 代码将一个个毛坯加工成合格的零件并装配成部件以至产品，完成设计和管理部门下达的任务；并将制造现场的各种信息实时地或经过初步处理后反馈到相应部门，以便及时地进行调度和控制。

MAS 的目标可归纳为：实现多品种、小批量产品制造的柔性自动化；实现优质、低成本、短周期及高效率生产，提高企业的市场竞争能力；为作业人员创造舒适而安全的劳动环境。

4. 计算机辅助质量控制系统（Computer Aided Quality System，CAQS）

在激烈的市场竞争中，质量是企业求得生存的关键。要赢得市场，必须在产品性能、价格、交货期、售后服务等方面满足顾客要求。因此需要一套完整的质量保证体系，CIMS 中的质量保证系统覆盖产品生命周期的各个阶段，主要包括四个子系统：

（1）质量计划子系统　该系统用来确定改进质量目标，建立质量标准和技术标准，计划可能达到的途径和预计可能实现的改进效果，并根据生产计划及质量要求制订检测计划及检测规程。

（2）质量检测管理子系统　该系统用来管理进厂材料、外购件和外协件的质量检验数据；采用自动或手动方式对零件进行检验，对产品进行试验，采集各项质量数据并进行校验和预处理；建立成品出厂档案，改善售后服务质量。

（3）质量分析评价子系统　该系统用来对产品设计质量、外购外协件质量、工序控制点质量、质量成本等进行分析，评价各种因素对造成质量问题的影响，查明主要原因。

（4）质量信息综合管理与反馈控制子系统。该系统包括质量报表生成、质量综合查询、产品使用过程质量综合管理，以及针对各类质量问题所采取的各种措施及信息反馈。

5. 数据库系统（Database System，DBS）

DBS 是一个支撑系统，它是 CIMS 信息集成的关键之一。CIMS 环境下的经营管理信息、工程技术、制造自动化、质量保证 4 个功能系统的信息数据都要在一个结构合理的数据库系统里进行存储和调用，以满足各系统信息的交换和共享。

CIMS 的数据库系统通常是采用集中与分布相结合的体系结构，以保证数据的安全性、一致性和易维护性。此外，CIMS 数据库系统往往还建立一个专用的工程数据库系统，用来处理大量的工程数据。工程数据类型复杂，它包含有图形、NC 代码等各种类型的数据。工程数据库系统中的数据与生产管理、经营管理等系统的数据均按统一规范进行交换，从而实现整个 CIMS 中数据的集成和共享。

6. 计算机网络系统（Computer Network System，NETS）

这是 CIMS 的一个支撑系统，通过计算机通信网络将物理上分布的 CIMS 各个功能分系

of various factors on the quality to finds out the main reasons.

(4) The integrated management of quality information and feedback control subsystem　It includes the quality report generation, quality comprehensive enquiry, quality integrated management during product service, as well as various measures and information feedback for all kinds of quality problems.

5. Database System (DBS)

DBS is a support system to achieve CIMS information integration. The information data of management information system, engineering design automation system, manufacturing automation system and quality assurance system in CIMS environment will be stored and called in a database system with reasonable structure so as to meet the system information exchange and sharing.

The database system of CIMS usually adopts the architecture of combination of centralization and distribution to ensure the security, consistency and maintainability of the data. In addition, the database system of CIMS is also established a dedicated engineering database system which is used to deal with a large number of engineering data. The types of engineering data are complex which contain various types of data such as graphics, NC code, etc. The data of the engineering database system and the data of production management and operation management are exchanged according to the unified standard, thereby achieving data integration and shared for CIMS.

6. Computer Network System (NETS)

This is a support system in CIMS. All information of each physically distributed functional subsystem of CIMS is connected by computer communication network, which achieves to share information. According to the size of geographical coverage of the enterprise, there are two kinds of computer networks available for CIMS, one is called local area network (LAN), another is called wide area network (WAN). At present, CIMS is mainly the interconnection of LAN. If the factory area is quite large, the LAN may be interconnected by the remote network which makes CIMS have the characteristics of both local area network and wide area network.

Under the support of the database and the computer network, the CIMS can easily realize the communication among the each functional subsystem, thus fulfilling effectively the integration of the whole system. The information exchange among the subsystems is shown in Fig. 6-3. In Fig. 6-3, FME stands for the flexible manufacturing equipment, QIS stands for the quality information system, EIS stands for the engineering information system, MIS stands for the management information system.

6.2.3　Technological Superiorities of CIMS

From the perspective of technology development, CIMS has experienced three stages, namely information integration (as early as the representative of the CIM), process integration (represented by concurrent engineering) and the integration in enterprises (represented by agile manufacturing). The former is the base of the latter, and the three kinds of integration technologies are also being developed.

1. The Information Integration

The information integration mainly solves the information exchange and sharing between the isolated islands of automation in enterprise, which mainly focus on the following aspects:

(1) Enterprise modeling, system design methods, software tools and standards　These are the

统的信息联系起来，以达到共享的目的。依照企业覆盖地理范围的大小，有两种计算机网络可供 CIMS 采用，一种为局域网，另一种为广域网。目前，CIMS 一般以互联的局域网为主，如果工厂厂区的地理范围相当大，局域网可能要通过远程网进行互联，从而使 CIMS 同时兼有局域网和广域网的特点。

CIMS 在数据库和计算机网络的支持下，可方便地实现各个功能分系统之间的通信，从而有效地完成全系统的集成。CIMS 分系统之间的通信如图 6-3 所示。在图 6-3 中，FME 代表柔性制造设备，QIS 代表质量信息系统，EIS 代表工程信息系统，MIS 代表管理信息系统。

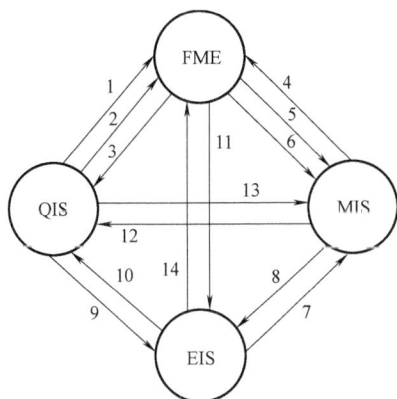

Fig. 6-3　Communication between CIMS subsystem CIMS 分系统之间的通信

1—Product quality control information 产品质量控制信息　2—Product quality report 产品质量报告

3—Product quality tracking information 产品质量跟踪信息　4—Manufacturing plan 生产计划

5—Inventory information 库存信息　6—Manufacturing planning tracking information 生产计划跟踪信息

7—Product process information 产品工艺信息　8—Design production plan 产品设计计划

9—Quality feedback information 质量反馈信息　11/14—Feedback of engineering and

process improvement information 工程反馈和工艺改进信息　12—Product quality information

产品质量信息　13—Product quality report 产品质量报告

6.2.3　CIMS 的技术优势

从 CIMS 技术发展的角度看，它共经历了三个阶段，即信息集成（以早期计算机集成制造为代表）、过程集成（以并行工程为代表）和企业间集成（以敏捷制造为代表）。前者是后者的基础，同时，三类集成技术也还在不断地发展之中。

1. 信息集成

信息集成主要解决企业中各个自动化孤岛之间的信息交换与共享问题，其主要内容有：

（1）企业建模、系统设计方法、软件工具和规范　这是系统总体设计的基础。企业建模及设计方法解决了一个制造企业的物流、信息流，以及资金流、决策流的关系，这是信息集成的基础。

（2）异构环境和子系统的信息集成　所谓异构是指系统中包含了不同的操作系统、控制系统、数据库及应用软件。如果各个部分的信息不能自动地进行交换，则很难保证信息传送和交换的效率和质量。早期的信息集成主要通过局域网和数据库来实现。近期采用企业网、外联网、产品数据管理（PDM）、集成平台和框架技术来实施。值得指出，

basis of system overall design. The modeling and design methods of the enterprise solve the relationship between logistics, information flow, capital flow and decision flow in a manufacturing enterprise, which are the basis of information integration.

(2) Information integration in heterogeneous environment and subsystem　Heterogeneous refers to system includes the different operating system, control system, database and application software. It is difficult to guarantee the efficiency and quality of information transmission and exchange if the information of each part cannot be exchanged automatically. The method of early information integration is realized mainly by local area network and database. Recently the method of information integration is realized mainly by Intranet, Extranet, product data management (PDM), integration platform and framework technology. It is worth pointing out that the integration framework based on object-oriented technology, soft component technology and web technology has become important supporting tools of system information integration.

2. The Process Integration

The traditional mode of product development is employing the serial product development process, in which the design and the machining are two independent functional departments. This mode is short of digital product definition and product data management, and is lack of computer and network environment to support group cooperative work, which undoubtedly prolongs product development cycle and increases the cost. The use of concurrent engineering can be a good solution to these problems.

3. The Integration Between Enterprises

The integration between enterprises is to optimize the use of internal and external resources to achieve agile manufacturing and adapt to the new situation of knowledge economy, global economy, and global manufacturing. From a management perspective, to establish dynamic enterprise alliance and to form flat type organization structure and the "dumbbell enterprise" will help to overcome the "small but independent" and "big and independent", to enhance the ability to develop new products and to open new markets, and to play the important role of human being in the system. The key technologies of the integration between enterprises include information integration technology, key technology of concurrent engineering, virtual manufacturing, enabling technology supporting agile engineering, agile manufacturing based on the network, as well as the resource optimization technology (such as MRP, supply chain, electronic commerce).

6. 2. 4　Development Tendency of Contemporary Integrated Manufacturing

(1) Integration　The trend of integration is that the modern integrated manufacturing technology is stepping into the stage of the integration in enterprises (represented by agile manufacturing) following the current enterprise internal information integration and functional integration to process integration (represented by concurrent engineering).

(2) Digitalization/Virtualization　The trend is that the digitization of all kinds of activities, equipment and entities in the whole life cycle of the product is being developed from the digital design of the product. On the basis of digitization, the virtualization technology is being developed

基于面向对象技术、软构件技术和 Web 技术的集成框架已成为系统信息集成的重要支撑工具。

2. 过程集成

传统的产品开发采用串行产品开发流程，设计与加工是两个独立的功能部门，缺乏数字化产品定义和产品数据管理，缺乏支持群组协同工作的计算机与网络环境，这无疑使产品开发周期延长，成本增加，而采用"并行工程"可以很好地解决这些问题。

3. 企业间集成

企业间集成优化是指企业内外部资源的优化利用，实现敏捷制造，以适应知识经济、全球经济、全球制造的新形势。从管理的角度，企业间实现企业动态联盟（Virtual Enterprise，VE），形成扁平式企业的组织管理结构和"哑铃型企业"，克服"小而全""大而全"，实现产品型企业，增强新产品的设计开发能力和市场开拓能力，发挥人在系统中的重要作用等。企业间集成的关键技术包括信息集成技术、并行工程的关键技术、虚拟制造、支持敏捷工程的使能技术系统、基于网络（如互联网、内联网、外联网）的敏捷制造，以及资源优化（如 ERP、供应链、电子商务）等。

6.2.4 现代集成制造技术的发展趋势

（1）集成化 从当前的企业内部的信息集成和功能集成，发展到过程集成（以并行工程为代表）、并正在步入实现企业间集成的阶段（以敏捷制造为代表）。

（2）数字化/虚拟化 从产品的数字化设计开始，发展到产品全生命周期中各类活动、设备及实体的数字化。在数字化基础上，虚拟化技术正在迅速发展，主要包括虚拟现实（Virtual Reality，VR）应用、虚拟产品开发（Virtual Product Development，VPD）和虚拟制造（Virtual Manufacturing，VM）。

（3）网络化 从基于局域网发展到基于互联网、内联网、外联网的分布网络制造，以支持全球制造策略的实施。

（4）柔性化 正积极研究发展企业间动态联盟技术、敏捷设计生产技术、柔性可重组机器技术等，以实现敏捷制造。

（5）智能化 智能化是制造系统在柔性化和集成化基础上进一步的发展与延伸，引入各类人工智能和智能控制技术，实现具有自律、分布、智能、仿生、敏捷、分形等特点的新一代制造系统。

（6）绿色化 绿色化包括绿色制造、环境意识的设计与制造、生态工厂、清洁化生产等。它是全球可持续发展战略在制造业中的体现，同时也是摆在现代制造业面前的一个崭新课题。

6.3 大批量定制

6.3.1 大批量定制及其分类

随着现代科学技术的迅猛发展和人们生活水平的日益提高，用户需求日趋多样化、个性化以及企业竞争的日趋激烈等许多原因，使得原先传统的大批量生产方式已不能适应快速多

rapidly, mainly including the virtual reality (VR) application virtual product development (VPD) and virtual manufacturing (VM).

(3) Networking　It means that the distribution network based on Intranet/Internet/Extranet manufacturing is to be developed from based on local area network (LAN), which the purpose is to support the implementation of the global manufacturing strategy.

(4) Flexibility　The enterprise dynamic alliance technology, agile design and production technology, flexible reconfigurable machine technology, etc. are being studied actively to realize agile manufacturing.

(5) Intellectualization　Intellectualization is the further development and extension of the manufacturing system based on flexibility and integration. All kinds of artificial intelligence and intelligent control technology are introduced to achieve a new generation of manufacturing system with characteristics of autonomy, distribution, intelligence, bionics, agility, fractal, and so on.

(6) Greenization　It includes green manufacturing, design and manufacturing of environmental awareness, ecological plant, and clean production, etc. Not only is it the embodiment of global sustainable development strategy in the manufacturing industry, but also it is a new topic in the modern manufacturing industry.

6.3　Mass Customization

6.3.1　Mass Customization (MC) and Its Classifications

With the rapid development of modern science and technology and the increasing improvement of people's living standard, the user requirements has become gradually diversified and personalized, and competition among enterprises becomes increasingly fiercer. For these reasons, the original traditional mass production mode cannot meet the fast-changing needs of the market. In the new market environment, enterprises urgently need to be a new mass production mode. So, the mass customization production mode arises at the historic moment.

Mass customization is to the individual mass production for customizing products and services. It integrates mass production into complete customization which organically combining make the advantages of mass production and complete customization. Its ultimate goal or desired objective is to design and produce customized products based on the efficiency and speed of the mass production. For customers, the obtained product is customized and personalized. For factory, products are mainly produced in mass production way.

Mass customization combines enterprise, customers, suppliers and environment as a whole. Under the guiding of system thought and with overall optimization idea it makes full use of enterprises existing resources to provide customized products and services through low cost, high quality and high efficiency of mass production with the support of information technology and advanced manufacturing technology and so on, according to customer's individual requirements.

Aiming at different custom market demands, the enterprises can take four different customiza-

变的市场需要。在新的市场环境中，企业迫切需要一种新的大批量生产模式，由此大批量定制（Mass Customization，MC）生产方式应运而生。

大批量定制是对定制产品和服务进行个别的大批量生产，它把大批量生产融入完全定制中，使大批量生产和完全定制的优势有机地结合起来。其最终目标或理想目标是以大批量生产的效率和速度来设计和生产定制产品。对客户而言，所得到的产品是定制的、个性化的，对厂家而言，则产品主要是以大批量生产方式生产的。

大批量定制是一种集企业、客户、供应商和环境于一体，在系统思想指导下，用整体优化的思想，充分利用企业已有的各种资源，在标准化技术、信息技术和先进制造技术等的支持下，根据客户的个性化需求，以大批量生产的低成本、高质量和高效率提供定制产品和服务的生产方式。

针对不同的定制市场需求，企业可以采取协同定制、装饰定制、调整定制和预测定制等4种不同的定制方式，也可以采用这4种方式的组合定制模式。

通常所指的大批量定制主要是协同定制。协同定制是客户参与的定制，通过企业与客户的协同共同确定定制的产品和服务，因此能够满足客户的特定需求。由于供应链被需求链所代替，"推"的生产方式被"拉"的生产方式所替代，从而消除了成品库存，减少了中间环节。对于那些客户必须进行大量选择才能确定功能和性能的产品定制，协同定制是一种正确的选择，如个人计算机、工业汽轮机的定制等。

6.3.2　大批量定制的基本原理

1. 相似性原理

大批量定制的关键是识别和利用大量不同产品和过程中的相似性。通过充分识别和挖掘存在于产品和过程中的几何相似性、结构相似性、功能相似性和过程相似性，利用标准化、模块化和系列化等方法，减少产品的内部多样化，提高零部件和生产过程的可重用性。

2. 重用性原理

在定制产品和服务中存在着大量可重新组合和可重复使用的单元（包括可重复使用的零部件和可重复使用的生产过程）。通过采用标准化、模块化和系列化等方法，充分挖掘和利用这些单元，将定制产品的生产问题通过产品重组和过程重组，全部或部分转化为批量生产问题，从而以较低的成本、较高的质量和较快的速度生产出个性化的产品。

3. 全局性原理

实施大批量定制，不仅与制造技术和管理技术有关，还与人们的思维方式和价值观念有关。除了从精益生产、敏捷制造、现代集成制造和成组技术等方式中吸取有益的思想以外，还要吸取一些特别重要的基本思想和方法，即定制点后移方法、总成本思想和产品全生命周期管理等。

6.3.3　大批量定制的关键技术

1. 面向大批量定制的开发设计技术

为了获得全面实施大批量定制的综合经济效益，首先应该在开发设计阶段应用大批量定制的原理。面向大批量定制的开发设计技术包括产品的开发设计技术与过程（制造与装配过程）的开发设计技术。完整的面向大批量定制的开发设计过程由面向大批量定制的开发

tion modes, which is collaborative customization, decorative customization, adjust customization and predict customization, and also can use the combining customization mode of four customizations.

Mass customization is usually referred to as collaborative customization. Collaborative customization is a kind of customer participation customization which the customized products and services are determined through the cooperation of enterprises and customers. So, it can meet the specific needs of customers. Because the supply chain is replaced by the demand chain, the "push" mode of production is replaced by the "pull" mode of production, thus eliminating the finished product inventory, reducing the intermediate link. For those customers who must make a lot of choices to determine the functionality and performance of product customization, the collaborative customization is the right choice, such as the customization of personal computers, the customization of industrial steam turbine, etc.

6.3.2　Rationale of MC

1. Similarity Principle

The key to mass customization is to identify and use the similarity in a number of different product and process. By fully identifying and mining geometrical similarity, structural similarity, functional similarity and process similarity that exist in product and process, using the standardization, modularization and serialization methods could reduce the internal product diversification and improve the reusability of components and production process.

2. Reusability Principle

There are many reconfigurable and reusable units (including reusable component and reusable production process) in customized product and service. By using standardization, modularization and serialization methods, these units are fully mined and used to make the production problems of customized products transform in whole or in part into the problems of mass production through product recombination and process recombination, so as to produce personalized products in lower cost, higher quality and faster speed.

3. Global Principle

The implementation of mass customization is not only related to manufacturing technology and management technology, but also is related to people's thinking mode and values. Beyond drawing useful ideas from lean production, agile manufacturing, modern integrated manufacturing and group technology etc, it also learn some very important basic idea and method, namely customizing time be moved backward method, the total cost concept and product lifecycle management, etc.

6.3.3　Key Technologies of MC

1. Development and Design Technology for Mass Customization

In order to obtain the comprehensive economic benefits of making full use of mass customization, the principle of mass customization should be firstly applied in the design and development stage. The development and design technology for mass customization includes the design and development of product as well as process (manufacturing and assembly process). The complete devel-

和面向大批量定制的设计两个过程组成。

2. 面向大批量定制的管理技术

面向大批量定制的管理技术是实现大批量定制的关键技术。为此，应该针对大批量定制在管理方面的特点，采用相应的管理技术，包括各种客户需求获取技术、面向大批量定制的生产管理技术、企业协同技术、知识管理和企业文化等。这些技术形成了一个完整的体系，分别在不同的阶段和从不同的层次支持企业实现大批量定制。

3. 面向大批量定制的制造技术

面向大批量定制的制造技术应该具有足够的物理和逻辑的灵活性，能够根据被加工对象的特点，方便、高效、低成本地改变系统的布局、控制结构、制造过程及生产批量等，有效地支持大批量定制。另外，为了有效地实现面向大批量定制的制造，在产品设计及工艺设计方面必须做到标准化、规范化及通用化，便于在制造过程中可以利用标准的制造方法和标准的制造工具（刀具和夹具等），优质、高效、快速地制造出客户定制的产品。

6.3.4　大批量定制的应用实例

针对不同的客户要求，存在两种不同的定制生产，即完全定制和大批量定制生产。当生产产品的流程及产品本身的变化较大时，宜采用完全定制生产法。当产品生产流程相对稳定，而产品相对变化较大时，宜采用大批量定制。完全定制的主要缺陷是敏捷性差和成本高，这对采用成本领先策略的企业来说是无法接受的。大批量定制则可以避免这一问题。其基本思想是采取主动的反应策略。首先是大大压缩修改或改变标准设计与工艺的比例，即通常所说的尽可能压缩非标件或非标工艺的比例。与此同时最大限度地增加标准件或标准流程的比重。

有些大的公司在大批量定制方面做得非常好，如戴尔公司的订单定制和网上直销。顾客在向戴尔下订单的时候，可根据需要自由组合各种配置，如 CPU、内存、硬盘、光驱、显示器等单元，从而装配出完全符合自己需要的计算机系统。通常戴尔系统中的原材料和零部件库存量大概只能维持 4 天的生产，而其同行业竞争者的库存量大多在 30~40 天。

在我国，大批量定制的代表当属海尔集团。海尔集团从 1999 年开始将定制化的思想引入家电产品的生产中，始终根据订单实施大量定制。海尔不但实现了家电产品的按需生产，同时还保证低成本和快速交货。海尔建立了一个可供顾客进行个性化定制的电子商务平台，把研制开发出的冰箱、洗衣机、空调等 58 个门类 9200 多种基本产品类型放到平台上，顾客可以在这些平台上进行模块化操作。

6.4　并行工程

6.4.1　并行工程的产生与内涵

20 世纪 80 年代中期以来，制造业商品市场发生了根本性的变化。同类商品日益增多，企业之间的竞争越来越激烈，而且越来越具有全球性。顾客对产品质量、成本和种类的要求越来越高，产品的生命周期越来越短。因此，企业为了赢得市场竞争的胜利，就不得不解决加速新产品开发、提高产品质量、降低成本和提供优质服务等一连串的问题。其中，迅速开

opment and design process for mass customization consists of two processes: development for mass customization and design for mass customization.

2. Management Technology for Mass Customization

The management technology for mass customization is the key technology of mass customization. Therefore, based on the characteristics of mass customization in management, the corresponding management technology should be used, including various acquirement technology of customer requirement, production management technology for mass customization, enterprise collaboration technology, knowledge management and enterprise culture etc. These technologies form a complete system to support enterprises to achieve mass customization at different stages and different levels respectively.

3. Manufacturing Technology for Mass Customization

The manufacturing technology for mass customization should have enough physical and logical flexibility. According to the characteristics of machined object, it can conveniently, efficiently and in low cost change the system layout, control structure, manufacturing process mass production and so on to support the mass customization effectively. In addition, in order to effectively implement manufacturing for mass customization, the product design and process design had to be standardization, standardization and generalization. Thus, it is easy to use standard manufacturing methods and standard manufacturing tools (cutter and clamp, etc.) in the manufacturing process to manufacture efficiently and fast the customized product with high-quality.

6.3.4　Application Example of MC

For different customer requirements, there are two different customization production, namely complete customization and mass customization production. When the process of production and the product itself change greatly, it is necessary to adopt the complete customization method. When the process of product production is relatively stable, but the relative change of the product is relatively large, it is appropriate to use mass customization. The main drawbacks of complete customization are poor agility and high cost, which is unacceptable for companies using cost leadership strategy. Mass customization can avoid this problem. The basic idea is to take the active response strategy. The first is to greatly compress or modify the proportion of standard design and process, that is, as far as possible to compress the proportion of non-standard parts or non-standard process, and maximizing the proportion of standard parts or standard processes at the same time.

Some big companies do very well in mass customization, for instance, the order customization and online direct selling of DELL. When making an order to DELL, the customers can freely combine a variety of configurations according to their needs, such as CPU, memory, hard drives, optical drives, monitors and other units, thus assembling computer system in full compliance with their need. Usually raw materials and spare parts inventory of DELL system can only be maintained for production of four days, and inventory of its competitors is mostly be within 30-40 days.

In China, Haier Group is the representative of mass customization. From 1999 Haier Group began to introduce the idea of customization into the production of household electrical appliances,

发出新产品，使其尽早进入市场成为赢得竞争胜利的关键。

要解决这一问题，必须改变传统的产品开发模式——串行生产模式，即概念设计—详细设计—过程设计—加工制造—试验验证—设计修改—工艺设计—正式投产—营销。传统的产品开发过程就像接力赛一样，产品总是从一个部门递交到下一个部门，每次都根据各自需要进行修改。这种"扔过墙"式的产品开发模式如图6-4所示。

Fig. 6-4　"Throw It Over the Wall" type development mode "扔过墙" 式的产品开发模式

在这种开发模式中，错误将沿着设计链一直传播下去，而且这些错误通常在需要昂贵的代价或在不便修正的情况下才被发现。由于开发者不能在产品设计阶段及早地考虑后续的工艺设计、制造、装配和质量保证等问题，致使各个生产环节前后脱节、设计改动量大、成本高，更主要的是产品开发周期长，难以满足激烈的市场竞争要求。人们不得不开始寻求更为有效的新产品开发方法。1988年，美国国防部防御分析研究所（The Institute for Defense Analyses，IDA）发表了非常著名的R-338报告，提出了"并行工程"（Concurrent Engineering，CE）的概念，并在美国西弗吉尼亚大学投资4~5亿美元建立了并行工程研究中心（Concurrent Engineering Research Center，CERC）。进入20世纪90年代后，美国许多大公司开始了并行工程实践的尝试，取得了实效。并行工程开始成为全球制造业关注的热点问题。

CE是一种对产品及其相关过程（包括制造过程和支持过程）进行并行的、一体化设计的工作模式。这种工作模式可使产品开发人员一开始就能考虑从产品概念设计到消亡的整个产品生命周期中的所有因素，包括质量、成本、进度和用户要求。CE的核心是实现产品及其相关过程设计的集成。图6-5所示为并行工程示意图。可见，并行工程可以使新产品开发时间大大缩短。

CE要求各有关部门人员在产品开发的早期阶段就要介入，而且参与每一个有关环节，强调各个部门的"协同工作"，即产品设计师、工艺师、财务分析人员、生产计划人员以及采购供应、市场营销人员在开始时就集合在一起，变成各部门之间的"同室协调"，如图6-6所示。可以充分利用集体智慧，获取有效的知识和经验，使产品设计能更好地满足用户要求，力争使产品开发能够一次获得成功，从而缩短产品开发周期，降低产品成本。

and always operate mass customization based on the order form. Haier not only complete the on-demand production for household electrical appliances, but also guarantee low cost and fast delivery. Haier Group has established a personalized e-commerce platform for customers to customize. The refrigerator, washing machine, air conditioning and other 58 categories of 9,200 kinds of basic types of products are put on the platform, so customers can make modular operation on the platform.

6.4　Concurrent Engineering

6.4.1　Emergence and Connotation of Concurrent Engineering (CE)

The goods market of manufacturing industry has taken place a fundamental change since the mid of 1980s. The identical goods is increased day by day, the competition between enterprises becomes more and more fierce, and being global increasingly. The demands of customers on product quality, cost and types are much higher, but product life cycle becomes much shorter. Therefore, in order to win in the market competition, enterprises have to resolve a series problems including accelerating new product development, improving product quality, reducing costs, providing quality services. Among them, the key to winning the competition is to develop new products rapidly and enter the market as early as possible.

To solve this problem, the traditional product development mode that is sequential production must be changed. The processes of sequential production sequentially include conceptual design, detailed design, process design, manufacturing, experimental verification, design modification, process design, formally put into production, and marketing. The traditional development process for products just like a relay race that products are always submitted from one department to next department and the modification every time need to be done according to their demands. This kind of development mode is known as "Throw It Over the Wall", as shown in Fig. 6-4.

In this mode, mistake in every process will be passed on following the design chains, and it is usually found at a high price or in case of inconvenient correction. Meanwhile, in product design stage this development mode could not early consider the follow-up process design, manufacturing, assembly, quality assurance, etc. Thereby, it will cause disconnect between each production link, a large quantity design changes and high cost. The more important is that product development cycle is long, being difficult to meet the fierce market competition. So, a more effective method of new product development has to be searched. In 1988, the Institute for Defense Analyses (IDA) of U.S. Ministry of National Defense published a very famous R-338 report, put forward concept of "Concurrent Engineering" (CE), and invested $400-$500 million to set up the Concurrent Engineering Research Center (CERC) in West Virginia university. After entering the 1990s, many large companies in the United States began to practice the concurrent engineering, and achieved some actual effect. Concurrent engineering has become the hot topic in the global manufacturing industry.

CE is a kind of operating mode which executes concurrent and integrated design for products and related processes (including manufacturing process and support process). This operating mode

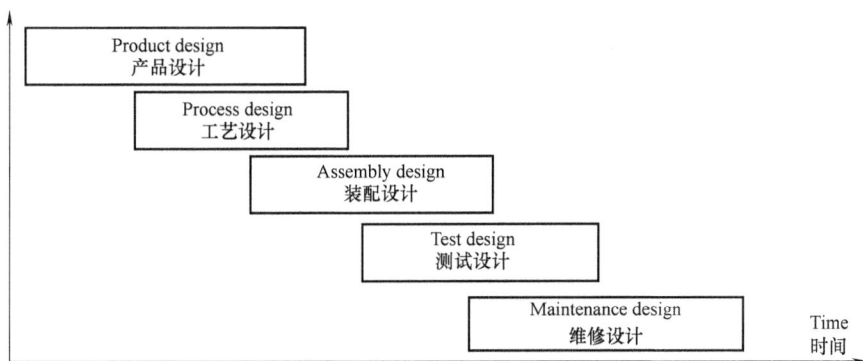

Fig 6-5　Schematic diagram of concurrent engineering 并行工程示意图

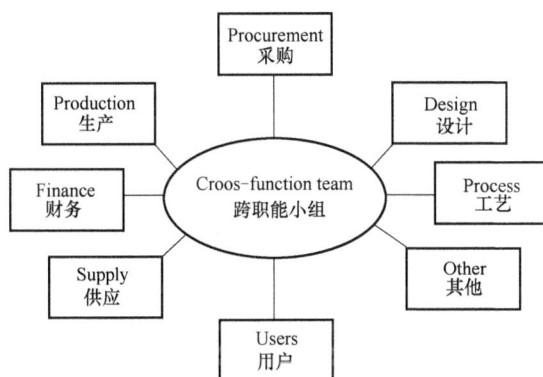

Fig. 6-6　"Room coordination" between departments 各部门 "同室协调"

6.4.2　并行工程的特性

1. 并行特性

把时间上有先后的作业活动转变为同时考虑或尽可能同时或并行处理的活动。

2. 整体特性

制造系统（包括制造过程）是一个有机的整体，在空间中似乎互相独立的各个制造过程和知识处理单元之间，实质上都存在着不可分割的内在联系，特别是丰富的双向信息联系。图6-7反映了产品开发过程中主要作业环节之间的内在联系。

并行工程强调全局性地考虑问题，即产品研制者从一开始就考虑到产品整个生命周期中的所有因素。并行工程追求的是整体最优，把产品开发的各种活动作为一个集成的过程进行管理和控制，以达到整体最优的目的。

3. 协同特性

并行工程特别强调设计群体的协同工作。现代产品的功能和特性越来越复杂，产品开发过程涉及的学科门类和专业人员也越来越多，要取得产品开发过程的整体最优，其关键是如何很好地发挥人们的群体作用。

（1）多功能的协同组织机构　并行工程根据任务和项目的需要组织多功能工作小组，小组成员由设计、工艺、制造和支持（质量、销售、采购服务等）的不同部门、不同学科

enable designer to consider all factors of the whole product life cycle from conceptual design to the product vanishing. These factors include quality, cost, schedule and user requirements. The core of CE is to realize the integration of the product and related process design. Fig. 6-5 is the schematic diagram of concurrent engineering. It is seen that CE can shorten the development time of new products greatly.

CE requires person in each relevant department to get involve in product development at the early stage, and participate in each related link. It emphasis on "cooperative work", namely, product designer, technologist, financial analysis personnel, production planning personnel, supply and market purchasing personnel in gather together from the beginning, becoming "coordination in one room" between departments, as shown in Fig. 6-6. The collective wisdom can be made full use of to obtain effective knowledge and experience, which making product design meet user requirements better, striving to make product development success, thus shortening product development cycle and reducing product cost.

6.4.2　Characteristics of CE

1. Concurrent Characteristic

The operating activities in chronological order are turned into activities that are considered concurrently or are operated simultaneously or concurrently as possible.

2. Overall Characteristic

The manufacturing system (including the manufacturing process) is an organic entirety.

There are inseparable inner links essentially, especially ample two-way information link, through it seems to be independent of each other in space between each manufacturing process and knowledge process unit. Fig. 6-7 reflects the inner links between the key operating segments in the process of product development.

Considering problems overall is emphasized in concurrent engineering, namely product developing personnel should consider all factors in the entire product life cycle from the start. The aim of concurrent engineering is to search for overall optimization, which product development activities are an integrated process to manage and control, so as to achieve the goal of overall optimization.

3. Collaborative Characteristic

It specially emphasized the team work of design group in concurrent engineering. The function and characteristics of modern products become more and more complex, and the product development process involved much more disciplines and professionals. To obtain the overall optimization for product development process, the key is how to play the group role greatly.

(1) Multifunctional collaborative organization　Multifunctional working groups are organized according to requirement of tasks and projects in concurrent engineering, in which group members are comprise of representatives of different departments and disciplines including design, technological process, manufacturing and support (quality, sales, procurement and services, etc.). Working groups have their own work plan and goal with their own responsibility, right and benefit. The group members complete collaboratively common tasks with the same terms and common information re-

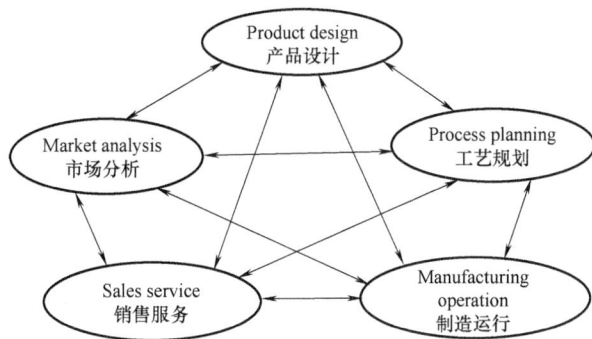

Fig. 6-7 The inner link of various segments of the manufacturing system

制造系统各环节之间的内在联系

的代表组成。工作小组有自己的责、权、利,有自身的工作计划和目标,小组成员之间用相同的术语和共同的信息资源工具,协同完成共同的任务。

(2) 协同的设计思想 并行工程强调一体化、并行地进行产品及其相关过程的协同设计,尤其注意早期概念设计阶段的并行和协调。

(3) 协同的效率 并行工程特别强调"1+1>2"的思想,力求排除传统串行模式中各个部门间的壁垒,使各个相关部门协调一致地工作,利用群体的力量提高整体效益。

这种途径生产出来的产品不仅有良好的性能,而且产品研制的周期也将显著缩短。

4. 集成特性

并行工程是一种系统集成方法,主要表现为:①管理者、设计者、制造者、支持者乃至用户集成为一个协调的整体;②产品全生命周期中的各类信息的获取、表示、表现和操作工具的集成和统一管理;③产品全生命周期中企业内各部门功能集成,以及产品开发企业与外部协作企业间功能的集成。

6.4.3 并行工程实施的关键技术

实践表明,并行工程的实施并不是轻而易举的,需要改变企业的文化、管理及各种用于设计、制造和支持的方法与技术。下面简单介绍几种实施并行工程的关键技术:

1. 过程管理与集成

并行工程与传统生产模式的本质区别在于它把产品开发的各个活动作为一个集成的、并行的产品开发过程,强调下游过程在产品开发早期参与设计过程,对产品开发过程进行管理和控制,不断改善产品开发过程。它主要包括建模技术、管理技术、评估技术、分析技术和过程集成技术。串行工程和并行工程在产品创新、质量、生产成本和柔性上的比较见表 6-1。

2. 集成产品开发团队

产品开发由传统的部门制或专业组变成以产品(型号)为主线的多功能集成产品开发团队(Integrated Product Team, IPT)。IPT 包括了市场设计、工艺、生产技术准备、制造、采购、销售、维修服务等各部门人员,有时甚至还包括用户、供应商或协作厂的代表。采用这种团队工作方式能大大增强产品生命周期各阶段人员之间的信息交流和合作,在产品设计时及早地考虑产品的可制造性、可装配性、可检验性等。

source tools.

（2）Collaborative design idea　Concurrent engineering emphasizes to conduct integrated and concurrent collaborative design of product and related process, especially pay attention to concurrent and coordination in the early concept design phase.

（3）Collaborative efficiency　Concurrent engineering places particularly emphasis on the thought of "1+1>2", striving to eliminate barriers between each department in traditional sequential mode. It makes each relevant department work collaboratively and applies group power to improve overall benefit.

The product produced by this way not only has good performance, but the product development cycle will be shortened significantly.

4. Integration Characteristic

Concurrent engineering is a kind of system integration method, which has three features: ①Managers, designers, producers, supporters and users are integrated as a collaborative entirety; ②The acquisition, representation, performance of all kinds of information and operating tools during the whole life cycle of products are integrated and managed uniformly; ③During the whole life cycle of products, the functional integration of each department in enterprise as well as the functional integration between product development enterprise and external cooperation enterprise are carried out.

6.4.3　Key Technologies of Implementing CE

Practice indicates that the implementation of CE is not easy, and it is need to change the enterprise culture, management as well as all kinds of method and technology used in design, manufacturing and support. Several key technologies of implementing concurrent engineering are to be introduced briefly as follows:

1. Process Management and Integration

The essential difference between concurrent engineering and traditional production mode is that each activity of product development is taken as an integrated and concurrent process, emphasizing the downstream process be involved in design process in the early product development. The product development process is managed and controlled, to constantly improve the product development process. Specific technologies mainly include modeling, management, evaluation, analysis and process integration technology. The comparison of sequential engineering and concurrent engineering in product innovation, quality, production cost, and flexible is shown in Tab. 6-1.

2. Integrated Product Development Team

The traditional department system or professional group for product development are changed into the multifunctional Integrated Product Team (IPT) taking product (types) as the main line. IPT includes each department personnel such as for design, process, production technology preparation, manufacturing, purchasing, sales, maintenance services, and sometimes even including the user, the representative of supplier or cooperation factory. Applying this team work mode can greatly improve the information exchanges and cooperation between persons at all stages of product life cycle, and during product design early consider the product's manufacturability, assemble ability, de-

Tab. 6-1 The comparison of sequential engineering and concurrent engineering

串行工程和并行工程的比较

Competitive advantage 竞争优势	Concurrent engineering 并行工程	Sequential engineering 串行工程
Product quality 产品质量	Better. The problems of product manufacturing have been noticed before production. 较好。在生产前即已注意到产品的制造问题	Lack of communication between design and manufacture results in product quality cannot be optimized. 设计和制造之间沟通不足,致使产品质量无法达到最优化
Product cost 产品成本	Due to the ease of manufacture of products is improved, the production cost is lower. 由于产品的易制造性提高,生产成本较低	The development cost of new product is lower, but manufacturing costs may be higher. 新产品开发成本较低,但制造成本可能较高
Production flexible 生产柔性	Suitable for product of small batch and multi-species. Suitable for products of new high-tech industry. 适于小批量、多品种生产;适于高新技术产业的产品	Suitable for production of mass and single variety. Suitable for low-tech products. 适于大批量、单一品种生产;适于低技术产品
Product innovation 产品创新	Be able to learn the methods of timely revision and innovation awareness from product development, new products put into the market faster and have strong competition ability. 能从产品开发中学习及时修正的方法及创新意识,新产品投放市场快,竞争能力强	It is difficult to obtain the latest technology and the changing trend of market demand, which is not conducive to product innovation. 不易获得最新技术以及市场需求变化趋势,不利于产品创新

3. 协同工作环境

在并行工程产品开发模式下，产品开发是由分布在异地的采用异种计算机软件工作的多学科小组完成的。多学科小组之间及多学科小组内部各组成人员之间存在着大量相互依赖的关系，协同工作环境支持 IPT 的异地协同工作。协调系统用于各类设计人员协调和修改设计，传递设计信息，以便做出有效的群体决策，解决各小组间的矛盾。基于 PDM 系统构造的 IPT 产品数据共享平台，在正确的时间将正确的信息以正确的方式传递给正确的人；基于 Client/Server 结构的计算机系统和广域的网络环境，使异地分布的产品开发队伍能够通过 PDM 和群组协同工作系统进行并行协作产品开发。

4. 数字化产品建模与 CAX/DFX 使能工具

基于一定的数据标准，建立产品生命周期中的数字化产品模型，特别是基于 STEP 标准的特征模型。产品设计主模型是产品开发过程中唯一的数据源，用于定义覆盖产品开发各个环节的信息模型，各环节的信息接口采用标准数据交换接口进行信息交换。数字化工具定义是指广义的计算机辅助工具集，最典型的有 CAD、CAE、CAPP、CAM、计算机辅助工装设计（Computer-Aided Tooling Design，CATD）、面向装配的设计（Design for Assembly，DFA）、面向制造的设计（Design for Manufacturing，DFM）、加工过程仿真（Manufacturing Process Simulation，MPS）等。

5. 全面质量管理技术方法和工具

全面质量管理工具用于收集用户需求，将这些需求转变成具体的时间、成本和性能值，

tectability, etc.

3. Collaborative Work Environment

The development of product is fulfilled by multidisciplinary team distributed in different place and adopting heterogeneous computer software in product development mode for concurrent engineering. There are a large number of interdependent relationships among the multidisciplinary teams and among the group staffs of multidisciplinary team. The collaborative work environment supports the collaborative work of IPT at different places. Collaborative system is used for coordinating all kinds of design personnel and modifying design, transmitting design information, in order to make effective group decision and solve the contradiction between each team. The IPT product data sharing platform built by PDM system is for passing the correct information to the right person in the right way at the right time. The computer system and the wider network environment based on Client/Server structure enable the product development teams distributed geographically to develop product concurrently and collaboratively through PDM and groups collaborative work system.

4. Digital Product Modeling and CAX/DFX Can Make Tools

Based on certain data standards, the digital product model in product life cycle, especially the characteristic model based on STEP standard, is established. The main model of product design is unique data source, used for defining information model throughout every stage during product development. The information interface of every stage exchanges information applying standard data interchange interface. The definition on digital tools refers to the generalized set of computer aided tools. The most typical tool includes CAD, CAE, CAPP, CAM, computer-aided tooling design (CATD), design for assembly (DFA), design for manufacturing (DFM), manufacturing process simulation (MPS), and so on.

5. Total Quality Management Techniques and Tools

Total quality management tool is used to gather user requirements, and these requirements are turned into specific time, cost and performance value. Meanwhile, the process of building the whole system is monitored, in order to maximize meet user requirements (including internal users and external users). These tools include Taguchi, quality function deployment (QFD), statistical process control (SPC), cost estimation, value engineering (VE), etc.

6. 4. 4 Application Examples of CE

Concurrent engineering has been widely used in the United States, Germany, Japan and other countries, involving automobile, airplane, computer, machinery, electronics and other industries. For example, Boeing Aircraft Company invested more than $4 billion to develop Boeing 777 jet airliner, using large computer network to support concurrent design and network manufacturing. It only took 3 years and 2 months that began to design in October 1990 until got successful trial-production in June 1994, and Boeing 777 jet airliner was carried out trial flight and succeeded at a time to come into service immediately. When the AT&T Company conducted the production of printing group component for computer, as the original design did not consider the production craft, the product quality was low, and the percent of pass was only 5%. After using concurrent design, com-

并监控整个系统建立的过程，以便最大限度地满足用户（包括内部用户和外部用户）需求。这些工具有：田口（Taguchi）方法、质量功能配置（Quality Function Deployment，QFD）方法、统计过程控制（Statistical Process Control，SPC）、成本分析（Cost Estimation）和价值工程（Value Engineering，VE）等。

6.4.4　并行工程应用实例

并行工程在美国、德国、日本等一些国家已得到广泛应用，其应用领域包括汽车、飞机、计算机、机械、电子等行业。例如，美国波音飞机制造公司投资 40 多亿美元，研制波音 777 型喷气客机，采用庞大的计算机网络来支持并行设计和网络制造。从 1990 年 10 月开始设计到 1994 年 6 月，仅花了 3 年零 2 个月就试制成功进行试飞，试飞一次成功即投入运营。AT&T 公司在生产计算机配套印刷组产品时，原来的设计中未考虑生产工艺性问题，致使产品质量低下，合格率仅为 5%。采用并行设计后，利用计算机虚拟检测，找出设计中的缺陷，使产品合格率达到 90%。美国 Mercury 计算机联合开发公司在开发 40-MHz Intel i860 微处理芯片时，运用并行工程方法，使产品从开始设计到被消费者检验合格由原来的 125 天减少到 90 天。

国内对 CE 的研究也已发展到了一定的高度，在航空、航天、铁路等领域均有广泛的应用。例如，西安飞机工业（集团）有限责任公司开发出支持飞机内装饰并行工程的系统工具，包括适用于飞机内装饰的 CAID 系统、DEA 系统和模具的 CAD/CAE/CAM 系统。在 Y7-700A 飞机内装饰工程中，应用过程建模与 PDM 实施、并行工程环境下的模具 CAD/CAM、飞机客舱内装饰数字化定义等技术手段，使研制周期从 1.5 年缩短到 1 年，减少设计更改 60% 以上，降低产品研制成本 20% 以上。

6.5　精益生产

6.5.1　精益生产的概念

1990 年，詹姆斯 P. 沃麦克、丹尼尔 T. 琼斯和丹尼尔·鲁斯在他们的《改变世界的机器》一书中创造了"精益生产"一词，描述了由丰田生产系统确立的生产模式。在 20 世纪 50 年代，丰田汽车公司率先推出一种先进制造方法，它旨在使单个产品在整个生产过程中流动占用的资源最少。它的灵感来自于 20 世纪初的享利·福特消除浪费的概念，丰田建立了关注生产工艺流程并系统识别和消除所有浪费的组织文化。在精益的情况下，他们认为任何不直接创造产品或服务客户需求的活动都是浪费。在许多企业流程中，像等待、产品不必要"接触"、生产过剩、浪费和原材料、能源的低效使用以及其他这样的"非增值活动"可能超过总活动的 90%。丰田先进制造方法的成功实施使在众多行业的其他数百家公司也用这些先进的生产方法来处理他们的业务。在这份报告中，"精益"一词用来描述广泛实施的几种先进制造方式。

精益生产通常代表了从均衡大规模生产下传统"批处理队列"模式转变为"一个流"的拉式产品排列模式。这种转变需要对操作过程的高度控制，以及良好的维护、排序和清洁的操作环境，并结合准时生产的原则和员工参与、系统范围、持续改进。要做到这一点，公

puter virtual test was carried out to find out the defects in the design, which resulted in the product qualification rate reach 90%. When the Mercury Computer Joint Development Company in the United States developed 40-MHz Intel i860 microprocessor chip, the CE method was applied to lead to the consumers of inspection from 125 days to 90 days using the method of concurrent engineering. The product developing period from the product design at the beginning to pass the consumers inspection was reduced from 125 days to 90 days by using the method of concurrent engineering.

The domestic research on CE has also been highly developed, and CE has been widely used in the railway, aviation, aerospace and other fields. For example, Xi'an Aircraft Industry (group) co. Ltd. has developed CE system tools which support the interior decoration of aircraft, including CAID system suitable for interior decoration of aircraft, DEA system and CAD/CAE/CAM system for mold. During the interior decoration engineering of Y7-200A, the process modeling and implementation of product data management (PDM), mold CAD/CAM based on concurrent engineering, digital definition of inner decoration of plane cabin project and so on are applied, resulting in shortening the developing period from 1.5 years to 1 year, reducing the design change by more than 60%, and decreasing product development costs by more than 20%.

6.5　Lean Production

6.5.1　The Concept of Lean Production

James Womack, Daniel Jones, and Daniel Roos coined the term "lean production" (LP) in their 1990 book *The Machine that Changed the World* to describe the manufacturing paradigm established by the Toyota Production System. In the 1950s, the Toyota Motor Company pioneered a collection of advanced manufacturing methods that aimed to minimize the resources it takes for a single product to flow through the entire production process. Inspired by the waste elimination concepts developed by Henry Ford in the early 1900s, Toyota created an organizational culture focused on the systematic identification and elimination of all waste from the production process. In the lean context, waste was viewed as any activity that does not lead directly to creating the product or service a customer wants when they want it. In many industrial processes, such "non-value added" activity can comprise more than 90% of the total activity as a result of time spent waiting, unnecessary "touches" of the product, overproduction, wasted movement, and inefficient use of raw materials, energy, and other factors. Toyota's success from implementing advanced manufacturing methods has lead hundreds of other companies across numerous industry sectors to tailor these advanced production methods to address their operations. Throughout this report, the term "lean" is used to describe broadly the implementation of several advanced manufacturing methods.

Lean production typically represents a paradigm shift from conventional "batch and queue", functionally aligned mass production to "one-piece flow", product-aligned pull production. This shift requires highly controlled processes operated in a well maintained, ordered, and clean operational setting that incorporates principles of just-in-time production and employee-involved, system-

司需采用各种先进的制造工具以降低生产的时间强度、材料强度和资本密集度。

丰田生产系统（Toyota Production System，TPS）是丰田独特的制造方法。这是许多精益生产运动的基础，多年来一直主导着制造业的发展。

6.5.2　精益生产的发展

在 20 世纪 30 年代，丰田的领导人参观了福特和通用汽车公司，研究他们的装配线，仔细阅读了亨利·福特的书《今天和明天》。在"二战"前，丰田就意识到日本市场太小，需求过于分散，不能使用美国的大量生产（一条美国的汽车生产线每月可生产 9000 台汽车，而丰田每月仅生产约 900 台，福特的生产力大约是丰田的 10 倍），但丰田需要在采用福特制造过程的同时实现高品质、低成本、短的提前期和灵活性。

1. 单件流——一个核心原则

大野耐一（TPS 的创始人之一）通过进一步走访美国树立了竞争标杆。大野耐一认为丰田需要掌握的主要构件之一是连续流动，而且在当时最好的例子是福特的流水装配线。他确定可以使用福特的连续物料流的理念（如装配线）开发一种能够根据客户需求灵活变化同时效率较高的"一个流"系统。灵活性要求整合工人的智慧，不断改进流程。

2. 创造改变世界的制造系统

在 20 世纪 50 年代，大野耐一开始了他在一些丰田工厂应用自働化和单件流原则的亲身实践。经过多年中几十次的实践，他提出了新的丰田生产系统。连同福特的教训，TPS 借用了许多来自美国的观念。一个非常重要的理念是拉式系统，它的灵感来自于美国超市。没有这种拉动系统，准时制（Just in time，JIT）——TPS 的两大支柱之一（另一个是"自働化"）就不会有发展。

JIT 是一套原则、工具和技术，可以使公司以最短的时间小批量生产和交付产品，满足特定客户的需求。简而言之，JIT 是在正确的时间提供正确数量的正确产品。JIT 的优势在于它能够响应每天变化的客户需求，这正是丰田自始至终所需要的。

丰田也秉承了美国质量先锋——戴明的教海。他在日本举办了美国的质量和生产力研讨会，并教导说，在一个典型的业务系统中，满足和超越客户的要求是组织中每个人的任务。他大大扩大了客户的定义，包括内部客户和外部客户，生产线或业务流程中的每一个人或每一个步骤都被视为一个客户，并在需要的确切时间提供所需的东西。它是 JIT 最显著的表现之一，因为在拉式系统中，它意味着前面的流程必须做后续流程要求的事情。否则，JIT 不工作。

戴明还鼓励日本采取解决问题的系统方法，即后来被称为戴明环或 PDCA 循环、持续改进的基石。持续改进是一个渐进改进的过程，不管改进多么小，它是实现消除全部增加成本不增加价值的浪费的精益目标。持续改进是一个总的理念，要求在日常的基础上追求精益求精和持续执行 TPS。

当大野耐一和他的团队在车间形成一个新的制造系统时，它不只是为了在一个特定的市场和文化中的某个公司，他们所创造的是制造业一种新的模式，一种新的看待、理解和解释生产过程中发生的事情的方式，可以推动他们超越大规模生产系统。

到了 20 世纪 60 年代，TPS 成为一个强大的理念，所有类型的业务和流程都能学会并使用它，但这需要一段时间。丰田推广精益的第一步是把 TPS 的原则努力地教给他们的关键

wide, continual improvement. To accomplish this, companies employ a variety of advanced manufacturing tools to lower the time intensity, material intensity, and capital intensity of production.

The Toyota Production System (TPS) is Toyota's unique approach to manufacturing. It is the basis for much of the lean production movement that has dominated manufacturing trends for many years.

6. 5. 2　The Development of LP

In the 1930s, Toyota's leaders visited Ford and GM to study their assembly lines and carefully read Henry Ford's book, *Today and Tomorrow*. Before World War Ⅱ, Toyota realized that the Japanese market was too small and demand too fragmented to support the high production volumes in the U. S. (A U. S. auto line might produce 9, 000 units per month, while Toyota would produce only about 900 units per month, and Ford was about 10 times as productive). But it needed to adapt Ford's manufacturing process to achieve simultaneously high quality, low cost, short lead times, and flexibility.

1. One-Piece Flow, a Core Principle

Ohno (one of the founders of TPS) benchmarked the competition through further visits to the U.S. One of the major components that Ohno believed Toyota needed to master was continuous flow and the best example of that at the time was Ford's moving assembly line. It determined to use Ford's original idea of continuous material flow (as illustrated by the assembly line) to develop a system of one-piece flow that flexibly changed according to customer demand and was efficient at the same time. Flexibility required marshaling the ingenuity of the workers to continually improve processes.

2. Creating the Manufacturing System That Changed the World

In the 1950s, Ohno began his many hands-on journeys through Toyota's few factories, applying the principles of jidoka and one-piece flow. Over years and then decades of practice, he had come up with the new Toyota Production System. Along with the lessons of Henry Ford, TPS borrowed many ideas from the U.S. One very important idea was the concept of the pull system, which was inspired by American supermarkets. Without this pull system, just-in-time (JIT), one of two pillars of TPS (the other is jidoka, built-in quality), would never have evolved.

JIT is a set of principles, tools, and techniques that allows a company to produce and deliver products in small quantities, with short lead times, to meet specific customer needs. Simply put, JIT delivers the right items at the right time in the right amounts. The power of JIT is that it allows you to be responsive to the day-by-day shifts in customer demand, which was exactly what Toyota needed all along.

Toyota also took to heart the teachings of the American quality pioneer, W. Edwards Deming. He gave U. S. quality and productivity seminars in Japan and taught that, in a typical business system, meeting and exceeding the customers requirements is the task of everyone within an organization. And he dramatically broadened the definition of customer to include both internal and external customers. Each person or step in a production line or business process was to be treated as a customer and to be supplied with exactly what was needed, at the exact time needed. It became one of

供应商，这使孤立的精益制造厂转向全面精益扩展型企业，使供应链上的每一个企业都使用相同的 TPS 原则。

最后，在 20 世纪 90 年代，通过美国麻省理工学院的国际汽车计划组织和基于其研究成果的畅销书《改变世界的机器》，世界制造业才发现精益生产，这是作者对丰田几十年前通过专注于供应链的速度而学到的东西的称谓：通过消除每一流程的浪费来缩短提前期，达到最好的质量、最低的成本，同时提高安全性和士气。

6.5.3 TPS 的核心：消除浪费

理解消除浪费的原则时，应当结合大野耐一在车间的经历，他在那里花费了大量的时间，试着规划增加产品的价值并去掉非增值活动。这是非常重要的，因为许多 TPS 的工具和丰田方式的原则都来源于这种行为。

TPS 是丰田方式的原则可以达到的最系统的、最高度发达的例子。丰田方式由让 TPS 功能有效的丰田义化的基本原则组成。虽然它们是不同的，但 TPS 的发展及其惊人的成功与丰田方式的演变和发展紧密相连。应用 TPS 的时候，应当从客户的角度审视制造过程。TPS 的第一个问题应该是"客户想从这个流程得到什么？"（在生产线下一步的内部客户和最后的外部客户），这定义了价值。通过客户的眼睛，可以观察到一个过程，并从非增值步骤分离增值步骤，可以将此应用于任何制造、信息或服务流程。

丰田确定了在业务或制造过程中的八种主要类型的非增值浪费，如下所述。人们可以将这些应用到产品开发、订单处理和办公室，而不仅仅是生产线。

1）过量生产的浪费。没有订单就生产产品，造成人员过多的浪费以及大量库存造成存储和运输的浪费。

2）等待的浪费（时间）。工人只照看一台自动化设备或等待下一工序、工具、供应、零件等，或只是因为缺货、很多处理的延误、设备停机和产能瓶颈而无工作可做。

3）不必要的运输或搬运的浪费。在流程之间长距离搬运在制品（WIP，Work in Progress）、移动物料和零件，或成品的出入库，造成运输效率低下。

4）过度加工或不正确的加工。采取不必要的步骤处理零件；糟糕的工具和产品设计造成的低效率的处理，引起不必要的运动和生产缺陷；提供比必要的质量更高质量的产品产生的浪费。

5）库存过剩。过量的原材料、在制品和成品会引起较长的生产准备时间、报废、货物损坏、运输和储存成本、延迟。此外，多余的库存还会掩盖如生产不平衡、供应商推迟交货、缺陷、设备停机、调整时间长等问题。

6）不必要的动作。员工在工作过程中必须进行很多浪费动作，例如寻找或堆放零件、工具等。步行也是浪费。

7）缺陷。即生产瑕疵的零件或纠正。修理或返工、报废、替换产品和检查意味着处置、时间和精力的浪费。

8）未发挥员工创造力。管理人员不倾听员工的建议或意见，而失去改进和学习的机会。

大野耐一认为生产过剩是根本的浪费，因为它导致大多数其他浪费。在制造过程中超过客户需求的任何生产操作必然导致库存在下游某处堆积，即物料滞留等待下一道工序加工。

大规模或大批量制造商可能会问，这是什么问题，只要人和设备生产零件？问题是巨大

the most significant expressions in JIT, because in a pull system it means the preceding process must always do what the subsequent process says. Otherwise JIT won't work.

Deming also encouraged Japanese to adopt a systematic approach to solve problem, which later became known as Deming Cycle or PDCA Cycle, a cornerstone of continuous improvement. The continuous improvement is the process of making incremental improvements, no matter how small, and achieving the lean goal of eliminating all waste that adds cost without adding to value. It is a total philosophy that strives for perfection and sustains TPS on a daily basis.

When Ohno and his team emerged from the shop floor with a new manufacturing system, it wasn't just for one company in a particular market and culture. What they had created was a new paradigm in manufacturing delivery a new way of seeing, understanding, and interpreting what is happening in a production process, that could propel them beyond the mass production system.

By the 1960s, TPS was a powerful philosophy that all types of businesses and processes could learn to use, but this would take a while. Toyota did take the first steps to spread lean by diligently teaching the principles of TPS to their key suppliers. This moved its isolated lean manufacturing plants toward a total lean extended enterprise when everyone in the supply chain is practicing the same TPS principles.

Finally, in the 1990s, through the work of MIT's Auto Industry Program and the bestseller based on its research, *The Machine That Changed the World*, the world manufacturing community discovered lean production the authors term for what Toyota had learned decades earlier through focusing on speed in the supply chain: Shortening lead time by eliminating waste in each step of a process leads to best quality and lowest cost, while improving safety and morale.

6.5.3 The Heart of TPS: Eliminating Waste

We touched on the philosophy of eliminating waste with Ohno's journey through the shop floor. He spent a great deal of time there, learning to map the activities that added value to the product and getting rid of non-value-adding activity. It's important to take a closer look at this, because many of the tools of TPS and principles of the Toyota Way derive from this behavior.

TPS is the most systematic and highly developed example of what the principles of the Toyota Way can accomplish. The Toyota Way consists of the foundational principles of the Toyota culture, which allow TPS to function so effectively. Though they are different, the development of TPS and its stunning success are intimately connected with the evolution and development of the Toyota Way. When applying TPS, you start with examining the manufacturing process from the customer's perspective. The first question in TPS is always "What does the customer want from this process?" (Both the internal customer at the next steps in the production line and the final, external customer.) This defines value. Through the customer's eyes, you can observe a process and separate the value-added steps from the non-value-added steps. You can apply this to any process manufacturing, information, or service.

Toyota has identified eight major types of non-value-adding waste in business or manufacturing processes, which are described below. You can apply these to product development, order taking,

的缓冲导致其他不理想的行为，如减少不断改善操作的动机。当丢弃有缺陷的零件时，为什么要过分担心一些质量问题？因为当一个有缺陷的工件流动到后续的操作，操作员试图装配的时候，在流程和缓冲中，可能存有几个星期的不合格零件。

图 6-8 通过铸造、加工和装配这一简单的时间线来表示这种浪费。在大多数传统的管理操作中，大部分花费在材料上的时间实际上是被浪费掉的。从精益的角度来看，在接近任何流程的过程中，首先应该根据材料的绕行路线绘制价值流图。最好是按照实际路径，获得充分的经验。在设计图上画出这个路径，计算花费的时间和通过的距离，然后给出路线图的技术名称。即使是那些在工厂里工作了一生的人也会惊讶于这个活动的结果。图 6-8 采用了非常简单的转化过程，使对该点的价值增加几乎无法辨认。

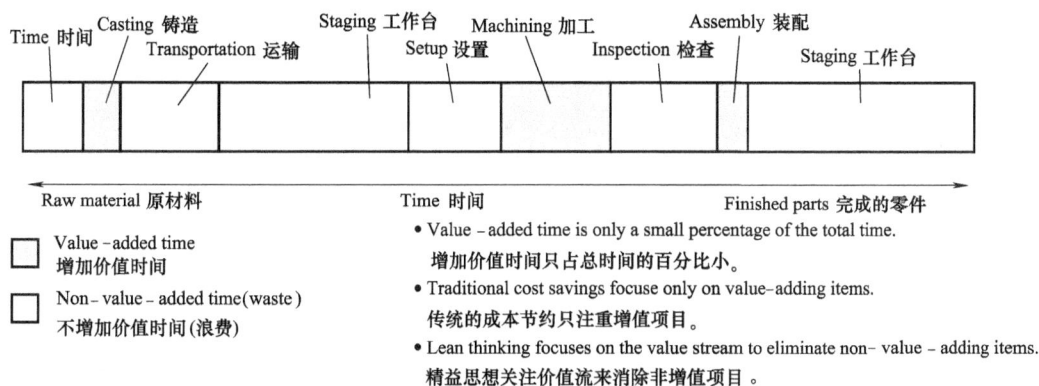

Fig. 6-8 Waste in a value system 价值体系中的浪费

6.5.4 丰田屋

几十年来，丰田每天在车间应用并改进 TPS，而没有记录 TPS 理论。工人和经理们在车间通过实践不断地学习新方法和旧方法的变化。在一个相对较小的公司，信息传递得非常快，所以在丰田开发的好方法迅速蔓延到其他丰田工厂和最终供应商。但随着在丰田内部实践的成熟，非常明确的是向基层传授 TPS 的任务永远不会结束。所以大野耐一的弟子张富士夫创造了一个简单的表示方法——一个房子。

丰田屋图（见图 6-9）已成为现代制造业最知名的标志。为什么是房子？因为房子是一个结构系统。只有屋顶、柱子和地基坚固，房子才坚固。任何一个薄弱环节都会削弱整个系统。有不同版本的房子，但核心原则保持不变。它从最好的质量、最低的成本、最短的提前期作为屋顶开始。TPS 有两个外部的柱子——准时制和自働化，准时制是 TPS 最明显和最广为人知的特性，自働化在本质上意味着永远不让缺陷传递到下一站，将人们从与人打交道的机械自动化中解脱出来。在丰田屋系统中心的是人。最后，有各种各样的基本元素，包括标准化、稳定、可靠的进程的需要，也包括平准化，这意味着在数量和品种上平衡的生产计划。一个分层的时间表或平准化使系统稳定而且允许最低库存。在某些产品的生产过程中，如果没有大量的库存被添加到系统中，大的产量波动将会造成零件的短缺。

房子的每一个元素本身是至关重要的，但更重要的是元素相互加强的方式。准时制意味着尽可能减少那些在生产中因发生问题而进行缓冲操作需要的库存。单件流的理想情况是根

and the office, not just a production line.

1) Overproduction. It means producing items for which there are no orders, which generates such wastes as overstaffing and storage and transportation costs because of excess inventory.

2) Waiting (time on hand). Workers merely serve to watch an automated machine or have to stand around waiting for the next processing step, tool, supply, parts, etc., or workers just plain have no work because of stockouts, lot processing delays, equipment downtime, and capacity bottlenecks.

3) Unnecessary transport or conveyance. It means carrying work in process (WIP) long distances, creating inefficient transport, or moving materials, parts, or finished goods into or out of storage or between processes.

4) Overprocessing or incorrect processing. Taking unneeded steps are taken to process the parts. Inefficiently processing due to poor tool and product design, which cause unnecessary motion and producing defects. Waste is generated when providing higher-quality products than is necessary.

5) Excess inventory. Excess raw materials, WIP, or finished goods cause longer lead times, obsolescence, damaged goods, transportation and storage costs, and delay. Also, extra inventory hides problems such as production imbalances, late deliveries from suppliers, defects, equipment downtime, and long setup times.

6) Unnecessary movement. The employees have to do many waste motions during work, suoh as looking for, or stacking parts, tools, etc. Walking is also waste motion.

7) Defects. Production of defective parts or correction. Repair or rework, scrap, replacement production, and inspection mean wasteful handling, time, and effort.

8) Unused employee creativity. It means losing time, ideas, skills, improvements and learning opportunities by not engaging or listening to the employees.

Ohno considered the fundamental waste to be overproduction, since it causes most of the other wastes. Producing more than the customer wants by any operation in the manufacturing process necessarily leads to a build-up of inventory somewhere downstream, the material is just sitting around waiting to be processed in the next operation.

Mass or larger-batch manufacturers might ask, what's the problem with this, as long as people and equipment are producing parts? The problem is that big buffers lead to other suboptimal behavior, like reducing your motivation to continuously improve your operations. Why get overly concerned about a few quality errors when you can just toss out defective parts? Because by the time a defective piece works its way to the later operation where an operator tries to assemble that piece, there may be weeks of bad parts in process and sitting in buffers.

Fig. 6-8 shows this waste through a simple time line for the process of casting, machining, and assembling. As in most traditionally managed operations, most of the time spent on material is actually wasted. From a lean perspective, the first thing you should do in approaching any process is to map the value stream following the circuitous path of material through your process. It is best to walk the actual path to get the full experience. You can draw this path on a layout and calculate the time and distance traveled and then give it the highly technical name of spaghetti diagram. Even people who have worked inside a factory for most of their adult lives will be amazed at the results of

据客户需求或节拍在单位时间内仅生产单位产品。使用较小的缓冲库存意味着像质量缺陷这样的问题变得立即可见。这加强了自働化而暂停生产过程，意味着工人必须立即解决问题并尽快恢复生产。房子的地基是稳定性。讽刺的是，当出现问题时，少库存和停止生产的要求会导致工人之间的不稳定和紧迫感。在大规模生产中，当一台机器坏了就没有紧迫感：维修部门计划修复它并利用库存保持生产运行。

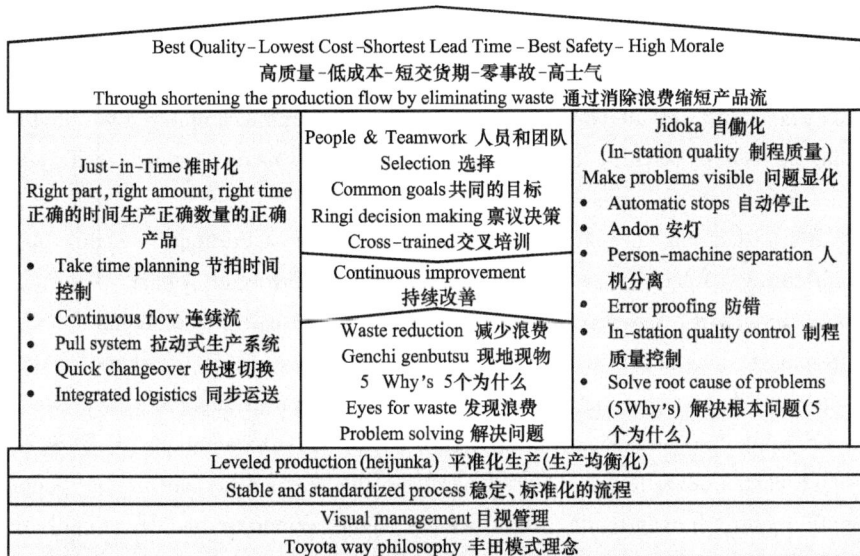

Fig. 6-9　The toyota production system 丰田生产系统

6.6　敏捷制造

6.6.1　敏捷制造的产生与含义

　　1991 年，美国里海大学在研究和总结美国制造业的现状和潜力后，发表了具有划时代意义的《21 世纪制造企业发展战略》报告，提出了敏捷制造（Agile Manufacturing，AM）的概念。敏捷制造是在具有创新精神的组织和管理结构、先进制造技术、有技术有知识的管理人员这三大支柱的支撑下得以实施的，通过所建立的共同基础结构，对迅速改变的市场需求和市场进度做出快速响应。敏捷制造比起其他制造方式具有更灵敏、更快捷的反应能力。

　　敏捷制造是以"竞争-合作（协同）"的方式，提高企业竞争能力，实现对市场需求做出灵活快速反应的一种新的制造模式。它要求企业采用现代通信技术，以敏捷动态优化的形式组织新产品开发，通过动态联盟、先进柔性生产技术和高素质人员的全面集成，迅速响应客户需求，及时交付新产品并投入市场，从而赢得竞争优势。下面从市场、企业能力和合作伙伴 3 个方面理解敏捷制造的内涵，如图 6-10 所示。

　　（1）敏捷制造的着眼点是快速响应市场/用户的需求　产品市场总的发展趋势是多样化和个性化，传统的大批量生产方式已不能满足瞬息万变的市场需求。敏捷制造思想的出发点是在对产品和市场进行综合分析时，首先明确用户是谁，用户的需求是什么，企业对市场做

this exercise. The point of Fig. 6-8 is that we have taken very simple transformation processes and stretched them to the point that the value added is barely recognizable.

6.5.4 TPS House

For decades Toyota was doing just fine in applying and improving TPS on the shop floor day in and day out without documenting TPS theory. Workers and managers were constantly learning new methods and variations on old methods through actual practice on the shop floor. Communication was strong in what was a relatively small company, so best practices developed within Toyota spread to other Toyota plants and ultimately to suppliers. But as the practices matured within Toyota, it became clear that the task of teaching TPS to the supply base was never ending. So Taiichi Ohno disciple Fujio Cho developed a simple representation a house.

The TPS house diagram (see Fig. 6-9) has become one of the most recognizable symbols in modern manufacturing. Why a house? Because a house is a structural system. The house is strong only if the roof, the pillars, and the foundation are strong. A weak link weakens the whole system. There are different versions of the house, but the core principles remain the same. It starts with the goals of best quality, lowest cost, and shortest lead time the roof. There are then two outer pillars just-in-time, probably the most visible and highly publicized characteristic of TPS, and jidoka, which in essence means never letting a defect pass into the next station and freeing people from machines automation with a human touch. In the center of the system are people. Finally there are various foundational elements, which include the need for standardized, stable, reliable processes, and also heijunka, which means leveling out the production schedule in both volume and variety. A leveled schedule or heijunka is necessary to keep the system stable and to allow for minimum inventory. Big spikes in the production of certain products to the exclusion of others will create part shortages unless lots of inventories are added into the system.

Each element of the house by itself is critical, but more important is the way the elements reinforce each other. JIT means removing, as much as possible, the inventory used to buffer operations against problems that may arise in production. The ideal of one-piece flow is to make one unit at a time at the rate of customer demand or the required beat. Using smaller buffers means that problems like quality defects become immediately visible. This reinforces jidoka, which halts the production process. This means workers must resolve the problems immediately and urgently to resume production. At the foundation of the house is stability. Ironically, the requirement for working with little inventory and stopping production when there is a problem causes instability and a sense of urgency among workers. In mass production, when a machine goes down, there is no sense of urgency: The maintenance department is scheduled to fix it while inventory keeps the operations running.

6.6 Agile Manufacturing

6.6.1 Emergence and Implication of Agile Manufacturing (AM)

In 1991, after researched and summarized the status and potential of U. S. manufacturing in-

出快速响应是否值得。只有这样，企业才能对市场/用户的需求做出响应，迅速设计和制造高质量的新产品，以满足用户的要求。

（2）敏捷制造的关键因素是企业的应变能力　企业要在激烈的市场竞争中生存和发展，必须具有"敏捷性"，即能够适时抓住各种机遇，把握各种变化的挑战，不断通过技术创新来领导市场潮流。敏捷企业能够以最快的速度、最好的质量和最低的成本，迅速、灵活地响应市场和用户需求，从而赢得竞争。

（3）敏捷制造强调"竞争-合作（协同）"　为了赢得竞争优势，必须采用灵活多变的动态组织结构，以最快的速度从企业内部某些部门和企业外部不同公司中选出设计、制造该产品的优势部分，组成一个单一的经营实体。在竞争-合作（协同）的前提下，企业需要考虑的问题包括：①哪些企业能成为合作伙伴？②怎样选择合作伙伴？③选择一家还是多家合作伙伴？④采取何种合作方式？⑤合作伙伴是否愿意共享数据和信息？⑥合作伙伴是否愿意持续不断地改进？

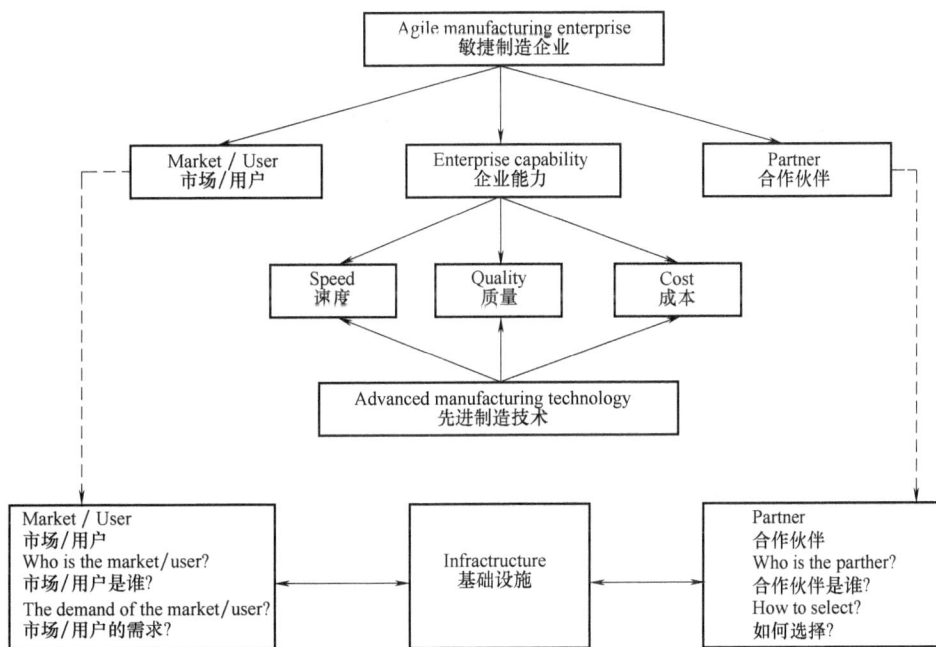

Fig. 6-10　Agile manufacturing concept diagram 敏捷制造概念示意图

6.6.2　敏捷制造的组成

敏捷制造主要由两个部分组成：敏捷制造的基础结构和敏捷的虚拟企业。基础结构为虚拟企业提供环境和条件，敏捷的虚拟企业来实现对市场不可预期变化的响应。

1. 敏捷制造的基础结构

物理基础结构、法律基础结构、社会基础结构和信息基础结构构成了敏捷制造的 4 个基础结构：

（1）物理基础结构　它是指虚拟企业运行所必需的厂房、设施、资源等必要的物理条件，是指一个国家乃至全球范围内的物理设施。当市场机会出现时，只需要添置少量必需的

dustry Lehigh University proposed the concept of agile manufacturing （AM） in the landmark report *21st Century Development Strategy of Manufacturing Enterprise*. Agile manufacturing is implemented under the support of three pillars of an innovative organization and management structure, advanced manufacturing technology, and skilled management personnel, and can respond to changing quickly market demands and market progress by establishing a common infrastructure. Agile manufacturing registers more sensitive and faster responding capacity compared with other manufacturing methods.

Agile manufacturing is a new manufacturing mode which can enhance the competitiveness of enterprises and achieve a flexible and rapid response to market demand by the way of "competition-cooperation （collaborative）". Agile manufacturing requires enterprises to adopt modern communication technology to organize new product development in the form of agile, dynamic and optimization. And it responds quickly to customer needs, deliveries new products timely and puts it on the market by the full integration of dynamic alliance, advanced flexible production technology and high-quality personnel, so as to win the competitive advantage. The following is to understand the meaning of agile manufacturing from three aspects of the market, business capabilities and partners, as shown in Fig. 6-10.

（1） Agile manufacturing focuses on responding to the needs of the market/users quickly   The general development trend of product market is diversification and personalization, so the traditional methods of mass production already cannot meet the market demand changing rapidly. The starting point of agile manufacturing is that when analyzing comprehensively the products and markets it should firstly make sure who is the user? What are the needs of the users? and whether it is worth for enterprises to make quickly response to the market? Only in this way, enterprises can quickly respond to the demands of the market/users, to design and produce high-quality new product, so as to meet the demands of users.

（2） The key factor for agile manufacturing is the enterprise's adaptability for change   If an enterprise wants to survive and develop in fierce market competition, it must has "agility". It means that enterprises can seize all kinds of opportunities, grasp various changing challenges, and continue to lead the market through technological innovation. Agile enterprise can quickly and agilely respond to the needs of the market and user with the fastest speed, best quality and lowest cost, so as to win the competitions.

（3） Agile Manufacturing emphasizes "competition-cooperation （collaborative）"   In order to win competitive advantages, it is necessary to adopt the flexible and dynamic organizational structure which can form a single operating entity at the fastest speed by selecting the advantage aspects of designing and manufacturing products from certain departments of inner-enterprise and other companies beyond enterprise.

Under the premise of competition-cooperation （collaborative）, some problems need to be considered by enterprises: ①Which companies can be partners? ②How to choose a partner? ③Choose one or more partners? ④What kind of cooperation mode? ⑤Are partners willing to share data and information? ⑥Are partners willing to keep on improving?

设备，集中优势开发关键设备，而多数设施可以通过选择合作伙伴得到，以实现敏捷制造。

（2）法律基础结构　它也称为规则基础结构，主要是指国家关于虚拟企业的法律和政策。具体来说，它应规定出如何组成一个法律上承认的虚拟企业。

（3）社会基础结构　虚拟企业要能生存和发展，还需要社会环境的支持。例如，人员需要不断地接受职业培训，不断地更换工作环境，这些都需要社会来提供职业培训、职业介绍的服务环境。

（4）信息基础结构　这包括能提供各种服务的网点、中介机构等一切为虚拟企业服务的信息手段。

敏捷制造的基本特征之一就是企业在信息集成基础上的合作与竞争。参加敏捷制造环境的企业可以分布在全国乃至世界各地。要建设敏捷制造环境，必须将各企业内部局域网络通过互联网或内联网连接起来，如图 6-11 所示。

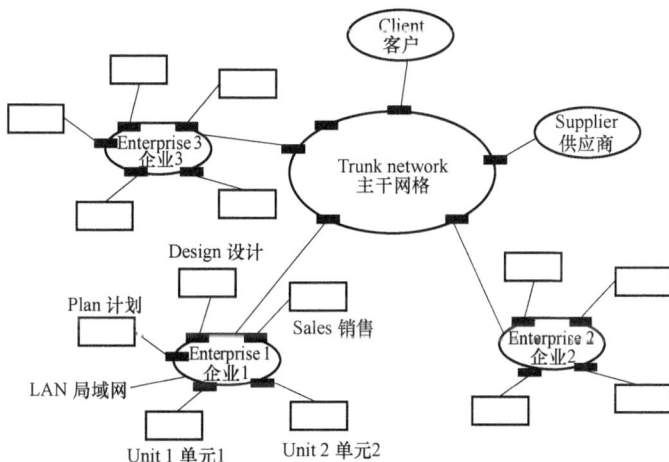

Fig. 6-11　Agile manufacturing computer network environment 敏捷制造计算机网络环境

图 6-12 所示是一个典型的信息集成基础结构框架，其中有 4 个层次：①网络通信层。连接异构设备和资源，进行结构和目标描述、定义节点在网络中的位置；②数据服务层。向计算机网络节点发送和从计算机网络节点请求信息，进行数据格式转换，在计算机网络节点间进行信息交换；③信息管理层。提供通用软件包和程序库，具有信息导航功能，支持电子邮件和超文本文件的传送；④应用服务层。提供支持企业经营、电子化贸易和加工制造活动的标准、协议、系统模型和接口等。

2. 虚拟企业

虚拟企业（Virtual Enterprises，VE）又称动态联盟（Virtual Orgnization，VO），是面向产品经营过程的一种动态组织结构和企业群体集成方式。它是一个依靠电子信息手段联系的、动态的合作竞争组织结构，将分布在不同地区、不同公司的人力资源和物质资源组织起来，实现市场需求的快速响应。参加虚拟制造环境的企业在通信网络上提供标准的、模块化的和柔性的设计与制造服务。

敏捷制造可以连接各种规模的生产资源，根据用户需求和虚拟制造环境中各企业现有能力，在合作竞争的基础上组成面向任务的虚拟企业。另外，在虚拟制造环境中，若干企业可以提供相同或类似的服务，系统可以从最优的目标出发，在竞争的基础上择优录用。

6.6.2　Components of AM

Agile manufacturing is mainly composed of two parts, namely infrastructure of AM and virtual enterprise of AM. The infrastructure provides a virtual enterprise with an environment and conditions, and agile virtual enterprise is for responding to unpredictable changes of market.

1. Infrastructure of AM

The physical infrastructure, legal infrastructure, social infrastructure and information infrastructure constitute the four basic structures of AM.

(1) Physical infrastructure　It refers to workshops, facilities, resources and other necessary physical conditions which are necessary for enterprise to operate, and also refers to the physical facilities of a country and even around the world. When market chance occurs, in order to achieve AM, the enterprise only need to purchase a small amount of necessary equipment and focuses on the development of key equipment, but most of the physical facilities can be got by selecting partners.

(2) Legal infrastructure　It is also called the rule infrastructure, mainly referring to national laws and policies on virtual enterprise. Specifically, it is should be regulated by law how to form a legal virtual business.

(3) Social infrastructure　The support of social environment is necessary for virtual enterprise to survive and develop. For example, personnel need to continually receive professional training, and constantly change the working environment, which requires the society to provide the service environment of professional training and introduction.

(4) Information infrastructure　It includes various information measures which service for the virtual enterprise, such as branch site, intermediary organization that can provide a variety of services.

One of the basic characteristics of AM is the cooperation and competition of enterprises on the basis of information integration. Companies that participate in the environments of AM can be distributed throughout the country and around the world. To build an AM environment, it is necessary to connect the internal LANs of the enterprises through the Internet (or Intranet), as shown in Fig. 6-11.

Fig. 6-12 shows a typical framework of information integration infrastructure with four levels: ①Network communication layer. It is in charge of connecting the heterogeneous devices and resources, describing the structure and target, and defining the location of nodes in the network; ②Data service layer. It sends information to a computer network node and requests information from a computer network node, performs data format conversion, and exchanges information between computer network nodes; ③Information management layer. It provides common software packages and program libraries, registers the function of information navigation, and supports the delivery of e-mail and hypertext documents; ④Application service layer. It provides standards, protocols, system models and interfaces that support business, e-trade and manufacturing activities.

2. Virtual Enterprises

Virtual enterprise (VE) is also known as virtual organization (VO). It is a kind of dynamic organization structure and enterprise group integration mode for product business process. It is a dynamic cooperative and competition organizational structure relying on the means of electronic infor-

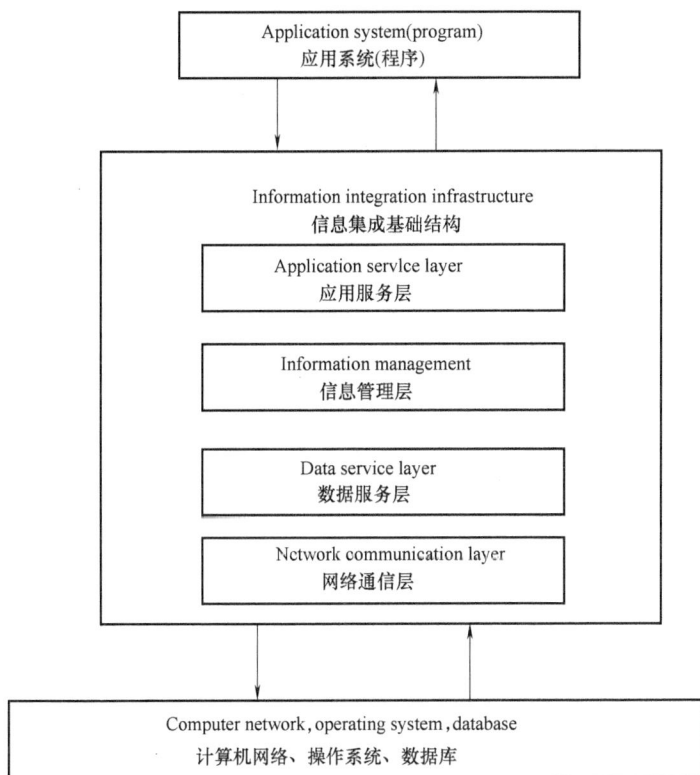

Fig. 6-12　Information integration infrastructure 信息集成基础

6.6.3　敏捷制造关键因素

敏捷制造的目的可概括为："将柔性生产技术，有技术、有知识的劳动力与能够促进企业内部和企业之间合作的灵活管理集成在一起，通过所建立的共同基础结构，对迅速改变的市场需求做出快速响应"。可见，敏捷制造主要包括 3 个要素：生产技术、管理技术和人力资源。

1. 敏捷制造的生产技术

敏捷制造的敏捷性是通过将技术、管理和人员 3 种资源集成为一个协调的、相互关联的系统来实现的。首先，具有高度柔性的生产设备是创建敏捷制造企业的必要条件（但不是充分条件）。其次，在产品开发和制造过程中，用数字计算方法设计复杂产品，可靠地模拟产品的特性，精确地模拟产品的制造过程。从用材料制造成品到产品最终报废的整个产品生命周期内，每个阶段的代表都要参加产品设计。再次，敏捷制造企业是一种高度集成的组织。信息在制造、工程、市场研究、采购、财务、销售等部门之间连续流动，而且还要在敏捷制造企业与其供应厂家之间连续流动。在敏捷制造系统中，用户和供应厂家在产品设计和开发中都应起到积极作用。最后，把分散的部门和人员集中在一起，靠的是严密的通用数据交换标准、坚固的"组件"（许多人能够同时使用同一文件的软件）、宽带通信通道（传递需要交换的大量信息）。

2. 敏捷制造的管理技术

敏捷制造在管理上所提出的最创新思想之一是"虚拟企业"。推出新产品最快的办法是利用不同公司的资源，使分布在不同公司内的人力资源和物资资源能随意互换，然后把它们

mation to contact. VE organizes human resources and material resources in different regions and different companies to achieve the rapid response to market demands. Companies participating in the virtual manufacturing environment provide standard, modular and flexible services of design and manufacturing on the communication network.

Applying agile manufacturing can connect to all kinds of production resources. The task-oriented virtual enterprise can be constituted on the basis of cooperative and competition according to the requirements of user and the existing capacity of the enterprises in virtual manufacturing environment. In addition, a number of enterprises can provide the same or similar services in a virtual manufacturing environment. The system can select the best one on the basis of competition considering the optimal target.

6.6.3 Key Factors of AM

The purpose of agile manufacturing can be summarized as that "integrates the flexible production technology, the labor force with technique and knowledge, and the agile management for collaboration of inner-enterprise and between enterprises together, responds rapidly to the market demands by the built common infrastructure". Thus, agile manufacturing mainly includes three elements: production technology, management and human resources.

1. Production Technology of AM

Agility is achieved by integrating the three resources of technology, management and personnel into a coordinated and interrelated system. Firstly, the production facilities with highly flexible is the necessary condition (but not the sufficient condition) to create an enterprise for agile manufacturing. Secondly, in the process of the product development and manufacturing, the method of digital calculation is used to design complex products, the characteristics of the product are simulated reliably, and the manufacturing process of product is also simulated accurately. During the entire life cycle of product, the representatives of each stage have to participate in product design from the finished product manufactured materials to the end of the product. Beyond that, the enterprise of agile manufacturing is a highly integrated organization. Information flows continuously not only among the departments of manufacturing, engineering, market research, procurement, finance, sales, but also flows continuously between agile manufacturing companies and their suppliers. In the system of agile manufacturing, users and suppliers should both play a positive role in the progress of product design and development. Finally, gathering the dispersive departments and personnel together relies on the strict exchange standards of common data, solid "components" (the software many people can use the same file at the same time), broadband communication channel (to send a lot of information needed to exchange).

2. Management Technology of AM

In management one of the most innovative ideas put forward by agile manufacturing is "virtual enterprise". The quickest way to launch new products is to use the resources of different companies to make the human resources and material resources distributed in different companies interchange freely, and then integrate them into a single entity—virtual enterprise relying on electronic means to

综合成单一的靠电子手段联系的经营实体——虚拟企业，以完成特定的任务。敏捷制造企业应具有组织上的柔性。产品的设计、制造、分配、服务将用分布在世界各地的资源（企业、人才、设备、物料等）来完成。根据工作任务的不同，有时可以采取内部多功能团队的形式，请供应者和用户参加团队；有时可以采用与其他企业合作的形式；有时可以采取虚拟企业形式。只要有效地运用这些手段，就能充分利用企业的资源。同时，应当把克服与其他企业合作的组织障碍作为首要任务，但需要解决因为合作而产生的知识产权问题，并开发虚拟企业管理技术。

3. 敏捷制造的人力资源

敏捷制造企业应能够最大限度地发挥人的主动性。有知识的人员是敏捷制造企业中唯一的最宝贵的财富。科学家和工程师参加战略规划和业务活动，对敏捷制造企业来说是决定性的因素。在制造过程的科技知识与产品研究开发的各个阶段，工程专家的协作是一种重要资源。

敏捷制造企业是连续发展的制造系统，该系统的能力仅受人员的想象力、创造性和技能的限制，而不受设备限制。敏捷制造企业中的每一个人都应该认识到柔性可以使企业转变为一种通用工具。柔性生产技术和柔性管理要使敏捷制造企业的人员能够实现他们自己提出的发明和合理化建议。应当提供必要的物质资源和组织资源，支持人员的创造性和主动性。

6.6.4　实施敏捷制造的技术

1994年美国能源部提出了"实施敏捷制造技术"（Technologies Enabling Agile Manufacturing，TEAM）的五年计划，并将其分为产品设计和企业并行工程、虚拟制造、制造计划与控制、智能闭环加工和企业集成五大类。

（1）产品设计和企业并行工程　其使命是按照客户需求进行产品设计、分析和优化，并在整个企业内实施并行工程。通过产品设计和企业并行工程，产品设计在概念优化阶段就可考虑产品整个生命周期的所有重要因素，诸如质量、成本、性能以及产品的可制造性、可装配性、可靠性与可维护性等。

（2）虚拟制造　虚拟制造就是"在计算机上模拟制造的全过程"。具体地说，虚拟制造将提供一个功能强大的模型和仿真工具集，并在制造过程分析和企业模型中使用这些工具。过程分析模型和仿真包括产品设计及性能仿真、工艺设计及加工仿真、装配设计及装配仿真等；而企业模型则考虑影响企业作业的各种因素。虚拟制造的仿真结果可以用于制订制造计划、优化制造过程、支持企业高层进行生产决策或重新组织虚拟企业。

（3）制造计划与控制　其任务是描述一个集成的宏观（企业的高层计划）和微观（详细的信息生产系统，包括制造路径、详细的数据以及支持各种制造操作的信息等）计划环境。该系统使用基本特征的技术、与CAD数据库的有效连接方法、具有知识处理能力的决策支持系统等。

（4）智能闭环加工　智能闭环加工是应用先进的控制和计算机系统以改进车间的控制过程的加工方式。当各种重要的参数在加工过程中能够得到监视和控制时，产品的质量就能够得到保证。智能闭环加工采用投资少、效益高、以微型计算机为基础的具有开放式结构的控制器，以达到改进车间生产的目的。

（5）企业集成　企业集成就是开发和推广各种集成方法，在适应市场多变的环境下运行虚拟的敏捷企业。TEAM计划建立了一个信息基础框架——制造资源信息网络，使得地理

complete a specific task. AM enterprise should have flexibility in organization. The design, manufacturing, allocation and service of the product will be finished by utilizing the resources (enterprises, talents, equipment, materials, etc.) distributed around the world. According to the different tasks, sometimes the enterprise can take the form of multi-functional team within the enterprise, and invite the suppliers and users to participate in the team; sometimes the enterprise can cooperate with other companies; sometimes the enterprise can adopt the virtual enterprise mode. As long as the enterprise takes use of these means effectively, it can make full use of the enterprise resources. Meanwhile, overcoming organizational obstacles to cooperation with enterprises should be taken as the first task. But, it needs to solve the intellectual property issues arising from cooperation, and develop the management technology for virtual enterprise.

3. Human Resources of AM

The enterprise of AM should enable people to utilize their initiative to the utmost extent. The knowledgeable people are the unique and most valuable treasure in the enterprise of AM. Scientists and engineers participate in strategic planning and operational activities, which is critical to the enterprise of AM. The collaboration of engineering experts is an important resource in each stage of the scientific and technical knowledge for manufacturing process and the product development.

The enterprise of AM is a manufacturing system being developed continually. This system capacity is only limited by imagination, creativity and skills of the staffs, but by the equipment. Each staff in the enterprise of AM should realize that flexibility can transform a enterprise into a general tool. The flexible production technology and flexible management must ensure the staffs in enterprise of AM to achieve the inventions and rationalization proposals put up by them. The necessary material resources and organizational resources should be provided to support the creativity and initiative of the staffs.

6.6.4 Technologies of Implementing AM

The Department of Energy in the U. S. has developed a five-year plan of "Technologies Enabling AM" (TEAM) in 1994, which was divided into five categories: product design and enterprise concurrent engineering, virtual manufacturing, manufacturing planning and control, intelligent closed-loop process and enterprise integration.

(1) Product design and enterprise concurrent engineering Its mission is to design, analyze and optimize the product in accordance with the demand of customer, and to implement concurrent engineering in the whole enterprise. Via product design and enterprise concurrent engineering, all important factors of the whole life cycle of product, such as quality, cost, performance, and manufacturability, assemblability, reliability and maintainability, can be taken into account during the conceptual optimization stage in product design.

(2) Virtual manufacturing Virtual manufacturing is "to simulate the whole process of manufacturing by the computer". Specifically, virtual manufacturing will provide a model and simulation tool sets with powerful function, and these tools will be used in the analysis of manufacturing process and enterprise models. Process analysis and simulation includes product design and performance simulation, process design and machining simulation, assembly design and assembly simulation,

上分散的各种设计、制造工作小组能够依靠这个制造资源信息网络进行合作，并能够依据市场变化而重组。

6.7 网络化制造

6.7.1 网络化制造的概念

1. 网络化制造的产生和内涵

21世纪以来，网络技术对制造业产生了重大影响，网络技术与制造技术相结合而形成的网络化制造技术应运而生。信息技术和网络技术向制造业的渗透早期是发生在设备层次的，如数控机床。到了计算机集成制造阶段，信息技术和网络技术开始深入到管理层上，通过企业内联网把企业中的各个部门紧密联系起来，实现企业内部信息化。网络化制造是为了适应当前经济全球化、区域和行业经济发展而采用的一种先进制造模式，也是实施敏捷制造和动态联盟的前提。

网络化制造尚处于不断发展的过程中，目前没有统一的定义。从发展过程可以看出，网络化制造是网络和制造的有机结合。这里的"制造"指的是大制造，包括产品整个生命周期过程及其所涉及的制造技术和制造系统。而"网络"具有广义性，可以是互联网、也可以是企业内联网或企业外联网。为此，网络化制造可以定义为：网络化制造是基于网络的制造企业的各种制造活动及其所涉及的制造技术和制造系统的总称。其中，网络包括互联网、企业内联网和企业外联网等各种网络，制造企业包括单个企业、企业集团以及面向某一市场机遇而组建的虚拟企业等各种制造企业及企业联盟；制造活动包括市场运作、产品设计与开发、物料资源组织、生产加工过程、产品运输与销售和售后服务等企业所涉及的所有相关活动。

2. 网络化制造的重要特性

网络制造与传统制造并不是对立的，网络制造不是对传统制造的取代，并不能代替传统制造中的许多功能，如产品的创新设计需要人的创造性劳动，零件的加工和装配需要相应的设备和人员，产品的销售需要物流系统等。网络化制造具有以下重要特性：

（1）敏捷性　网络化制造既能以最快的速度响应市场和客户的需求，也能根据市场需求的变化，灵活、快捷地对系统的功能和运行方式进行快速重构。

（2）协同性　网络化制造通过协同提高企业间合作的效率，缩短产品开发周期，降低制造成本，缩短整个供应链的交货周期。

（3）数字化　产品设计、制造、管理和控制等各种信息都是通过网络传递的，因此数字化是实施网络化制造的重要基础。

（4）直接性　企业通过网络化制造的方式，不仅可以直接与用户建立连接，从而减少消息传递过程造成的信息失真和时间上的延误，还可直接与供应商建立连接，降低零部件采购成本。

（5）远程化　企业利用网络化制造系统，可以对远程的资源和过程进行控制和管理，也可以与远方的合作伙伴、供应商进行协同工作。

（6）多样性　可以针对企业的具体需求，设计各种基于网络化的制造系统。如网络化产品定制系统、网络化协同制造系统、网络化营销系统、网络化售后服务系统等。

etc. All kinds of factors that affect the enterprise operations should be considered in the enterprise model. The simulation results of virtual manufacturing can be used to make the plan of manufacturing, optimize the manufacturing processes, to support enterprise executives to make the production decisions or reorganize virtual enterprises.

(3) Manufacturing planning and control   The task of manufacturing planning and control is to describe an integrated planning environment including macro (enterprise's high-level plan) and micro (detailed information production system, including manufacturing path, detailed data and information supporting various manufacturing operations, etc) environment. The system will take use of the technology with the basic characteristics, the method of connecting effectively with the CAD database, and the decision supporting system with the ability of knowledge processing.

(4) Intelligent closed-loop process   Intelligent closed-loop machining is a process that the advanced control and computer systems are used to improve the workshop control. The product quality can be guaranteed as all the various important parameters can be monitored and controlled during the processing. During the intelligent closed-loop process, the controllers with open structure based on microcomputer, less investment and high efficiency will be adopted to achieve the purpose of improving the workshop production.

(5) Enterprise integration   Enterprise integration is to develop and spread a variety of integration methods, then to operate the virtual agile enterprise in the changing market. The TEAM plan built an information infrastructure framework—an information network of manufacturing resource, enabling design and manufacturing groups dispersed geographically to collaborate each other relying on this network, can reorganize according to market change.

6.7   Network Manufacturing

6.7.1   Concept of Network Manufacturing (NM)

1. Emergence and Connotation of Network Manufacturing

Since 21st century, network technology has a major impact on the manufacturing industry. The combination of network technology and manufacturing technology makes the network manufacturing emerge at the right moment. In the early time, information technology and network technology which penetrated into the manufacturing industry occurred in the equipment level, for instance, CNC machine tool. In the computer-integrated manufacturing stage, information technology and network technology started to step deeply into the management level, various departments in the enterprise can be connected tightly together through Intranet to realize the informatization of inter-enterprise. Network manufacturing is an advanced manufacturing mode which is applied to adapt to the current economic globalization and the development of the regional and industrial economy. It is also the demand for conducting agile manufacturing and dynamic alliance.

Being in the process of continuous development, there is no uniform definition at present. It can be seen from the development process that network manufacturing is the significant combination of the network and the manufacturing. The "manufacturing" herein means the broad manufacturing,

6.7.2　网络化制造的关键技术

网络化制造的关键技术大致可以分为总体技术、基础技术、集成技术与应用实施技术。

（1）总体技术　总体技术主要是指从系统的角度，研究网络化制造系统的结构、组织与运行等方面的技术，包括网络化制造的模式、网络化制造系统的体系结构、网络化制造系统的构建与组织实施方法、网络化制造系统的运行管理、产品全生命周期管理和协同产品商务技术等。

（2）基础技术　基础技术是指网络化制造中应用的共性与基础性技术，这些技术不完全是网络化制造所特有的技术，包括网络化制造的基础理论与方法、网络化制造系统的协议与规范技术、网络化制造系统的标准化技术、产品建模和企业建模技术、工作流技术、多代理系统技术、虚拟企业与动态联盟技术和知识管理与知识集成技术等。

（3）集成技术　集成技术主要是指网络化制造系统设计、开发与实施中需要的系统集成与使能技术，包括设计制造资源库与知识库开发技术、企业应用集成技术、ASP 服务平台技术、集成平台与集成框架技术、电子商务与 EDI 技术、Web Service 技术，以及 COM+、CORBA、J2EE 技术、XML、PDML 技术、信息智能搜索技术等。

（4）应用实施技术　应用实施技术是支持网络化制造系统应用的技术，包括网络化制造实施途径、资源共享与优化配置技术、区域动态联盟与企业协同技术、资源（设备）封装与接口技术、数据中心与数据管理（安全）技术和网络安全技术等。

6.7.3　网络化制造系统的结构

网络化制造系统是一个运行在异构分布环境下的制造系统。在网络化制造集成平台的支持下，企业在网络环境下开展业务活动和实现不同企业之间的协作，包括协同设计制造、协同商务、网上采购与销售、资源共享和供应链管理等。

图 6-13 给出了网络化制造系统的结构。由图中可见，网络化制造系统分为 4 个层次：

（1）基础层　它主要为实施区域网络化制造提供基础性的支持，包括基础数据库（产品资源库、制造资源库、基础数据库等）、相关的技术基础（标准、规范、系统体系结构和网络化制造系统实施指南等）、网络化制造相关标准与协议等。

（2）应用与使能工具层　它主要包括各种实施网络化制造所需的应用软件系统（CAD、CAPP、CAM、PDM、ERP 等）和使能工具（项目管理、企业建模与诊断、设备互联等）。

（3）应用系统层　它是企业实施网络化制造最主要的功能支持层。

（4）企业用户层　它是通过互联网实现企业互联，在项目管理和过程管理系统的支持下开展企业网络化制造的实际应用。

6.7.4　网络化制造系统的实施

1. 网络化制造系统的实施过程

（1）需求分析　需求分析是网络化系统设计的出发点和依据。概括地说，网络化制造系统的需求分析就是要根据企业的具体情况，明确企业需要什么样的网络化制造系统，需要什么样的功能和性能，为什么需要，以及各种需求的紧迫程度如何。通过需求分析建立起来

including the whole lifecycle process of production as well as involved manufacturing technology and manufacturing system. But network has the generalized characteristics, which may be Internet or may be Intranet or Extranet of enterprise. For this reason, it can be defined that network manufacturing is the umbrella term for the various manufacturing activities and involved manufacturing technology and manufacturing system based on network manufacturing enterprises. Among them, network consists of various networks, such as Internet, enterprise Intranet and enterprise Extranet. The manufacturing enterprises include various manufacturing enterprises and enterprise alliance, such as single enterprise, enterprise groups and the virtual enterprise built by responding to a market chance. Manufacturing activities consists of all the correlative activities that contain the market operation, product design and development, manipulation for material resources, process of production, transportation and sale of products and after-sale service.

2. Importance Characteristics of Network Manufacturing

Network manufacturing is not opposite to traditional manufacturing, which is not the substitution of traditional manufacturing, and cannot replace many functions of traditional manufacturing, such as product innovation design need for person's creative work, process and assembly of parts for the corresponding equipment and personnel, product sales for logistics system, etc. The network manufacturing possesses the following important characteristics:

(1) Agility　Network manufacturing can respond to the needs of market and customers at the fastest speed. And it can also rebuild the systematical function and operating mode in a flexible and prom.

(2) Cooperativity　The network manufacturing just focuses on the coordination to enhance the efficiency of the enterprise corporation, shorten the development cycle of product, decrease the manufacturing cost, reduce the delivery cycle in the whole supply chain.

(3) Digitization　All the various information for the product design, manufacturing, management and control is transmitted by network. Therefore, digitalization is the critical base to implement network manufacturing.

(4) Directness　By the way of network manufacturing the enterprise can not only connect directly with users to reduce the information distortion and the delay in time in the process of information transmission, but also connect with suppliers to reduce the purchasing cost of components.

(5) Long distance　Enterprise can control and manage the remote resource and process, also can also work cooperatively with partners and suppliers by using the network manufacturing system.

(6) Diversity　According to the specific needs of enterprise, various network manufacturing systems can be designed, such as customization system of network product, network cooperative manufacturing system of product, network sale system, network after-sale service system, etc.

6. 7. 2　Key Technologies of Network Manufacturing

The key technologies of network manufacturing can be divided into overall technology, basic technology, integration technology, application and implementation technology.

(1) Overall technology　The overall technology mainly refers to research the structure of network manufacturing system, the technology of organization and operation from a system perspective,

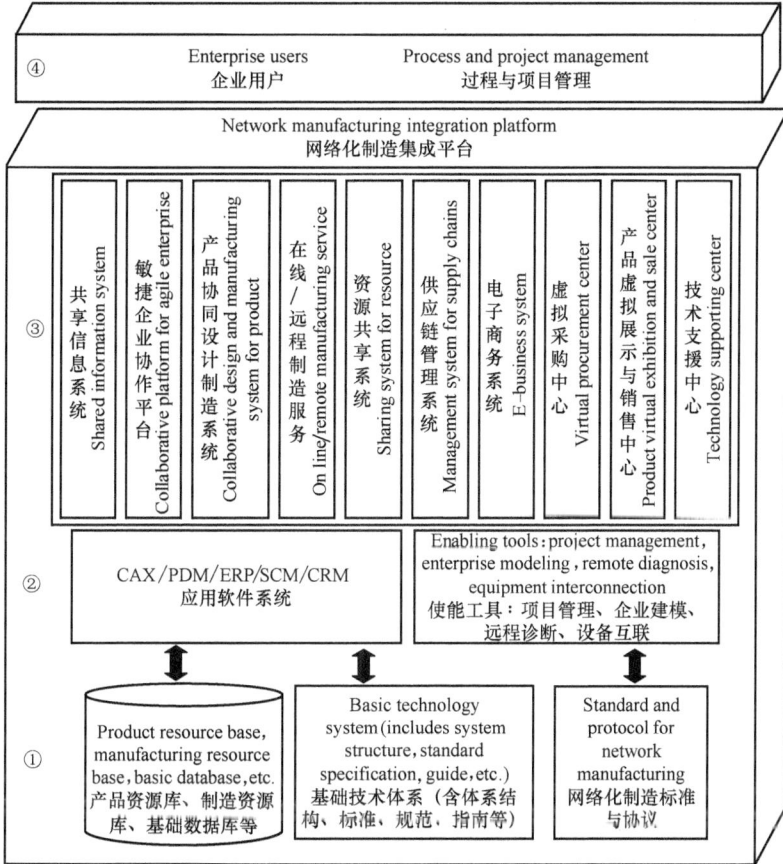

Fig. 6-13　Structure of network manufacturing system 网络化制造系统的结构

的网络化制造系统才能达到预期的目标，取得预定的经济效益。

（2）总体设计　总体设计的主要任务是确定企业网络化制造系统的系统需求、建立目标系统的功能模型、确定信息模型的实体和联系（信息模型建模的初期阶段）、提出网络化制造系统实施主要技术方案。在系统需求分析和主要技术方案设计方面，应深入到各子系统内部的功能需求，并产生相应的系统需求说明。

（3）网络化制造系统的实施、运行和维护　主要任务是实现总体设计的内容，产生一个可运行的系统。为此要完成应用软件编码、安装、调试，计算机硬件和生产设备的安装调试，全局数据库和局部数据库、网络的安装调试，以及组织机构落实和人员定岗等。各项工作最终都要达到可行的程度，并最终被用户接受。

（4）项目管理　网络化制造系统是一个复杂的大系统，其实施过程复杂、周期长、费用高、涉及面宽、风险大，因此，需要引进网络化制造系统应用工程的项目管理。项目管理的主要内容有：项目计划、组织、调度、资源分配、项目监控等。高效率的项目管理还需要有计算机辅助工具，如最常用的辅助制订项目计划等。

（5）搭建软件支持平台　平台软件包括工具软件、应用软件和集成平台支持软件等，平台支持复杂信息环境下网络化制造系统的应用开发、应用集成和系统运行。

（6）采用应用服务提供商（Application Service Provider，ASP）模式　ASP是指在共同签署的外包协议或合同的基础上，客户将其部分或全部与业务流程相关的应用委托给服务

including network manufacturing mode, architecture of network manufacturing system, establishment of network manufacturing system and organizational implementation method, operation management of network manufacturing system, whole lifecycle management of product and business technology of collaborative product, etc.

(2) Basic technology　It is the universal and basic technology in the application of network manufacturing. These technologies are not the specific technology for network manufacturing, involving basic theory and method of network manufacturing, protocol and normal technology of network manufacturing system, standard technology of network manufacturing system, technology of product modeling and enterprise modeling, workflow technology, multi-agent technology, technology of virtual enterprise and dynamic alliance, technology of knowledge integration, and so on.

(3) Integration technology　Integration technology mainly refers to the system integration and enabling technology needed for network manufacturing system design, development and implementation, including the development technology of resource base and knowledge base for design and manufacturing, technology of enterprise application and integration, ASP service platform technology, technology of integration platform and integration framework, e-commerce and EDI technology, Web service technology, as well as COM+, CORBA, J2EE technology, XML, PDML technology, information intelligent search technology, etc.

(4) Application and implementation technology　Application and implementation technology is the technology of supporting the application of network manufacturing system, involving implementation pathway of network manufacturing, technology of resource sharing and optimal configuration, technology of regional dynamic alliance and enterprise collaboration, technology of resource (equipment) encapsulation and interface, data center and date management (safety) technology and network security technology, etc.

6.7.3　Structure of Network Manufacturing System

Network manufacturing system is a kind of manufacturing system which operates in the distributed heterogeneous environment. Under the support of network manufacturing integration platform, the enterprises promote business activities and realize cooperation among different enterprises, including collaborative design and manufacturing, collaborative commerce, purchasing and sale online, resource sharing and supply chains management, etc.

Fig. 6-13 shows the system structure of network manufacturing. As can be seen from Fig. 6-13, the network manufacturing system is divided into four levels:

(1) Foundation layer　It is mainly to provide basic support for the implementation of regional network manufacturing, including the basic database (product resource base, manufacturing resource base, basic database, etc.), the relevant technological fundamentals (standard, specification, system structure, guide for implementing network manufacturing, etc.), standard and protocol for network manufacturing, and son on.

(2) Application and enabling tools layer　It mainly includes application software systems and enabling tools needed for implementing network manufacturing (CAD, CAPP, CAM, PDM,

商，服务商将保证这些业务流程的平滑运转。

2. 实现网络化制造要解决的问题

（1）互联网基础设施建设　互联网是网络化制造的基础。网络化制造已从基于企业内联网走向了基于内联网/互联网或内联网/外联网/互联网的集成，从企业内部走向了企业外部，并在迅速走向全球。

（2）网络化制造技术平台建设　网络化制造涉及多种高技术的集成，以及企业内部与外部、产品生命周期全过程、大量硬件和软件、技术和管理的集成，因此具有多种功能的网络化制造技术平台将成为网络化制造的技术支持工具，能有效地支持企业实施网络化制造。

（3）网络安全性　网络化制造的数据对企业十分重要。由于使用了 ASP 技术在网上传递信息，必须高度重视网络安全性问题。要加大数据传输安全的技术研究与应用，建立认证、编码等安全传输体系。

（4）网络化制造的可持续性　需建立健全道德和信用体系，否则，即使技术方面再先进，如果存在欺诈行为，网络化制造也难以生存。对开发产品的知识产权保护应制定明晰的法规，以保证网络化制造系统的健康运行。网络化制造服务提供商的运营和盈利模式也是值得关注的问题。

6.8　智能制造

6.8.1　智能制造技术的兴起和内涵

智能制造（Intelligent Manufacturing，IM）思想起源于 20 世纪 80 年代的美国。1988 年，美国的 P. K. Wright 和 D. A. Bourno 在其所著的《智能制造》一书中认为："智能制造的目的是通过集成知识工程、制造软件系统、机器人视觉和机器控制，对制造领域的技能和专家知识进行建模，以使机器人在没有人工干预的情况下进行小批量生产。"智能制造技术是制造技术、自动化技术、系统工程与人工智能等学科互相渗透、互相交叉而形成的一门综合技术。

目前比较通行的一种定义是：智能制造技术（Intelligent Manufacturing Technology，IMT）是指利用计算机模拟制造业人类专家的分析、判断、推理、构思和决策等智能活动，并将这些智能活动与智能机器有机地融合起来，将其贯穿应用于整个制造企业的各个子系统（经营决策、采购、产品设计、生产计划、制造装配等），以实现整个制造企业经营运作的高度柔性化和高度集成化，从而取代或延伸制造环境中人类专家的部分脑力劳动，并对制造业人类专家的智能信息进行搜集、存储、完善、共享、继承与发展。

自智能制造概念提出以来，智能制造系统一直受到众多国家的重视和关注。从 20 世纪 90 年代开始，美国国家科学基金会就着重资助有关智能制造的诸项研究。2005 年，美国国家标准与技术研究院提出了"智能加工系统（Smart Machining System，SMS）"研究计划。2009 年，美国提出和实施了"再工业化"计划，目标是重振实体经济，增强国内企业竞争力；发展先进制造业，实现制造业的智能化；保持美国制造业价值链上的高端位置和全球控制者地位。

日本于 1990 年提出为期 10 年的智能制造系统（Intelligent Manufacturing System，IMS）的国际合作计划，其目的是把日本工厂和车间的专业技术、欧盟的精密工程技术和美国的系统技术组合起来，开发出能使人和智能设备都不受生产操作和国界限制，且能彼此合作的高

ERP, etc.).

(3) Application system layer　It is the main functional support layer for enterprises to implement network manufacturing.

(4) Enterprise user layer　It develops the practical application of enterprise network manufacturing under the support of project management and process management system by realization of enterprise interconnection through Internet.

6.7.4　Implement of Network Manufacturing System

1. Implementation Process of Network Manufacturing

(1) Requirement analysis　Requirement analysis is the starting point and basis of network system design. Generally speaking, the requirement analysis of network manufacturing system, according to the specific situation of enterprises, aims just to clear what kind of network manufacturing system is needed, what kind of function and performance and why is needed, as well as what the urgent degree of various requirements is. The network manufacturing system can reach the expected objective and obtain the predetermined economic benefits by requirement analysis.

(2) Overall design　The main task of overall design is to determine requirements of the enterprise network manufacturing system, establish functional model of objective system and confirm the entity and relationship of information model (initial stage of information modeling), put forward the main technical scheme for implementing the network manufacturing system. The requirement analysis of system and design of main technical scheme should be carried out into the requirements of each subsystem internals, and to produce the corresponding of specification of system requirements.

(3) Implementation, operation and maintenance of network manufacturing system　Its main task is to realize the content of the overall design, and to generate a operational system. For this purpose, the application software coding, installation and debugging, installation and adjustment of computer hardware and manufacturing facility, installation and debugging of global database and local database and network, as well as organization implementation and staffs position, are to be accomplished. Various tasks should be reach the practicable degree ultimately and be accepted by users.

(4) Project management　The network manufacturing system is a complex and large scale system, which the implement of the system registers the following features, such as complex process, long cycle, high cost, wide range involved, great risk, etc. Therefore, it needs to introduce the project management for application engineering of network manufacturing system. The main contents of project management include project planning, organization, scheduling, resource allocation, monitoring, and so on. Efficient project management also needs computer-aided tools, such as the assistant project plan most commonly used, etc.

(5) Construction of software support platform　Software involves the tools software, application software and software for supporting the integration platform, etc. The platform can support the application development, application integration and system operation of network manufacturing system under the complex information environment.

(6) Application service provider (ASP) mode On the basis of cosigned outsourcing agreement

技术生产系统。2006 年 10 月，日本提出了《创新 25 战略》，其目的是在全球大竞争时代，通过科技和服务创造新价值，提高生产力，促进日本经济的持续增长。"智能制造系统"是该计划中的核心理念之一。欧盟于 2010 年启动了第七框架计划（FP7）的制造云项目，特别是作为制造业强国的德国，继实施智能工厂之后，2013 年 4 月在汉诺威工业博览会上正式推出了投入达 2 亿欧元的"工业 4.0"战略，旨在奠定德国在关键工业技术上的国际领先地位。另外，以英国为代表的老牌工业国家以及以韩国为代表的后发工业国家在其最新的经济发展计划中都对智能制造概念尤为重视。表 6-2 列出了一些国家的具体计划。

Tab. 6-2 Intelligent manufacturing planning of major countries 主要国家的智能制造计划

Country 国家	Planning name 计划名称	Proposed time 提出时间	Implementation goal 实施目标
Japan 日本	"Innovation 25 Strategy" 《创新 25 战略》	2006	To create new value through science and technology and services, taking "Intelligent Manufacturing System" as the core idea of the program, to promote the sustainable growth of the Japanese economy and to deal with the era of big competition around the world. 通过科技和服务创造新价值，以"智能制造系统"作为该计划核心理念，促进日本经济的持续增长，应对全球大竞争时代
The U. S. A 美国	"Re-industrialization" program "再工业化"计划	2009	To develop the advanced manufacturing industry, achieve the intelligence of the manufacturing industry, and keep the U. S. the high-end status on the value chains and global control position in manufacturing industry. 发展先进制造业，实现制造业的智能化，保持美国制造业价值链上的高端位置和全球控制者地位
South Korea 韩国	"New Growth Power Planning and Development Strategy" "新增长动力规划及发展战略"	2009	To determine 17 industries in the three major fields to be the developing tasks, promote the digital industrial design and the digital collaborative construction for manufacturing industry, strengthen the policy support for the basic development of intelligent manufacturing. 确定三大领域 17 个产业为发展重点，推进数字化工业设计和制造业数字化协作建设，加强对智能制造基础开发的政策支持
Germany 德国	"Industry 4.0" strategy "工业 4.0"战略	2013	By the distributed and combined industrial manufacturing unit module, to form the multiple combined, intelligent industrial manufacturing system, to deal with the "Fourth Industrial Revolution" in which the intelligent manufacturing is the leading position. 由分布式、组合式的工业制造单元模块，通过组建多组合、智能化的工业制造系统，应对以智能制造为主导的"第四次工业革命"
Britain 英国	"High Value Manufacturing" strategy "高价值制造"战略	2014	To apply the intelligent technology and professional knowledge, produce product, production process and related service with sustainable growth and high economic value potential by creativity, so as to achieve the goals of reviving the UK's manufacturing industry. 应用智能化技术和专业知识，以创造力带来持续增长和高经济价值潜力的产品、生产过程和相关服务，达到重振英国制造业的目标

or contract, ASP refers to the customers entrust the partial or full applications related with business procedures to the service providers. The service providers will ensure the smooth operation of these business procedures.

2. Problems of Realizing Network Manufacturing

(1) Construction of network basic infrastructure   Network is the basis of network manufacturing. The network manufacturing has stepped into the integration of Intranet/Internet or Intranet/Extranet/Internet form the integration based on Intranet, and into the inner-enterprise, even will become globalization quickly.

(2) Construction of network manufacturing technology platform   Network manufacturing involves the integration of various high technologies, as well as the integration of inside and outside the enterprise, the whole process of product lifecycle, lots of hardware and software, technology and management. Therefore, the multifunctional network manufacturing technology platform will become the technical supporting tools, which can effectively support the enterprise to implement the network manufacturing.

(3) Network security   The data of network manufacturing is important to enterprise. Due to using the ASP technology to convey information online, so the security of network must be paid high attention. The research and application on technologies of date transmission security should be reinforced, and the secure transmission system of authentication and coding is to be built.

(4) Sustainability of network manufacturing   It needs to establish a sound system of moral and credit. Otherwise, even if the technology is progressive, if there are fraud behaviors, it is difficult to survive for the network manufacturing.

6.8   Intelligent Manufacturing

6.8.1   Emergence and Connotation of IM Technology

The thought of intelligent manufacturing (IM) originated from America in the 1980s. In 1988, professor Wright and Bourne published a their book *Intelligent Manufacturing*, in which "the purpose of IM is to model the skills and expertise knowledge of the manufacturing field by integrating knowledge engineering, manufacturing software system, robot vision and machine control, so that robots can perform small batches production without human intervention". Intelligent manufacturing technology is a comprehensive technology which is formed by mutual infiltration and interpenetration of manufacturing technology, automation technology, system engineering, artificial intelligence and other disciplinary technologies.

At present, one common definition is that intelligent manufacturing technology (IMT) refers to using computer to simulate the analysis, judgment, reasoning, conception and decision-making and other intelligent activities of human experts in manufacturing industry, in the meantime making these intelligent activities blend with intelligent machines, and applying throughout each subsystem in the entire manufacturing enterprise (such as business decisions, purchasing, product design, produc-

6.8.2 智能制造的技术体系

下面介绍一种以提高制造系统智能为目标，以机器人、智能体或全能体为手段的 Holonic 制造系统（Holonic Manufacturing System，HMS），也称全能制造系统的结构。图 6-14 所示为 Holonic 制造系统的结构示意图。

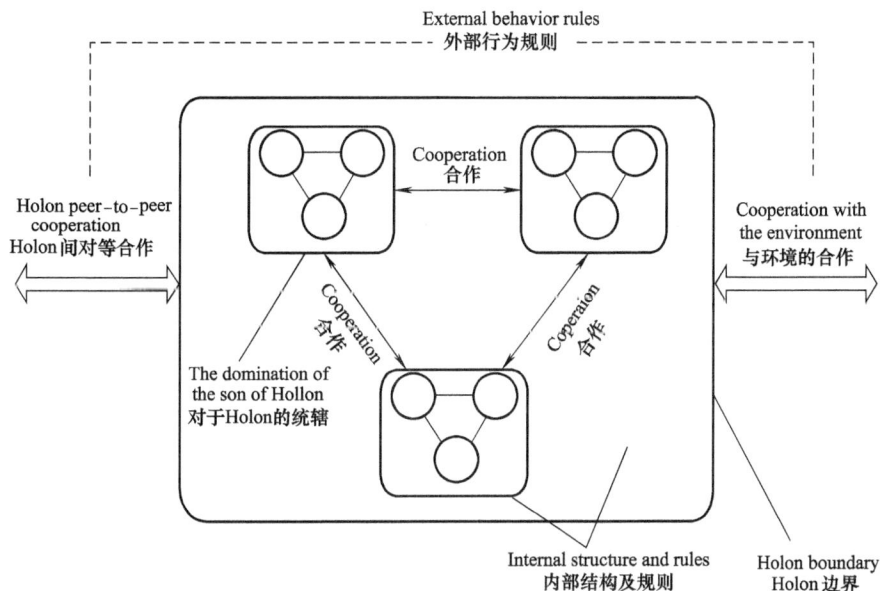

Fig. 6-14 Schematic diagram of Holonic manufacturing system structure Holonic 制造系统结构示意图

HMS 是由若干全能体（Holon）组成的，它强调每个全能体都能独立自主和协同工作。每个全能体都具有自治性、协同性和柔性，它们是既相互独立又相互协作的系统构造块。HMS 不强调全面自动化，更不是无人化工厂，其独立的含义是自主地制订和执行作业计划以及运行策略。它的组织结构不是固定不变的，而是具有暂时的递阶性以适应外部环境的变化。

Holon（全能体）一词是由 A. Koestler 在他的《机器中的灵魂》一书中提出，其含义是全能的整体是由局部的模块组成的。Holon 是一种具有自律特性和协作特性的制造系统结构单元。它在制造系统中用于对信息或物理对象进行转换、运输、储存和确认。Holon 通常包括一个信息处理部分和一个物理处理部分。Holonic 结构是指由能够协作达到一个目的或目标的 Holon 组成的 Holonic 系统。Holonic 结构定义了 Holon 协作的基本规则。

HMS 是一个集成了包括从订单预订到设计、生产和打入市场在内的所有制造活动的、以实现敏捷制造为目的的 Holonic 结构。图 6-15 所示是一个 Holonic 制造系统的示例。

6.8.3 智能制造的支撑技术

智能制造技术是以知识信息处理技术为核心的面向 21 世纪的制造技术，其主要支撑技术如下：

（1）人工智能技术 智能制造技术利用计算机模拟制造业人类专家的智能活动，旨在取代或延伸人的部分脑力劳动，而这些正是人工智能（Artificial Intelligent，AI）技术的研究内容。人工智能研究的是利用机器来模拟人类的某些智能活动的相关理论和技术。

tion planning, manufacturing assembly, etc.) to achieve high flexibility and high integration in the whole operations of manufacturing enterprise. Thus, it can replace or extend part of mental work of human beings in the manufacturing environment, which fulfills in the collection, storage, improvement, sharing, succession and development of intelligent information of human experts in manufacturing industry.

Since the concept of IM was put forward, the intelligent manufacturing system has been paid much more attention and caution by many countries. Since 1990s, the National Science Foundation of America has funded lots of researches into IM. In 2005, the National Institute of Standards and Technology put forward the "Smart Machining System" (SMS) research program. In 2009, the United States proposed and implemented the "Re-industrialization" program, which the goal was to revive the real economy, enhance the competitiveness of domestic enterprises, develop advanced manufacturing industry, achieve the intelligence of the manufacturing industry, and keep the U. S. the high-end status on the value chains and global control position in manufacturing industry.

In 1990, Japan proposed a 10-year international cooperation program for intelligent manufacturing system (IMS), which aimed to combine the professional technology of Japanese factories and workshops, the precision engineering technology of the European Union and the system technology of American, to develop a high-tech production system that enables people and intelligent devices to cooperate each other without the limits of production operations and national boundaries. In October 2006, the "Innovation 25 Strategy" was proposed. The aim of the program is to enhance the productivity and promote the sustainable growth in the Japanese economy by creating new value through technology and services in the era of global competition. "Intelligent Manufacturing System" is one of the core concepts in this program. The European Union launched the manufacturing cloud project of the Seventh Framework Program (FP7) in 2010. Especially, the Germany, as a powerful manufacturing country, aiming to establish the leading international position in the key industrial technology, officially launched the Industry 4. 0 project with 200 million Euro investment on the Hannover Messe in April 2013, followed the implementation of the Smart Factory. In addition, the old-line industrial countries represented by the United Kingdom, as well as the post-industrial countries represented by Korea, all attach particular importance to the concept of IM in the latest economy development plan. Tab. 6-2 lists the specific planning of different countries.

6.8.2 Technological System of IM

The following is to introduce the Holonic manufacturing system (HMS) using robots, intelligent agents or Holon as its means, which aims to enhance the intelligence of manufacturing system, is also called the almighty manufacturing system structure. Fig. 6-14 shows the structure sketch of the Holonic system.

HMS is composed of a number of omnipotent units (Holon), emphasizing that each Holon can be independent and collaborative work. Each Holon records autonomy, cooperativity and flexibility, but the Holons are the system constructive modules mutually independent and mutually cooperative each other. HMS does not emphasize total automation, but not be unmanned factory, which its

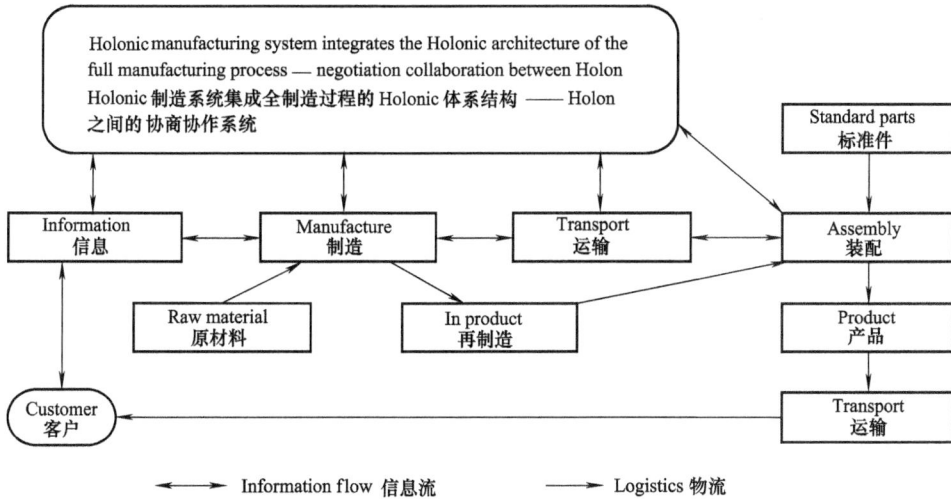

Fig. 6-15 An example for Holonic manufacturing system Holonic 制造系统示例

（2）并行工程　并行工程通过组织多学科产品开发小组、改进产品开发流程和利用各种计算机辅助工具等手段，使多科学小组在产品开发初始阶段就能及早考虑下游的可制造性、可装配性、质量保证等因素，从而达到缩短产品开发周期、提高产品质量、降低产品成本，增强企业竞争力的目标。

（3）虚拟制造技术　虚拟制造是建立在利用计算机完成产品整个开发过程这一构想基础之上的产品开发技术，它综合应用建模、仿真和虚拟现实等技术，提供三维可视交互环境，对从产品概念到制造全过程进行统一建模，并实时、并行地模拟出产品未来制造的全过程，以期在真实执行制造之前，预测产品的性能和可制造性等。

（4）计算机网络与数据库技术　计算机网络与数据库的主要任务是采集 IMS 中的各种数据，以合理的结构存储数据，并以最佳的方式、最少的冗余、最快的存取响应为多种应用服务，同时为应用共享数据创造良好的条件，从而使整个制造系统中的各个子系统实现智能集成。

6.8.4 工业 4.0 与制造业的未来

1. 工业 4.0 的诞生及内涵

为保持并提高其在全球的优势，德国政府于 2010 年发布了《德国 2020 高技术战略》报告，并在 2013 年 4 月举办的汉诺威工业博览会上发布了《实施"工业 4.0"战略建议书》。同年 12 月，德国又发布了"工业 4.0"标准化路线图。自此，"工业 4.0"成为德国政府确定的面向 2020 年的国家战略，引领工业制造业朝高度信息化、自动化、智能化方向发展。

"工业 4.0"的第一个内涵是智能化、绿色化和人性化。由于每个客户的需求不同，个性化或定制化的产品不可能大批量生产。"工业 4.0"必须解决的第一个问题就是单件小批量生产要能够达到大批量生产同样的效率和成本，构建能生产高品质、个性化智能产品的"智能工厂"。绿色化包括产品全生命周期，以实现可持续制造。正是由于绿色化生产，工厂可以建在城市里、甚至靠近员工的住处，大大改善生产与环境和人的关系，如图 6-16 所示。

"工业 4.0"的第二个内涵是实现资源、信息、物品和人相互关联的"虚拟网络-实体物理系统（Cyber-Physical System，CPS）"，也称信息物理融合生产系统。借助移动终端和无

meaning for independence is to develop and implement independently the operating plan and operation strategy. Its organizational structure is not fixed, but has a temporary hierarchy to adapt to changes in the external environment.

Holon (omnipotent unit) is a term coined by A. Koestler in his book *The Soul of Machinery*, which means the almighty ensemble is constituted by the local modules. Holon is a kind of manufacturing system structure unit with the characteristics of autonomy and cooperation. It is used to convert, transport, store and confirm the information or physical objects in manufacturing system. Holon usually includes an information processing section and a physical processing section. Holonic structure refers to the Holonic system consisting of many Holons which can achieve one aim or goal. The Holonic structure defines the basic rules for Holon collaboration.

HMS is a Holonic structure that integrates all manufacturing activities including from order booking to design, production and access to market for achieving the agile manufacturing. Fig. 6-15 is an example of a Holonic manufacturing system.

6.8.3 Support Technologies of IM

Intelligent manufacturing is a kind of manufacturing technology for the 21st century in which the kernel is the processing technology of knowledge and information. The main support technologie of intelligent manufacturing is introduced as follows:

(1) Artificial intelligence technology Intelligent manufacturing technology is to use computer to simulate the intelligent activities of human experts, which aims to replace or extend part of mental work of human beings. Exactly, it is the research contents of the artificial intelligence (AI) technology. Artificial intelligence focuses the research on the theory and technology for using machines to simulate some intelligent activities of human beings.

(2) Concurrent engineering By means of organizing a multidisciplinary team for product development, improving the product development procedure and using a variety of computer-aided tools, using concurrent engineering can make the multidisciplinary team in the initial stage of product development early consider the factors such as the downstream manufacturability, assemble ability, quality guarantee, etc. Thus, the objective to shorten the product development period, enhance product quality, reduce product cost and strengthen the competitiveness of enterprises is achieved.

(3) Virtual manufacturing technology Virtual manufacturing is a kind of product development technology built on the conception of using computer to accomplish the whole development process of product. It applies synthetically modeling, simulation, and virtual reality technologies to provide three-dimensional visual interactive environment, uniformly modeling throughout the product concept and the manufacturing process, and real-time and concurrently simulating the whole process of product manufacturing in future. It is hoped to predict product performance, manufacturability, and so on, prior to the real implementation of manufacturing.

(4) Computer network and database technology The main task is to collect various data in the IMS and store data by means of a reasonable structure, and serve for a variety of applications in the optimal way, the least redundancy, the fastest access response. At the same time, computer network

线通信,虚拟世界和现实世界能够无障碍沟通,使设备和人在空间和时间上可以分离,机器与机器相互之间可以通信,处于不同地点的生产设施可以集成。可以设想,在统一的生产计划系统指挥调度下实行柔性工作制,由若干相对独立的 CPS 制造岛组成的分散网络化制造将成为一种高效率、省资源、宜人化的先进生产模式,如图 6-17 所示。

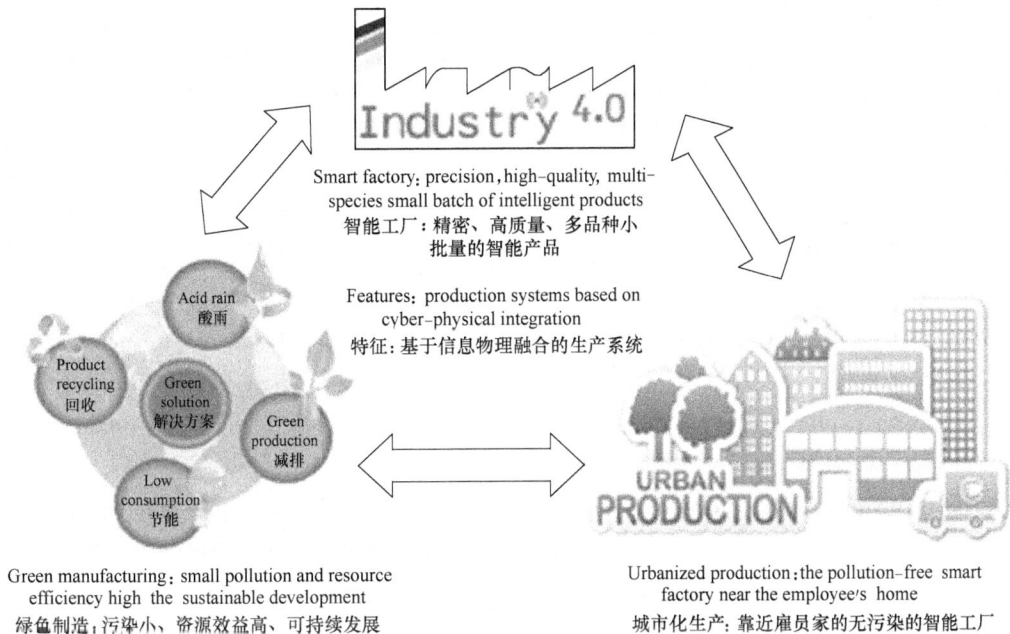

Fig. 6-16 Intelligence,green and humanity factory 智能化、绿色化和人性化工厂

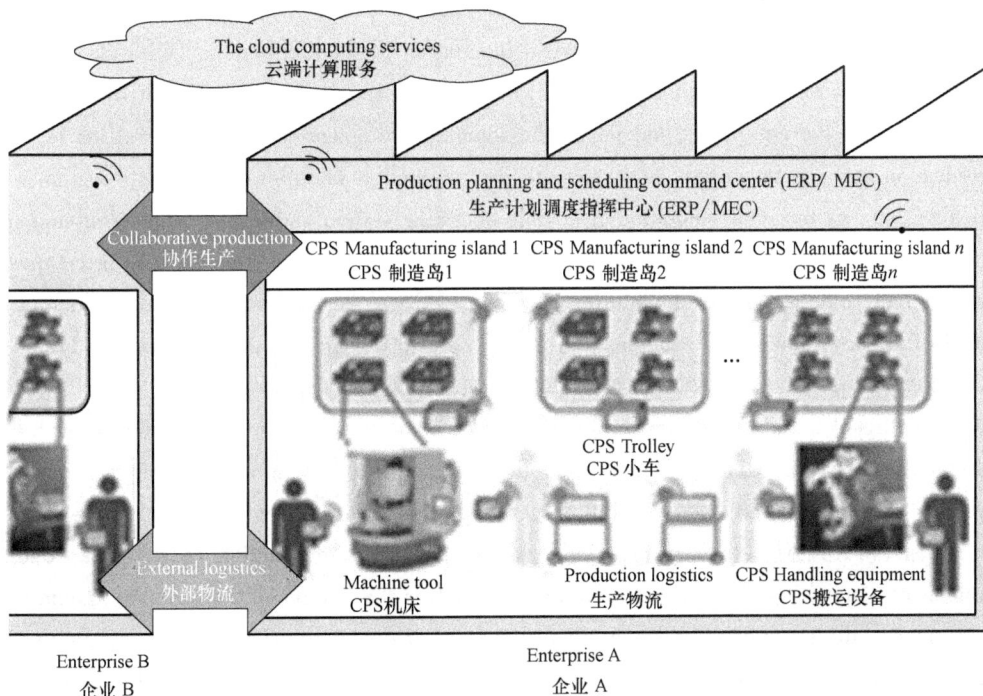

Fig. 6-17 Decentralized networked manufacturing 分散网络化制造

and database technology creates good conditions for the application of shared data, so as to make the intelligent integration of each subsystem in the entire manufacturing system be implemented.

6. 8. 4　Industry 4. 0 and the Future of Manufacturing Industry

1. Birth and Connotation of Industry 4. 0

In order to maintain its global advantages, the German government in 2010 formulated the "High Technology Strategy 2020" program, and issued the *Proposal of Implementation of Industry 4. 0 Strategy* on the Hannover Messe held in April. In December of the same year, Germany has issued a standardized route map for "Industry 4. 0". Since then, the "Industry 4. 0" has become the national strategy determined by the German government for the year 2020, leading the manufacturing industry towards the development of high informatization, automation, and intelligence.

The first connotation of "Industry 4. 0" is intelligence, greenization and humanity. Because of the different needs of each customer, the personalized or customized products are not likely to be produced in large quantities. The first problem to be solved for "Industry 4. 0" is that a single part and small batch production can achieve the same efficiency and cost as the mass production, and build an "intelligent factory" which can produce high-quality and personalized intelligent products. Greenization involves in the whole lifecycle of product to achieve sustainable manufacturing. Because of the greenization production, factories can be built in cities, or even close to employees' residence, greatly improve the relationships among the production, environment and human beings, as shown in Fig. 6-16.

The second connotation of "Industry 4. 0" is to realize the "Virtual Network-Real Physical System" related to resources, information, goods and people, namely cyber-physical system (CPS), which is also called the information and physical integration production system. With the help of mobile terminal and wireless communication, the virtual world and the real world can freely communicate, which makes the equipment and people be separated in space and time, as well as realize the communication between machines, and integrate the production facilities at different locations. It is conceivable that, by the implementation of flexible operating system under the commanding and scheduling of the uniform production planning system, the distributed network manufacturing composed of several relatively independent manufacturing island of CPS will become an advanced production mode with high efficiency, saving on resources and comfortable to human, as shown in Fig. 6-17.

2. The Future of Automated Manufacturing: Intelligent Factory

The intelligent factory based on the conception of "Industry 4. 0" marks the future of the automated manufacturing industry. It is composed of the physical system and the virtual information system, called the information and physical integration production system, its skeleton structure is shown in Fig. 6-18. As seen from Fig. 6-18, there is a virtual information system corresponding to the system for conducting the material production, which is the "soul" of the physical system to control and manage the production and operation of the physical system. The cooperative interaction of the physical system and the information system is carried out by mobile Internet and Internet of things. Intelligent factory is realized by providing the production system with the CPS. This kind of

2. 自动化制造业的未来：智能工厂

基于工业 4.0 构思的智能工厂标志着自动化制造业的未来。它由物理系统和虚拟的信息系统组成，称之为信息物理融合生产系统，其框架结构如图 6-18 所示。从图中可见，对应于进行物质生产的系统有一个虚拟的信息系统，它是物理系统的"灵魂"，控制和管理物理系统的生产和运作。物理系统与信息系统通过移动互联网和物联网协同交互。智能工厂是通过在生产系统中配备 CPS 来实现的。这种工厂未必是一个有围墙的实体车间，它也可以借助网络利用分散在各地的设备，这是"全球本地化"的工厂。相对于传统生产系统，智能工厂的产品、资源及处理过程因 CPS 的存在，将具有非常高水平的实时性，同时在资源和成本的节约中也颇具优势。

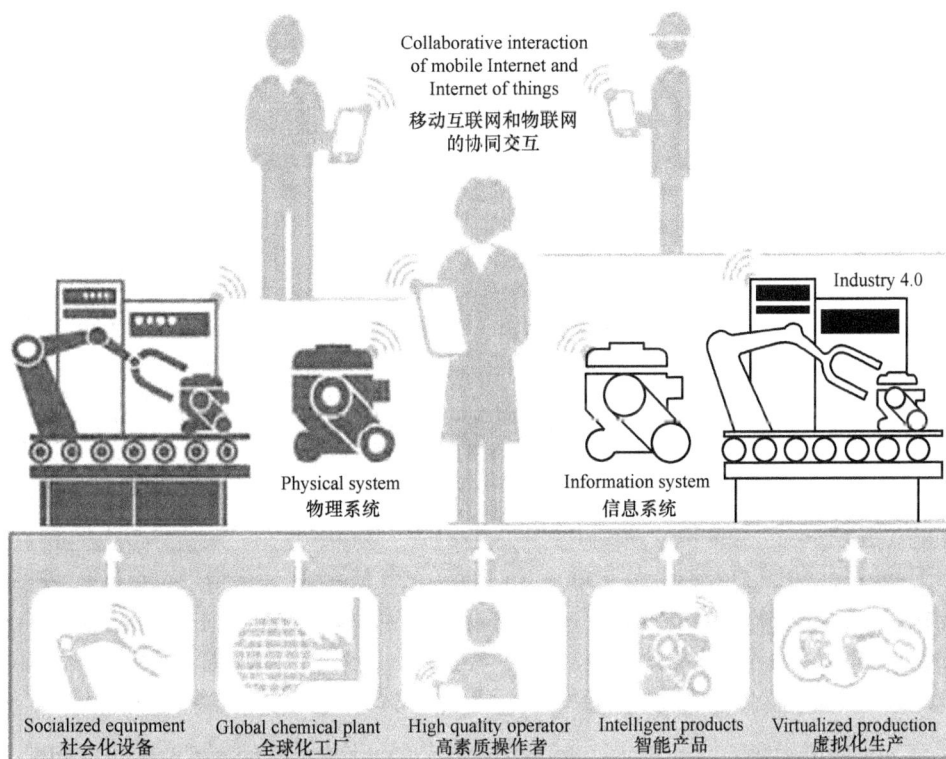

Fig. 6-18　Physical information integration production system 信息物理融合生产系统

3. 未来制造：云制造

制造业已经进入大数据时代，智能制造需要高性能的计算机和网络基础设施，传统的设备控制和信息处理方式已经不能满足需要，基于云计算的云制造已经指日可待。云计算提供计算资源的共享池（网络、服务器、应用程序和存储），本地计算机安装 SCADA 数据采集和监控系统后，可将数据发送给云进行处理、存储和分配，并在需要时从云接收指令。一群机器人的云端控制概念如图 6-19 所示。

6.8.5　智能制造的主要发展趋势

智能制造已成为新型工业应用的标志性概念，一些工业发达国家将发展智能制造作为打

factory which is not necessarily a walled entity workshop, can use the equipment distributed different places by means of the network, so it is a factory characterized by "global localization". Contrasted to the traditional production system, the products, resources and processing of the intelligent factory because of the existence of CPS will have a rather high real-time performance and also present advantages in resources and cost saving.

3. Future Manufacturing: Cloud Manufacturing

Manufacturing has entered the era of the big data. Intelligent manufacturing requires high-performance computer and network infrastructure, and the traditional facility control and information processing methods cannot meet the needs. Thus, the cloud manufacturing based on the cloud computing can be expected soon. The cloud computing provides a shared pool of computing resources (network, server, application and storage). While the local computer is installed with SCADA data acquisition and monitoring system, the data can be sent to the cloud for processing, storage and distribution, and receive instructions from the cloud when needed. A cloud control concept for a swarm of robots is shown in Fig. 6-19.

6. 8. 5　Development Tendency of IM

The intelligent manufacturing has become a landmark concept for new industrial application. Some industrial developed countries focus on developing the intelligent manufacturing as the core content to create new advantages in international competition. Our country also takes the intelligent manufacturing as the important means to seize the commanding heights in a new round of industrial competition at present and in a period of future. The intelligent manufacturing is an important link for the implementation of "Made in China 2025". The development trend of intelligent manufacturing is represented in the following aspects:

1. New Industrial Revolution Focus on Intelligent Manufacturing Concerned Greatly

In 2016, the "New Industrial Revolution" which regards the intelligent manufacturing as the kernel becomes the attention focus of the international society once again. The theme of the 46th Davos World Economic Forum was "The Fourth Industrial Revolution". It is reported that on the G20 Summit held at Hangzhou, China in September 2016, the "New Industrial Revolution" is an important part of the meeting topic, aiming at "promote the new industrial revolution, make full use of the roles of new technologies, new elements and new industrial organization mode in promoting domestic production and create employment".

2. The Intelligent Manufacturing Going Forward Deeper and Broader by the Promotion of the Information and Network Production Mode

At present, Internet plus manufacturing industry is becoming a major trend. The network production mode firstly embodies in the intelligent allocation of global manufacturing resources, and is transformed from centralized production into coordinative production in different places by network. The information network technology realizes the information sharing between enterprises in different links, can quickly find and dynamically adjust cooperating partners around the global, to achieve global decentralization production in each industrial chain including research and development, man-

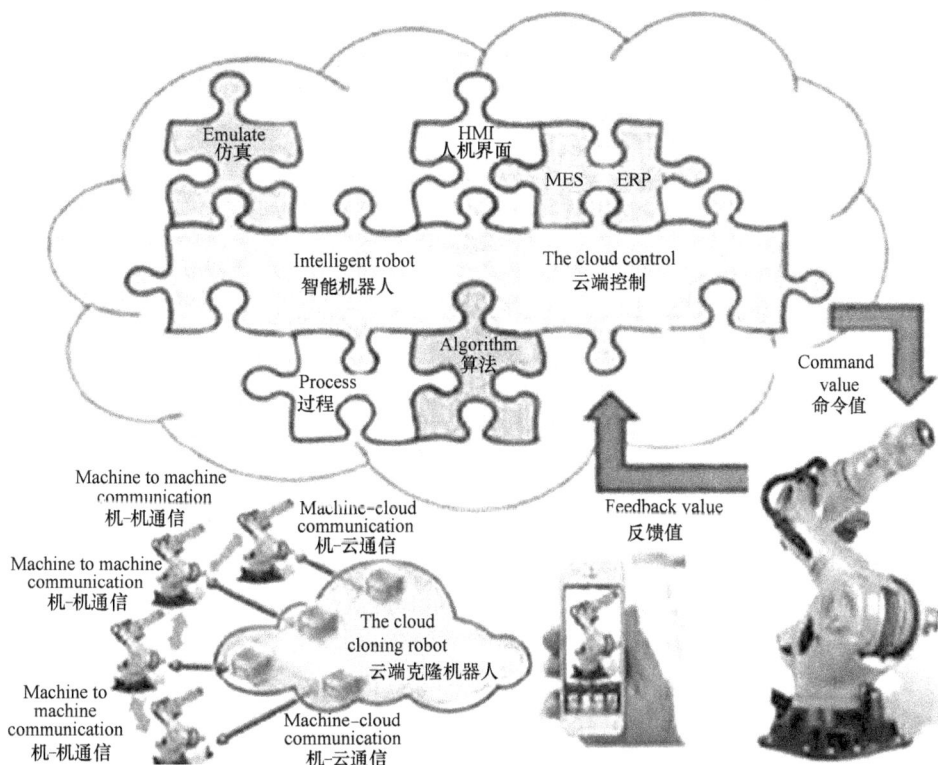

Fig. 6-19 Cloud manufacturing 云制造

造国际竞争新优势的核心内容。我国也将智能制造作为当前和今后一个时期抢占新一轮产业竞争制高点的重要手段,智能制造已成为实施《中国制造2025》的重要环节。智能制造的发展趋势表现在以下几个方面:

1. 以智能制造为核心的新工业革命再度引发国际社会高度关注

2016年,以智能制造为核心的新工业革命再度成为国际社会关注的焦点。第46届达沃斯世界经济论坛的主题为"第四次工业革命"。2016年9月在中国杭州举办的G20峰会上,"新工业革命"成为会议议题的重要组成部分,旨在"推动新工业革命,充分发挥新技术、新要素和新工业组织模式在促进国内生产和创造就业中的作用"。

2. 信息网络化生产方式进一步推动智能制造向深度和广度进军

当前,互联网+制造业正成为一种大趋势。网络化生产方式首先体现在全球制造资源的智能化配置上,由集中生产向网络化异地协同生产转变。信息网络技术使不同环节的企业间实现信息共享,能够在全球范围内迅速发现和动态调整合作对象,整合企业间的优势资源,在研发、制造、物流等各产业链环节实现全球分散化生产。其次,大规模定制生产模式的兴起也催生了如众包设计、个性化定制等新模式,这从需求端推动了企业采用信息网络技术集成度更高的智能制造方式。

3. 国际社会竞相打造智能制造系统平台

近年来,国际社会竞相打造智能制造系统平台,如图6-20所示。以物联网、移动互联网、大数据、云计算为代表的新一代信息技术,以3D打印、机器人、人机协作为代表的新型制造技术,与新能源、新材料及生物科技呈现多点突破、交叉融合,智能制造技术创新不

ufacturing, logistics, etc. Secondly, the rise of mass customization production mode has also spawned new modes such as the crowd sourcing design, customization etc. This suggests that the enterprise is promoted to use the intelligent manufacturing mode with highly integrated information network technology from the demand aspect.

3. International Society Competes to Build the Intelligent Manufacturing System Platform

In recent years, the international society is racing to build the intelligent manufacturing system platform, as shown in Fig. 6-20. The new generation of information technology representing Internet of things, mobile Internet, big data, cloud computing, and the new type manufacturing technology characterizing by 3D printing, robot, man-machine cooperation, present multi-point breakthrough overlapping and integration in new energy, new materials and bio-technology, which resulted in innovation breakthroughs of intelligent manufacturing technology.

Surrounding the construction of intelligent manufacturing system platform, the United States reinforces software development of the cyber-physical system (CPS) and industrial Internet platform construction through the implementation of "Partnership Program for Advanced Manufacturing Industry". German promotes the "Industry 4.0" strategy, and builds intelligent manufacturing system architecture with CPS. China combines the implementation of "Made in China 2025", and explores to set up intelligent manufacturing innovation center, and guides research institutions, manufacturing enterprises and information communication enterprises to strengthen the deep cooperation and build the intelligent manufacturing system platform jointly in accord with the actual development of China manufacturing industry.

4. To Re-construct the Basic Standardization and Promote the Systematization of Intelligent Manufacturing

The basic standardization system of intelligent manufacturing is the foundation of intelligent manufacturing. The re-construction of standardization procedure realizes the large-scale industrial application of intelligent manufacturing, especially the unified specifications for key smart components, equipment and system, the unified procedure for products, production process, management and service, etc., which will greatly improve the overall level of intelligent manufacturing. The establishment of the standard system of intelligent manufacturing also indicates that the intelligent manufacturing this turn is the re-construction and upgrading as to the traditional manufacturing mode.

5. The Concept of Internet of Things Changes Systematically the Overall Appearance of Intelligent Manufacturing

With the emergence of a large number of new concepts, such as industrial Internet of things, industrial cloud and so on, the intelligent manufacturing presents the overall characteristics by being promoted systematically. In recent years, a number of innovation achievements in Internet of things technology are obtained, especially using Internet of things technology brings about "the machine substituting for human" and Internet of things factory, promotes the subversive substitution of "green and safe" manufacturing mode to the traditional "pollution and dangerous" manufacturing mode.

断取得新突破。

围绕智能制造系统平台建设，美国借助实施"先进制造业伙伴计划"加强信息物理融合系统（CPS）软件开发和工业互联网平台建设。德国推行"工业 4.0"战略，搭建以 CPS 为核心的智能制造系统架构。我国结合"中国制造 2025"的实施，探索建立智能制造创新中心，引导科研机构、制造企业与信息通信企业加强深度合作，联合搭建符合中国制造业发展实际的智能制造系统平台。

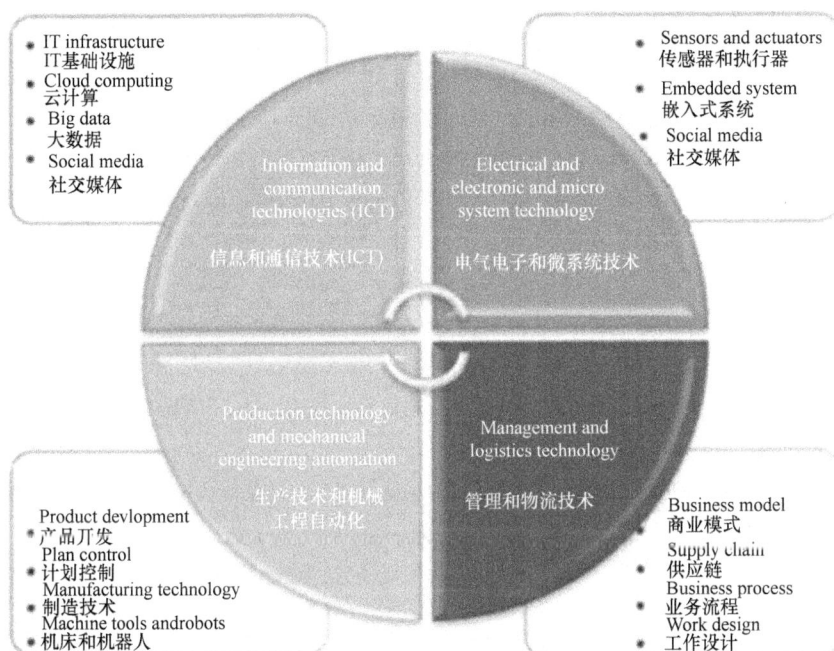

Fig. 6-20　Intelligent manufacturing system platform model 智能制造系统平台模型

4. 再造基础性标准化，推动智能制造的系统化

智能制造的基础性标准化体系是智能制造的根基。标准化流程再造使得工业智能制造的大规模应用得以实现，特别是关键智能部件、装备和系统的规格统一，产品、生产过程、管理、服务等流程统一，这将大大促进智能制造总体水平的提升。智能制造标准化体系的建立也表明本轮智能制造是从本质上对传统制造方式的重新架构与升级。

5. 物联网理念系统性改造智能制造的全局面貌

随着工业物联网、工业云等新理念的产生，智能制造呈现出系统性推进的整体特征。近年来，物联网技术取得了一批创新成果，特别是物联网技术带来的"机器换人"、物联网工厂，推动着"绿色、安全"制造方式对传统"污染、危险"制造方式的颠覆性替代。物联网制造是现代方式的制造，将逐步颠覆人工制造、半机械化制造与纯机械化制造等现有的制造方式。

The manufacturing of Internet of things is a modern manufacturing method, which will gradually subvert the existing manufacturing methods such as manual manufacturing, semi mechanical manufacturing and pure mechanical manufacturing, and so on.

Review Questions and Problems

6-1　What is the strategic goal of advanced manufacturing mode?

6-2　Which main stages has the development of the manufacturing industry production mode undergone generally?

6-3　What are the basic viewpoints and connotations of CIM?

6-4　What are the definition and connotation of CIMS?

6-5　Which systems are composed of CIMS? Briefly describe the architecture of CIMS.

6-6　What is mass customization? What are the differences between mass customization and mass production?

6-7　What kinds of mode does mass customization include? Which kind of mode is often applied by enterprises?

6-8　Briefly describe the basic principle of mass customization. What key technologies can be used to realize mass customization?

6-9　Compare the differences between concurrent engineering and sequential engineering. Analyze the operating mode and the functional characteristics of concurrent engineering.

6-10　What is the essence of concurrent engineering? What scopes are mainly embodied characteristics of concurrent engineering?

6-11　What do the key technologies of CE include?

6-12　What is lean production? What is JIT?

6-13　Briefly describe the eight major types of non-value-adding waste.

6-14　What background does agile manufacturing generate?

6-15　What is the main concept of agile manufacturing?

6-16　What characteristics does agile manufacturing enterprise have?

6-17　What is network manufacturing? What characteristics does network manufacturing have?

6-18　What scopes are embodied key technologies of network manufacturing?

复习题与习题

6-1　先进制造模式的战略目标是什么？

6-2　制造业的生产方式的发展大致经历了哪几个主要阶段？

6-3　CIM 的基本观点与内涵是什么？

6-4　CIMS 的定义与内涵是什么？

6-5　CIMS 由哪几个系统组成？简述 CIMS 的体系结构。

6-6　何谓大批量定制？大批量定制和大批量生产有何联系和区别？

6-7　大批量定制有哪几种方式？企业通常采用的是何种定制方式？

6-8　简述大批量定制的基本原理。实现大批量定制的关键技术有哪些？

6-9　比较并行工程（CE）与串行工程的区别，分析并行工程的运行模式和功能特点。

6-10　并行工程的实质是什么？其特点主要体现在哪些方面？

6-11　CE 的关键技术有哪些？

6-12　什么是精益生产？什么是 JIT？

6-13　简述八种主要的非增值浪费。

6-14　敏捷制造是在什么样的背景下产生的？

6-15　敏捷制造的主要概念是什么？

6-16　敏捷制造企业的特点有哪些？

6-17　何谓网络化制造？网络化制造有哪些重要特性？

6-18　网络化制造的关键技术体现在哪几方面？

References
参 考 文 献

[1] 任小中. 先进制造技术 [M]. 3 版. 武汉：华中科技大学出版社，2016.

[2] 宾鸿赞. 先进制造技术 [M]. 武汉：华中科技大学出版社，2010.

[3] 宾鸿赞，王润孝. 先进制造技术 [M]. 北京：高等教育出版社，2006.

[4] 张世昌. 先进制造技术 [M]. 天津：天津大学出版社，2004.

[5] 李蓓智. 先进制造技术 [M]. 北京：高等教育出版社，2007.

[6] 何涛，杨竞，范云. 先进制造技术 [M]. 北京：北京大学出版社，2006.

[7] WRIGHT P K. 21 世纪制造 [M]. 冯常学，等译. 北京：清华大学出版社，2004.

[8] TURNER W C, MIZE J H, CASE K E. Introduction to Industrial and Systems Engineering [M]. 3rd ed. 北京：清华大学出版社，2002.

[9] 唐一平. 先进制造技术 [M]. 英文版. 北京：机械工业出版社，2002.

[10] 袁哲俊，王先逵. 精密和超精密加工技术 [M]. 2 版. 北京：机械工业出版社，2007.

[11] 庞滔. 超精密加工技术 [M]. 北京：国防工业出版社，2000.

[12] 刘飞. 先进制造系统 [M]. 北京：中国科学技术出版社，2001.

[13] 艾兴. 高速切削加工技术 [M]. 北京：国防工业出版社，2003.

[14] 刘志峰，张崇高，任家隆. 干切削加工技术及应用 [M]. 北京：机械工业出版社，2005.

[15] 刘延林. 柔性制造自动化概论 [M]. 武汉：华中科技大学出版社，2001.

[16] 苑伟政，马炳和. 微机械与微细加工技术 [M]. 西安：西北工业大学出版社，2001.

[17] KALPAKJIAN S, SCHMID S R. Manufacturing Engineering and Technology [M]. Upper Saddle River, N. J.：Prentice Hall，2001.

[18] REGH J A, KRAEBBER H W. Computer-Integrated Manufacturing [M]. 2nd ed. Upper Saddle River, N. J.：Prentice Hall，2001.

[19] AYRES, ROBERT U. Computer Integrated Manufacturing [M]. London：Chapman and Hall，1991.

[20] 中国工程院《新世纪如何提高和发展我国制造业》课题组. 新世纪的中国制造业 [J]. 中国机械工程学会会讯，2002（10）：1-7.

[21] 杨叔子，吴波. 先进制造技术及其发展趋势 [J]. 机械工程学报，2003，39（10）：77-78.

[22] 梁福军，宁汝新. 可重构制造系统理论研究 [J]. 机械工程学报，2003，39（6）：36-43.

[23] 孙林岩，等. 先进制造模式的分类研究 [J]. 中国机械工程，2002（1）：84-88.

[24] 任小中，邓效忠，苏建新，等. 数控成形磨齿机计算机辅助模块化设计 [J]. 农业机械学报，2008，39（2）：144-146.

[25] 张绍国，任小中，段明德. 拖拉机发动机气道自由曲面的反求技术 [J]. 拖拉机与农用运输车，2007（4）：52-53.

[26] 张曙. 五轴加工机床：现状和趋势 [J]. 金属加工（冷加工），2015，（15）：1-5.

[27] 高晓平. 先进制造管理技术及其应用 [M]. 北京：机械工业出版社，2005.

[28] MEYER H, FUCHS F, THIEL K. Manufacturing Execution Systems-Optimal Design, Planning, and Deployment [M]. New York：McGraw-Hill Companies, Inc.，2009.

[29] GRABSKI S, LEECH S, SANGSTER A. Management Accounting in Enterprise Resource Planning Systems

[M]. London: CIMA Publishing, 2009.

[30] SHEHAB E M, SHARP M W, SUPRAMANIAM L, et al. Enterprise Resource Planning: An Integrative Review [J]. Business Process Management Journal, 2004, 10 (4): 359-386.

[31] COHEN S, ROUSSEL J. Strategic Supply Chain Management [M]. New York: McGraw-Hill Companies, Inc., 2005.

[32] CHOPRA S, MEINDL P. Supply Chain Management: Strategy, Planning, and Operation [M]. London: Pearson Education, Inc., 2013.

[33] ROSS D F. Introduction to Supply Chain Management Technologies [M]. London: Taylor and Francis Group, 2011.

[34] LIKER J K, MEIER D. The Toyota Way Fieldbook [M]. New York: McGraw-Hill Companies, Inc., 2006.

[35] DIXIT A, DAVE V. Lean Production System: A Future Approach [J]. International Journal of Engineering Sciences & Management Research, 2015, 2 (6): 69-74.

[36] LIKER J K. The Toyota Way: 14 Management Principles from the World's Greatest Manufacturer [M]. New York: McGraw-Hill Companies, Inc., 2004.

[37] 熊光楞. 并行工程的理论与实践 [M]. 北京: 清华大学出版社, 2001.

[38] 张申生. 敏捷制造的理论、技术与实践 [M]. 上海: 上海交通大学出版社, 2000.

[39] 童秉枢, 李建明. 产品数据管理 (PDM) 技术 [M]. 北京: 清华大学出版社, 2003.

[40] 柴跃廷, 刘义. 敏捷供需链管理 [M]. 北京: 清华大学出版社, 2001.

[41] 刘飞, 雷琦, 宋豫川. 网络化制造的内涵及研究发展趋势 [J]. 机械工程学报, 2003, 39 (8): 1-6.

[42] 祁国宁, 顾新建, 杨青海, 等. 大批量定制原理及关键技术研究 [J]. 计算机集成制造系统-CIMS, 2003, 9 (9): 776-782.

[43] 张曙. 工业4.0和智能制造 [J]. 机械设计与制造工程, 2014, 43 (8): 1-5.